Lives in the Balance

Lives in the Balance

The Ethics of Using Animals in Biomedical Research

The Report of a Working Party of the Institute of Medical Ethics

Edited by

JANE A. SMITH

and

KENNETH M. BOYD

Oxford New York Tokyo

OXFORD UNIVERSITY PRESS

1991

Oxford University Press, Walton Street, Oxford OX2 6DP
Oxford New York Toronto
Delhi Bombay Calcutta Madras Karachi
Petaling Jaya Singapore Hong Kong Tokyo
Nairobi Dar es Salaam Cape Town
Melbourne Auckland

and associated companies in
Berlin Ibadan

Oxford is a trade mark of Oxford University Press

Published in the United States
by Oxford University Press, New York

A catalogue record for this book is available from the British Library

Library of Congress Cataloging in Publication Data
Lives in the Balance: the report of a
working party of the Institute of Medical Ethics / edited by Jane A. Smith and Kenneth M. Boyd.
Includes index.
1. Animal experimentation–Moral and ethical aspects. I. Smith, Jane A. II. Boyd, Kenneth M. III. Institute of Medical Ethics (Great Britain)
[DNLM: 1. Animals, Laboratory. 2. Animal Welfare. 3. Bioethics. 4. Research–standards. QY 50 E84]
R853.A53E74 1991 179'.4–dc20 91–3945
ISBN 0–19–854744–7

Set by Footnote Graphics, Warminster, Wiltshire
Printed in Great Britain by
Bookcraft (Bath) Ltd
Midsomer Norton, Avon

Preface

The enactment of the Animals (Scientific Procedures) Act 1986 opened a new era of statutory control for the use of animals in research in the United Kingdom. The Act was a product of a century of experience gained in administrative control, since the Cruelty to Animals Act 1876, through the Home Office and its Inspectorate; of hard thinking in the Home Office Advisory Committee; and of Parliamentary activity and debate in which bodies committed to the welfare of animals and to scientific research played constructive parts. There is a general belief that the Act, in its intent and in its administrative provisions, is well founded. But it could not end the moral and philosophical debate on the research use of animals. Neither is it desirable that that debate should end: it is not the purpose of legislation to put consciences to sleep.

The Institute of Medical Ethics, accordingly, set up in September 1986 a Working Party to study the ethical issues in the use of animals in biomedical research, and so, it was hoped, to make a major contribution to the continuing national and international debate. It has been my privilege to chair that Working Party. Members were chosen for the knowledge and experience that they could bring from the range of disciplines involved, and for the strong convictions with which they would defend the legitimate interests of animals, of scientific research, and of the public conscience. Each member served in his personal capacity, and not as a representative of any organization. Tribute must be paid to their diligence and loyal service. They met eighteen times as a full Working Party. They met also in smaller groups, in which potentially conflicting interests were brought together, to work on particular subjects. They were generous in their writing of memoranda and drafts, and in the provision of information, out of which the final chapters were written. Above all, by looking behind the words of contention to discern the truths contended for, they broke out of the polarization in which these debates are often, and unrewardingly, conducted, and produced an agreed report.

This report is now published. Its aim is to promote good thinking in the continued debate; to raise standards of perception, and of sensitivity to the factors involved; to further the refinement, reduction, and replacement of the use of living and sentient creatures consistently with the essential needs of scientific research and safety evaluation. In full awareness of the past, and of a present in which scientific and technical developments and alert consciences are in full vigour, the report looks to a future in which there must be change; it points to directions in which change should go.

The Working Party and the Institute of Medical Ethics are deeply indebted to Dr Jane Smith and the Reverend Dr Kenneth Boyd, the Director of Research at the Institute, who together furnished both the research team and the secretariat for the task. They have served a group, formidable alike in knowledge, experience, and moral conviction. Readers in particular disciplines of science and philosophy will recognize the professional hand in some of the chapters; but Dr Smith and Dr Boyd, tireless and generous in their patience, have been responsible for the major part of the writing of successive drafts and of the final report.

The major funding of the exercise was by a grant from the Leverhulme Trust. To this were added donations from other bodies which, by their gifts, demonstrated their faith in the non-partisan character of the exercise and the potential value of the work. They were:

British Dental Association
British Veterinary Association Animal Welfare Foundation
CIBA–Geigy plc
Fisons plc
Fund for the Replacement of Animals in Medical Experiments
Glaxo Holdings plc
The Humane Research Trust
Imperial Chemical Industries plc
Merck Sharp & Dohme Limited
Pfizer Limited
Reckitt & Colman Pharmaceutical Division
Royal Society for the Prevention of Cruelty to Animals
St Andrew Animal Fund
Sandoz Institute for Medical Research
Smith Kline & French Laboratories Limited
Unilever plc
Universities Federation for Animal Welfare.

To all of these, thanks are returned.

We express our thanks also to experts outside the Working Party, who assisted the research staff by welcoming them into laboratory animal facilities, by providing information and stimulating discussion, and by reading drafts; to the Universities of Leicester and Birmingham which, through the kindness of Professor David Morton, provided office facilities for Dr Smith; and to the ever-hospitable staff of the Ciba Foundation, London, where our meetings were held.

The Report is written in standard English, with grammatical use of the common gender in personal pronouns. Where sexual difference is to be drawn, the masculine and feminine forms are in place.

It remains for me to commend the Report to its readers, not as an academic study only but as material for their own personal engagement with the issues of conscience which thinking and sensitive people must feel. Erasmus wrote, in *The handbook of a Christian soldier*,

> Man is a marvellous creature composed of two or three very diverse parts, a soul which is like a divinity and a body which is like a beast without understanding. Indeed, as far as the body is concerned, so far are we from surpassing the rest of unreasoning creation that we are actually found inferior to them in every physical endowment. But with regard to the soul we have such a capacity for divinity that we can soar past the minds of the angels and become one with God.

The nature of man might be delineated differently today. But whatever there is of divinity in humanity, it enshrines the concept of duty. How we discharge our duty to other creatures puts our humanity to the proof.

Exeter
1990

G. R. DUNSTAN
Chairman

Contents

List of members of the Working Party xv

1. Introduction 1

1.1 Establishing the facts 2
1.2 Moral issues: benefits 3
1.3 Animal awareness and sensitivity 4
1.4 'Weighing' costs and benefits 4
1.5 Control of laboratory animal use 5
1.6 Philosophical arguments and moral concerns 6

2. The researchers, their aims and methods 8

2.1 Aims of biomedical research 8
2.2 Methods in biomedical research 9
2.2.1 Clinical and animal research 9
2.2.2 *In vivo* and *in vitro* research 12
2.3 Constraints on potential biomedical researchers 13
2.4 Scale of animal use in biomedical research 17

3. Benefits of biomedical research involving animal subjects 25

3.1 Historical perspective 25
3.2 Changing moral sentiments 29
3.3 Two moral positions on benefits 31
3.4 The argument from necessity 36
3.5 Assessment of benefits 39

4. Pain, stress, and anxiety in animals 45

4.1 Introduction 45
4.2 Brain development 46
4.3 General cognitive abilities 50
4.4 Awareness of 'self' 54
4.5 Pain 58
4.6 Stress 67
4.7 Anxiety 70
4.8 General conclusions 73

5. Animals as experimental subjects 78

5.1 Recognition of adverse effects in animals 78
5.1.1 Ways of recognizing adverse effects in animals 79
5.1.2 Comments on schemes for the recognition of adverse effects 82
5.2 Examining the costs imposed on animals by experiments 86
5.2.1 Species of animal used 87
5.2.2 Animal supply 90
5.2.3 Animal housing and husbandry 101
5.2.4 Scientific procedures performed on the animals 110
5.2.5 Number of animals used 114
5.3 Strategies for assessing and reducing costs: a summary 115

6. Ethical considerations in the development and use of replacement alternatives to animal experiments 122

6.1 Definitions 122
6.2 The legal requirement to consider replacement alternatives 124
6.3 Advantages of replacement alternatives 125
6.4 Ethical considerations in the use of replacement alternatives 126
6.4.1 General ethical considerations 127
6.4.2 Ethical aspects of specific types of replacement alternative 128
6.5 Strategic use of replacement alternatives 134

7. The assessment and 'weighing' of costs and benefits 138

Introduction 138
A scheme for the assessment of potential and likely benefit in research involving animal subjects 141
A scheme for the assessment of likely cost to animals in research involving animal subjects 144

Case study 1: Autotransplantation of isolated islets of Langerhans in the Cynomolgus monkey 148
Case study 2: Evaluation of a new therapy for acute pancreatitis 153

Case study 3: Investigation of the development and nature of immune suppression accompanying malnutrition 158
Case study 4: Investigation of the route of water uptake in a Cephalopod 161
Case study 5: Localization of plant chemicals 166
Case study 6: A school biology practical exercise 173
Case study 7: Development of a pesticide (hypothetical example) 176

8. Ethical considerations in the use of animals in toxicology and toxicity testing 183

8.1 Introductory remarks 183
8.2 What is tested and why 184
8.2.1 Exposure to chemicals 184
8.2.2 Reasons for testing chemicals 184
8.2.3 Benefits of synthetic chemicals 186
8.3 The use of animals in toxicity testing 187
8.3.1 Regulations 187
8.3.2 Toxicity test procedures 188
8.3.3 Harm caused to animals in toxicity testing 194
8.4 Ethical analysis 198
8.4.1 Introduction 198
8.4.2 Necessity of chemical products 199
8.4.3 Reduction and refinement of practice 201
8.4.4 Use of information generated by the tests 205
8.4.5 Effects of regulations on testing practices 206
8.5 Alternative strategies: replacement of animals in tests 211
8.6 Conclusions 214

9. Ethical considerations in the use of animals in education and training 220

9.1 Use of animals in primary and secondary schools 220
9.1.1 Possible effects of using animals in schools 220
9.1.2 Permitted uses of animals in schools 221
9.1.3 Animals in primary schools 223
9.1.4 Animals in secondary schools 223
9.2 Use of animals in tertiary education 229
9.2.1 Permitted uses of animals in tertiary education 229
9.2.2 Scale of animal use in tertiary education 230
9.2.3 Variety of animal use in tertiary education 231

9.2.4 Alternatives to the use of animals 234
9.2.5 Manner in which animals are used 236
9.3 Use of animals in research for postgraduate degrees 238
9.4 Training for licensees 239
9.5 Conclusions 241

10. From theory to practice: control of biomedical research involving animal subjects 245

Part 1: Legal control of research involving animal subjects 245
10.1 Need for legislative control 245
10.2 Types and extent of legal controls: international comparisons 247
10.2.1 An historical perspective: development of legislation in the UK 247
10.2.2 Current UK legislation: Animals (Scientific Procedures) Act 1986 254
10.2.3 International comparisons 256
10.2.4 West European legislative controls 261
10.2.5 USA: Problems in implementing effective legislative control 264
10.3 Voluntary self regulation 268
10.4 Research review committees 271
10.4.1 Types of research review committee 271
10.4.2 Possible roles for research review committees 273
10.4.3 Some conclusions regarding research review committees 274
10.5 An overview 275
10.5.1 Ethics and the law 275
10.5.2 Law, education, and society 277
10.5.3 Public reassurance 278
10.5.4 When things go wrong 283
Part 2: Other controls on research involving animal subjects 284
10.6 Peer review within institutions 284
10.7 Review by grant-giving bodies 285
10.8 Role of learned societies 286
10.9 Editorial policies 287
Appendix 1 292
Appendix 2 293

11. The philosophical and moral debate on the use of animals in biomedical research 296

11.1 Philosophical arguments and moral concerns: an introduction 296
11.1.1 Philosophers on animals 298
11.1.2 Moral concerns and conflicts 306
11.2 The Working Party's conclusions: a philosophical analysis 310
11.2.1 The Working Party's general approach 310
11.2.2 The ethics of use and moral consensus 313
11.2.3 Conclusion 326

12. Summary and conclusions 329

12.1 General conclusion 329
12.2 Benefits of biomedical research involving animal subjects 329
12.3 Costs imposed on animals in biomedical research 330
12.3.1 Species of animal used 331
12.3.2 Animal supply 332
12.3.3 Animal housing and husbandry 332
12.3.4 Scientific procedures performed on the animals 333
12.3.5 Number of animals used 333
12.4 Replacement alternatives to animal experiments 333
12.5 'Weighing' costs and benefits 334
12.5.1 Use of animals in toxicity testing 335
12.5.2 Use of animals in education and training 338
12.6 Control of laboratory animal use 340
12.6.1 Statutory controls 340
12.6.2 Non-statutory controls 342
12.7 Philosophical arguments and moral concerns 344

Index 347

Members of the Working Party

G. R. Dunstan CBE (Chairman)
Professor Emeritus of Moral and Social Theology in the University of London.

W. Norman Aldridge OBE
Professor in Toxicology, Robens Institute of Health and Safety, University of Surrey.

Michael Balls
Chairman of the Trustees of FRAME, and Reader in Medical Cell Biology, University of Nottingham Medical School.

Patrick Bateson FRS
Professor of Ethology, University of Cambridge, and Provost of King's College, Cambridge.

G. W. Bisset
Professor Emeritus of Pharmacology, University of London; now working at the National Institute for Medical Research, Mill Hill.

Peter Byrne
Lecturer in the Philosophy of Religion, King's College, London.

Brian W. Cromie
Formerly chairman of Hoechst Pharmaceuticals and of Hoechst Animal Health; previously member of the Medicines Commission.

Gerald Dworkin
Herchell Smith Professor of Intellectual Property Law, Queen Mary Westfield College, University of London.

Roger Ewbank OBE
Director, Universities Federation for Animal Welfare.

R. G. Frey
Professor of Philosophy at Bowling Green State University, Ohio, USA; Senior Research Fellow, Social Policy Center, Bowling Green State University.

F. A. Harrison
Animal Health and Welfare Officer, Agricultural and Food Research Council Institute of Animal Physiology and Genetics, Cambridge.

Clive Hollands
Secretary, St Andrew Animal Fund, and Consultant to Advocates for Animals.

E. Stewart Johnson
Medical Director and Senior Vice-President, Clinical Research and Development SmithKline Beecham Pharmaceuticals. Reader Emeritus in Pharmacology, University of London.

David B. Morton
Professor of Biomedical Science and Biomedical Ethics; Director of Biomedical Services, University of Birmingham.

Iain F. H. Purchase
Director, Central Toxicology Laboratory Imperial Chemical Industries plc.

Derek Tavernor
Home Office Inspector.

Research staff

Kenneth M. Boyd
Director of Research, Institute of Medical Ethics; Honorary Fellow, Faculty of Medicine, University of Edinburgh.

Jane A. Smith
Lecturer in Biomedical Science and Biomedical Ethics, University of Birmingham.

1
Introduction

This book is the result of a three-year study undertaken by a multidisciplinary Working Party convened by the Institute of Medical Ethics. The book has been written primarily for those who are uncertain about the ethics of using animals in research because they feel less than fully and impartially informed about the facts and arguments which they should take into account when making up their own minds on the subject. Thus, although the Working Party's conclusions, in general and on specific topics, are summarized in the final chapter, these cannot be regarded as more than interim conclusions. In themselves, they are of less significance than whether the discussions which led to them, as reflected in other chapters, assist the reader in making his or her own informed moral judgements on the issues; and thus contribute to forming the moral judgements of society.

With such readers ultimately in mind, the Institute of Medical Ethics determined that the first aim of its study of this subject should be to establish the relevant facts about animal experiments. To achieve this most efficiently and comprehensively, it included among those invited to become members of its Working Party: basic biological and ethological scientists, toxicologists, physicians, and veterinary surgeons with particular experience of research using animals, a Home Office inspector, an expert in alternatives to the use of animals in research, and officers of animal welfare organizations. The Working Party, which was chaired by a moral theologian, also included philosophers and a lawyer, and its research staff comprised a biologist and an ethicist.

As an independent organization with educational goals, and existing to promote the impartial study of biomedical ethics, the Institute was at pains to ensure that the membership of the Working Party reflected as broad a diversity of moral views as was compatible with reasoned discussion of the Working Party's subject of study. This inevitably meant that some moral points of view were not represented. But the Institute was confident that these are represented elsewhere both adequately and eloquently, and that readers of this book's exploration of the middle ground will have no difficulty in finding numerous publications reflecting the views not represented here.

The Institute's intentions with respect to the membership of the Working Party were justified by the openness and productivity of its deliberations. Because the members of the Working Party were all well informed about their subject of study, those with divergent moral views on any aspect of it were able to respect one another's views, even when they were not able to agree with them; and where ethical evaluation turned on disputed interpretations of what was in fact the case, discussion was pursued vigorously until some agreed conclusion, or clear statement of what was in dispute, was reached. When gathering the information on which successive drafts of the report's chapters were based, the research staff occasionally found that those responding to enquiries were initially cautious, until they had been fully informed of the nature of the study. In the case of some research establishments, this was presumably attributable to concern about the violence used by some extreme animal liberation organizations. But apart from reticence attributable to this, the Working Party and its research staff encountered no evidence of any wish, among those using animals, to conceal any aspect of their work from those seeking to make a reasoned moral critique of it.

1.1 ESTABLISHING THE FACTS

The Working Party's study of the relevant facts about animal experiments began by clarifying the purposes of biomedical research and the procedures involved in it. Experiments on animals are described in literature dating from as early as 500 BC, and techniques for dissecting living animals were greatly improved by the ancient physician Galen of Pergamon (c.130–201 AD). While there appears to have been little animal experimentation during the Middle Ages, Renaissance anatomists and physiologists of the sixteenth century both repeated and extended Galen's original experiments and devised new ones. This work led to important new discoveries, including that of 'lacteals' in the gut of a living dog (1627), which led in turn to an understanding of the lymphatic absorption of food, and the discovery of the circulation of the blood (published by William Harvey in 1628). In the sixteenth and early seventeenth centuries, few experimenters saw any religious or moral objections to the dissection of living animals. But some, including Robert Boyle in the seventeenth century, began to regard the practice as requiring some moral defence, on the grounds of necessity; and by the nineteenth century, public criticism of work such as that of the French physiologist Claude Bernard and his pupil Magendie, led not only to legislative control, but also, in the present century, to study of the care and treatment of laboratory animals, and also of alternatives to using them.

Some further information about historical aspects of animal experimentation is given later, in Chapter 11, when philosophical arguments and moral concerns are discussed, and also in Chapter 3, in connection with the benefits of biomedical research. But in order to explain why animals are used today, Chapter 2 provides some background information, about the aims and methods of biomedical research, about the constraints and controls which safeguard its standards when using animals, and about the scale of animal use involved.

1.2 MORAL ISSUES: BENEFITS

The first aim of the Working Party was to establish the relevant facts about animal experiments. The second was to clarify the moral issues involved, and the third, to seek consensus (or a clear statement of conflicting views) on moral principles relevant to the area. Against the background of the discussion in Chapter 2 of the purposes and procedures of biomedical research, Chapter 3 identifies points of moral claim arising from the empirical facts. It points out, that is, where impressions of 'ought' and 'ought not', or 'should' and 'should not' emerge, and begins the process of moral reasoning about conflicting claims and how to work towards a resolution of these.

Chapter 3 does this specifically by examining moral claims related to the benefits, to humans and animals, which have accrued from the use of animals in research; and by asking what benefits might serve sufficiently serious purposes to be weighed against the cost to the animals used. The discussion involves an examination of the complex relationship between fundamental research and practical benefits. Partly because of this complexity, it argues, judgements about the likely benefits of particular research projects are best made case by case, in the light of agreed criteria developed through a dialogue between the scientific community and informed public opinion. A moral justification of this is offered, at this point, in terms of interim necessity, and its practical implementation looks forward to the *Scheme for the assessment of potential and likely benefit of research involving animal subjects*, set out in Chapter 7.

1.3 ANIMAL AWARENESS AND SENSITIVITY

The relevant facts to be established about animal experiments include not only those which are about experiments, but also those about animals—in particular, about their awareness of pain and psychological stress and their abilities to anticipate or to remember. While there is no certain way of assessing the thoughts and feelings of other species, Chapter 4 develops an approach based on extrapolation from human experience. This, the Working Party believes, can lead to plausible inferences about animals' capacities for experiencing adverse states, such as pain, distress, and anxiety. The evidence reviewed in this chapter relates specifically to brain development, cognitive abilities, awareness of 'self', pain, stress, and anxiety, in different animal species, both vertebrate and invertebrate. Its general conclusion, that experiments can impose costs in terms of suffering on many animals, leads to the moral claim that, even if it cannot be proved that some species are sentient, it is more appropriate morally to treat them as if they were.

The Working Party considered that any moral justification for imposing such costs must, at the very least, attempt to assess both their relative severity and how to reduce this as much as possible. To this end, Chapter 5 reviews empirical evidence about the recognition of adverse effects in animals, and identifies features of experimental contexts which are likely to contribute to harm or suffering. These features relate to both the numbers and species used, to animal supply, housing, and husbandry, and to the particular scientific procedures used. Chapter 5 also describes strategies for reducing costs in terms of the 'refinement' and 'reduction' of animal use—two of the 'Three Rs' of the concept of alternatives. The third R, that of 'replacement' alternatives, is discussed in Chapter 6, with particular reference to legal requirements, practical advantages, ethical considerations, and differing views on their potential contribution in the long term.

1.4 'WEIGHING' COSTS AND BENEFITS

Having explored alternatives, however, the Working Party acknowledged that, at present, the use of animals in some, if not all, areas of biomedical research inevitably imposed some costs, in terms of suffering, on the animals involved. But some use of animals in research, it had agreed earlier, could be justified in terms of interim necessity. How could it be determined whether or not this justification applied in the case of specific research proposals?

The Working Party's response to this takes the form of (a) two assessment schemes, set out in Chapter 7, designed to help make this judgement in particular cases; and (b) the Working Party's view on how the relatively different dimensions listed in these schemes for assessing benefit and cost may be put together to arrive at an overall assessment. While many empirical questions have to be answered before arriving at this overall assessment, it also inevitably involves making contestable moral, or value, judgements. The Working Party acknowledges this. What made this response persuasive to its own members was the moral adequacy, not so much of these judgements themselves, as of the procedures to be used in order to arrive at a final assessment. The Working Party's hope is that an approach of this kind, if developed through open dialogue between the scientific community and informed public opinion, will attract a wider consensus.

To illustrate how this dialogue might be developed, Chapter 7 also includes a number of case studies which show how the principles set out in the Working Party's assessment schemes might be applied. Chapters 8 and 9 then discuss two further areas, neither strictly 'research', to which the lines of moral reasoning developed by the Working Party were applied. The first of these is toxicity testing, which, unlike many other forms of research, requires adverse effects to be produced in animals. Chapter 8 explores and recommends a variety of practical ways in which the particularly acute moral conflicts arising from this requirement might be diminished. The second area is that of the use of animals in education and training. This is reviewed in Chapter 9, which again includes a number of practical recommendations.

It was not the Working Party's purpose to examine every scientific or experimental procedure which engages public attention from time to time, or receives inflammatory coverage in some newspapers. It would have been both unscientific and unethical to pronounce summarily on subjects like research related to bacteriological or chemical warfare or brain research without a detailed study of the empirical facts of each matter. Such studies would have been disproportionate to the Report. Rather, the intention was to establish methods of moral analysis and reasoning which can be applied to particular cases by those called upon and competent to do so.

1.5 CONTROL OF LABORATORY ANIMAL USE

When the Institute of Medical Ethics was convening its Working Party, the Animals (Scientific Procedures) Act 1986 had just become part of the law of the United Kingdom, and the Institute was asked whether further discussion of the subject really was necessary. The Institute, and subsequently the Working Party itself, believed that it was. While statutory controls are important, United Kingdom legislation is not the only model for them; and because biomedical research is an international enterprise, one effect of controls designed to protect animal subjects may be to drive the work to another country, where animals have less protection. The Working Party thus had to examine this international dimension; and Chapter 10 not only describes how legislation developed in the United Kingdom, and the provisions of its present law, but also includes a comparative account of relevant legislation in other countries, paying particular attention to voluntary self regulation and research review committees. Statutory controls, moreover, form only one aspect of safeguarding standards in biomedical research involving animal subjects. Chapter 10 includes, in addition, an account of various non-statutory forms of control and suggests how the Working Party's own assessment schemes may contribute usefully to the process of reviewing research projects using animals. The Working Party believes that the attitudes, both of the scientific community and of the public, are no less important than the laws, in safeguarding standards and protecting animal subjects. With this in mind, it wishes again to emphasize the need for a continuing open dialogue between the scientific community and informed public opinion.

1.6 PHILOSOPHICAL ARGUMENTS AND MORAL CONCERNS

An increasing part in this dialogue is now played by philosophical arguments. Chapter 11 discusses these in two distinct ways. The first part of the chapter reviews the contribution of philosophy both to the moral debate in general and to the development of thinking about the use of animals in biomedical research, taking particular note of the writings of some recent philosophers. It then relates this philosophical debate to some of the reasons why people may be morally concerned about the use of animals; and it concludes with a brief account of an argument, related to the notion of 'tragic conflict', which may help to illustrate some aspects of the Working Party's deliberations. The second part of the chapter then sets out a more detailed analysis and defence of the Working Party's general approach and conclusions. This pays particular attention to questions of justice and common morality, examining how a different research ethics for human and animals can be justified in terms of a morally relevant difference between them. The second part of Chapter 11 is especially concerned with the claims that the contrast between human and animal research ethics rests on no more than a species prejudice (or 'speciesism'), and to find rational arguments which support the Working Party's conclusions. The Working Party's conclusions are then summarized in the final chapter.

2

The researchers, their aims and methods

Discussion of the ethics of using animals in biomedical research should be based on adequate understanding of the purposes of the research and the procedures involved in it. In this chapter, we aim to provide some of the background information needed to embark on such discussion. The chapter includes consideration of the variety of research activities which might be classed as 'biomedical'; a brief description of the various methods used in such research (especially the involvement of animal procedures); an outline of the constraints and controls which aim to safeguard standards in research involving animal subjects; and an examination of the scale of animal use in these activities.

2.1 AIMS OF BIOMEDICAL RESEARCH

The heading 'biomedical research' covers a wide range of different types of research activity. In the most general terms, it can be said to be the study of living things and their interactions, usually for one of three main purposes:

(1) improving medical, veterinary or biological understanding, so as to enhance or maintain the well-being of humans, animals or the environment;
(2) testing the safety of chemicals or other products, so as to safeguard the welfare of humans, animals and the environment; and
(3) improving fundamental biological knowledge.

The first two categories may be said to be examples of *applied* research, being dedicated to the solution of practical problems, whilst the last category is *fundamental* (pure or basic) research. As Francis Bacon described, there are *experimenta fructifera*, or 'experiments yielding fruit', and *experimenta lucifera*, 'experiments shedding light' (see Paton 1984).

Investigators carrying out fundamental research may be aiming solely to increase knowledge, with no practical application in mind, or may have in view a practical problem, the solution of which would require further fundamental biological knowledge (Home Office 1988). The zoologist studying the ecology of a species of bat; the biochemist engaged in identifying receptor sites on the surface of mammalian cells; the pharmacologist investigating the distribution of chemical transmitters in the brain; and the geneticist studying the genetic control of limb development in insects, may all be engaged in fundamental biomedical research. It is difficult to classify fundamental research into particular categories, since such a wide variety of research activities comes under this heading. However, it is possible to recognize a number of distinct categories of applied biomedical research, divided according to the nature of the problem addressed. These categories are shown in Table 2.1. A further category of applied work, related to biomedical research, is education and training. This is considered further in Chapter 9.

In practice, fundamental and applied research go hand in hand. Fundamental research often leads to applied research whilst, conversely, much fundamental knowledge has been derived from applied research. As Paton (1984) has observed, 'research workers are constantly mixing these approaches or switching from one to another'. In pharmacology, for example, fundamental research on the definition of receptor sites for drugs (that is, the recognition sites in the cell with which the drugs interact to produce their effects) and of neurotransmitters (that is, naturally occurring substances which are released from nerve endings and transmit the nerve impulses to receptors on effector cells) has provided information which has led to the development of effective drug therapy for several debilitating disorders, such as peptic ulcers, and neurological or mental diseases including depression and schizophrenia. Indeed, there is an increasing tendency for pharmaceutical and other companies to carry out, or support, fundamental research. In addition, many clinical procedures depend on fundamental research. The artificial kidney, for example, would not be possible without knowledge of the physiology of renal function and of water and electrolyte balance. The problem of rejection of organ transplants arises because basic immunology lags behind advances in surgery.

2.2 METHODS IN BIOMEDICAL RESEARCH

2.2.1 Clinical and animal research

Sometimes, biomedical research involves the study of animals for their own sake (as in the zoological study of the ecology of bats), or

Table 2.1 Categories of applied biomedical research

Selection of new medical and veterinary pharmaceuticals, by a number of approaches.

Serendipity, for example, the chance observation that the excretion of sodium in the urine was increased in patients treated for typhoid with sulphonamides led to the development of new diuretics which have proved to be an outstanding success in the treatment of heart failure and hypertension.

An empirical approach, in which new synthetic products are submitted to a systematic series of screening tests in the hope of revealing useful pharmacological properties;

Computer design, on the basis of the structure–activity relations of molecules.

Imitating the structure of known endogenous substances such as hormones or compounds already marketed ('me-too' or congener drugs).

Modification of naturally occurring plant or animal products; for example, the development of semisynthetic derivatives of morphine, designed to retain the analgesic properties of naturally occurring morphine, but to remove the chemical's addictive properties.

Development of medical and veterinary pharmaceuticals, including tests for efficacy and mode of pharmacological action.

Toxicity testing (considered in detail in Chapter 8), including:

Mandatory or elective tests of new or existing drugs intended for clinical use (for example, in the UK tests are carried out to meet the requirements of the British Pharmacopoeia, and the Medicines Act 1968 and its subsequent statutory instruments).

Mandatory or elective safety tests of other chemical products, such as environmental pollutants, pesticides, substances used in industry, household substances, food additives, cosmetics and toiletries.

Testing batches of vaccines for undue toxicity.

Development and testing of medical and surgical materials, such as sutures, dressings, orthopaedic prostheses for joint replacement, and plastics, such as Dracon, for artificial replacement of parts of the oesophagus or blood vessels.

Development or improvement of surgical procedures, such as organ transplants.

Study of experimental disease and pathology, such as the induction of diabetes mellitus in dogs by the drug Streptozotocin to study the development and prevention of cardiovascular complications; the study of leprosy in the armadillo; the investigation of ways of improving healing in experimental burns; cancer studies, involving the induction and maintenance of tumours in animals; the study of the epidemiology of experimental nematode worm infections in sheep.

Development and production of antisera and vaccines for use in man and other animals.

Development of medical and veterinary diagnostic techniques.

Other studies dedicated to the solution of particular medical, veterinary or environmental problems.

---

for improvement of understanding in the field of veterinary science (as in the study of nematode infections in sheep). Most often, animals are used as 'models' for human subjects, in order to address fundamental or applied questions in medical science.

This latter research can also involve the use of volunteer human subjects and, since it is intended that the results should be applied towards the solution of medical problems in humans, humans might be regarded as the ideal subjects. The use of humans, however, is limited in two main ways. First, ethical and legal constraints limit the amount of harm, or risk of harm, which may be inflicted on humans in research. Subjects must be fully informed about the risks involved, and must have consented, voluntarily, to the proposed procedures. Second, clinical studies may be difficult to control (in the scientific sense) since, in addition to the experimental treatments, a large number of variables, including the age, life-style, past experience, and genetic constitution of the human subjects, may be involved. In contrast, experiments involving animals, although sometimes purely observational in nature, may also be 'invasive'; causing more harm, or involving a greater risk of harm, than would be acceptable in the case of human subjects. Studies using animals are also easier to control. It is possible, for example, to control all of the human variables mentioned above, by breeding colonies of genetically identical animals, under controlled conditions.

Scientific procedures involving animal subjects may be carried out:

(1) wholly under anaesthesia, the animals being killed under the anaesthetic at the end of the experiment;
(2) without anaesthesia, in cases where anaesthesia would frustrate the object of the experiment and/or where minor procedures are involved. Such experiments may involve, for example, taking blood samples, administering drugs or antigens or changing the animal's diet; or
(3) with initial anaesthesia followed by recovery and continued observation. Examples are the investigation of the function of endocrine organs by their extirpation; the induction of neoplasia by implantation of tumour cells; study of organ rejection after transplantation; cannulation of blood vessels under anaesthesia to allow subsequent blood sampling and recording of blood pressure in the conscious animal; study of recovery from experimental trauma.

Must of the rest of this book is devoted to discussion of the ethical issues raised by the use of animals in these ways, and whether such uses can be justified.

2.2.2 *In vivo* and *in vitro* research

Use of whole animals, that is *in vivo* work, is not the only way around the problems of invasiveness and controllability of experiments in biomedical research. Some experiments can be carried out on isolated parts (organs, tissues, cells or subcellular constituents) of animals or humans (see also Chapter 6). These are *in vitro* studies. *In vitro* studies, by definition, are more controllable than either clinical or animal studies and for this reason, in mechanistic studies at least, researchers often attempt to move to *in vitro* work as soon as possible. In practice, *in vivo* and *in vitro* methods are interdependent and complementary.

In vitro studies, along with physical, chemical and mathematical techniques, may all be considered as possible replacement alternatives for animals in biomedical research. The possibilities for employing such replacements, as well as the ethical problems raised by their use, are considered in Chapter 6. At present, however, it is true to say that many areas of biomedical research rely on whole animal studies. There may be two main reasons for this. First, living organisms are very complex. The different tissues and organs participate in an integrated physiology, so that it is difficult to address many biomedical questions using isolated cells, tissues or organs. It is, for example, impossible to study some diseases in *in vitro* systems, since the disease state may only be produced in living animals. Furthermore, experiments on isolated parts of animals may sometimes produce quite different results from similar experiments performed in whole organisms. For example, noradrenalin causes the isolated heart to beat at a much faster rate, but compensatory reflexes *in vivo* cause it to decrease heart rate.

Second, it can be argued that the use of whole animals enables researchers to observe the unexpected. *In vivo* experiments are used, for example, in pharmacokinetic studies (studies of the way in which drugs are distributed or broken down within the body and eventually eliminated from the body). In whole animals, drugs may be broken down into active or toxic compounds, which it is important to recognize, and they may occasionally interact in unpredictable and sometimes harmful ways within the body. Unfortunately, it has not yet been possible to develop *in vitro* experiments which are capable of mimicking the way in which the whole animal deals with drugs, so as to reveal such toxic breakdown products or interactions.

It should also be recognized, however, that, like *in vitro* studies, *in vivo* methods have their limitations. For example, in both cases, difficulties may be encountered in extrapolating from the 'model' to the real-life situation.

2.3 CONSTRAINTS ON POTENTIAL BIOMEDICAL RESEARCHERS

Workers undertaking biomedical research are mainly, but by no means exclusively, graduates in medicine, dentistry, veterinary science, basic medical sciences such as physiology, pharmacology, toxicology, biochemistry and anatomy, or in other biological sciences. Such researchers may have both graduate and non-graduate technical staff working with them. Some researchers may be registered for 'higher degrees', such as Master of Science (MSc), Doctor of Philosophy (PhD, DPhil), Doctor of Medicine (MD) or Master of Surgery (MS). In addition, undergraduate students may work on research projects as part of the requirement of their degrees, for example, medical students taking an 'intercalated' BSc degree between their preclinical and clinical studies.

In the UK, work involving the use of 'regulated procedures' on 'protected animals' is controlled by the Animals (Scientific Procedures) Act 1986. Living vertebrate animals (other than humans) are protected under the Act. The animals become protected when, in the case of mammals, birds and reptiles, they are halfway through their gestation or incubation period, and, in the case of fishes and amphibians, when they are capable of independent feeding. Every 'experimental or other scientific procedure applied to a protected animal which may have the effect of causing that animal pain, suffering, distress or lasting harm', is regulated under the Act. The use of animals in biomedical research is controlled by legislation in several other countries, and some of these laws are considered in Chapter 10, where the UK Act is also examined in more detail.

In the UK, researchers wishing to use animals in their work must first gain acceptance at an establishment 'designated', under section 6 of the Animals (Scientific Procedures) Act, as a place where regulated procedures may be performed. Such designated places may be:

(1) commercial concerns, including
   (a) laboratories in the pharmaceutical or chemical industry, for the development and testing of pharmaceuticals, or the safety testing of other chemical products (such as pesticides, industrial products, consumer products, and cosmetics);

- (b) industrial laboratories undertaking fundamental research; and
- (c) commercial laboratories undertaking contract work;

(2) science departments in universities and polytechnics and basic science departments in medical, dental and veterinary schools;

(3) government departments, such as the UK Ministry of Agriculture, Fisheries and Food, and Ministry of Defence Establishments;

(4) research institutes (including public bodies and non-profit-making organizations), such as the UK Medical and Agricultural Research Councils' (MRC and ARC) Institutes, the Royal College of Surgeons, the British Museum, the Flour Milling and Baking Research Association, the Imperial Cancer Research Fund and the British Heart Foundation laboratories;

(5) hospitals, in departments of pathology (including haematology, virology, and clinical pathology), clinical biochemistry and medical and surgical research units;

(6) public health laboratories, such as the UK Public Health Service Laboratory at Porton Down.

Before a Certificate of Designation is issued, Home Office Inspectors, appointed under the terms of the Act,* must be satisfied that the establishment's facilities are satisfactory for the maintenance of the animals. In particular, the design of the animal accommodation and standards for the husbandry and welfare of the animals must meet with the guidelines contained in the Home Office *Code of practice for the housing and care of animals used in scientific procedures* (1989). Applicants for certificates of designation are also required to nominate one or more persons responsible for the day-to-day care of the animals and one or more veterinary surgeons to provide advice on animal health and welfare. These people have a statutory duty to ensure that the animals in the establishment are properly cared for at all times. Their responsibilities are spelt out in the Home Office *Guidance on the operation of the Animals (Scientific Procedures) Act 1986* (1990*a*).

Before they can start work, researchers must obtain licences under the Act. Each researcher must obtain a 'personal' licence which, in effect, is a certificate of competence, qualifying him to perform certain, specified, regulated procedures. If necessary, a condition will be attached to the licence, requiring the researcher

* The Home Office is responsible for the administration of the Act in England, Scotland and Wales. In Northern Ireland, inspectors are appointed by the Northern Ireland Department of Health and Social Services.

to work under supervision, until competence is gained. The personal licence, however, does not, by itself, authorize a researcher to carry out the procedures. These may be performed only as part of a specified programme of work, authorized under a 'project' licence.

The scope of project licences can vary a great deal. At one end of the scale, a project licence might cover the work of one personal licensee using a few animals of one species. At the other end, a large project licence, involving, say, the testing of a group of chemical products, might cover a number of personal licensees using large numbers of animals of several species (Home Office 1990*a*). In all cases, applicants for project licences are required to describe the background, purpose and scientific justification for the proposed project, and to give details of the procedures they propose to use. They must include an assessment of the severity of the procedures, in terms of the pain, suffering, distress or lasting harm likely to be imposed on the animals involved. Under section 5(4) of the Act, in deciding whether or not to grant a project licence, the Secretary of State (in practice, the Home Office inspectors acting on his behalf) must 'weigh the likely adverse effects on the animals concerned against the benefit likely to accrue' from the project. Under section 5(6) of the Act, special justification is required for the use of primates, dogs, cats, or equidae (horses, donkeys, and mules).

Besides this weighing of benefits and severity, several other considerations are important in deciding whether or not to grant the licence (Home Office 1990*a*). As outlined in the guidance on the operation of the Act, the Home Office Inspectors also consider:

(1) whether it might be possible to reduce the number of animals used, refine the procedures so as to reduce animal suffering, or avoid using protected animals altogether in the project;
(2) whether the proposed anaesthetic techniques or other methods of mitigating pain and distress are appropriate to the animals and procedures;
(3) whether the design of the project is suitable in relation to its stated aims;
(4) whether the facilities in the designated establishment are adequate for the proposed procedures; and
(5) whether the facilities for animal housing and care are adequate for the species involved.

On the first point, section 5(5) of the Act requires that a project licence cannot be granted unless 'the applicant has given adequate consideration to the feasibility of achieving the purpose' of the project 'by means not involving the use of protected animals'.

The severity of the procedures which may be performed on the animals are limited by the project licence conditions. Researchers must prevent or attempt to alleviate any pain or distress suffered by the animals, by the administration of appropriate analgesics or anaesthetics, and to terminate the experiment if the animals appear to be suffering undue pain or distress, and/or if the severity limit is breached.

The Home Office Inspectors, as well as advising the Secretary of State on applications for personal and project licences and for certificates of designation, have a duty to visit designated places and check that the conditions of the licences and certificates are being complied with. The Inspectors therefore play an important role in ensuring the welfare of animals involved in experiments. They get to know the licensees involved in the work and the technicians who are caring for the animals; and they are available to discuss and give advice on the procedures employed at both the research proposal stage and whilst the research is being carried out.

In addition to gaining acceptance at a designated establishment, and obtaining licences, researchers must also obtain funding for their work. In the UK, potential non-industrial researchers may seek funds from four main areas:

(1) from their Institute's budget by agreement with senior management and senior scientists;
(2) from government via the Research Councils, such as the Medical Research Council or the Science and Engineering Research Council;
(3) from industry via contract;
(4) from charitable organisations, amongst the largest of which are the Wellcome Trust, the Leverhulme Trust, the Nuffield Foundation, and the Wolfson Foundation.

There can be great competition for grants for biomedical research, so that in most cases, only the best science is funded. Whether this competition for funds might exert a form of 'control' over the use of animals in biomedical research is considered in Chapter 10.

An additional *post hoc* constraint on biomedical research involving animals may be imposed by learned societies to whom the results of the work are communicated and by the editors and referees of journals, who may reject papers if they fail to conform with defined ethical practices or are not of a sufficiently high scientific standard. The question of whether such editorial policies can indeed influence practice is also taken up in Chapter 10.

2.4 SCALE OF ANIMAL USE IN BIOMEDICAL RESEARCH

In Great Britain, researchers holding project licences under the Animals (Scientific Procedures) Act are required to send to the Home Office details of the procedures which have been performed under the authority of their licences. Using this statistical return, the Home Office publishes annual *Statistics of scientific procedures on living animals in Great Britain.** Under previous legislation, the Cruelty to Animals Act, 1876, researchers were also required to make annual returns, so there is now a detailed and long-standing database on the use of animals in experiments in Great Britain. The Netherlands seems to be the only other country in the world which collects and publishes such detailed statistics (Veterinary Public Health Inspectorate in the Netherlands 1978–87).

The Home Office statistics for Great Britain record that 3 315 125 scientific procedures were performed on animal subjects during 1989 (the most recent statistics at the time of writing, Home Office 1990*b*). It should be noted that the statistics record the number of scientific procedures carried out, not the number of animals used. Each procedure or series of procedures (for example, the giving of an anaesthetic, followed by surgery with recovery, then monitoring of the animal with blood sampling), carried out for a specified purpose on a single animal, is counted as one scientific procedure. Most frequently, only one such scientific procedure is performed on a given animal. If this were always the case, the number of scientific procedures would indicate the number of animals used. Sometimes, however, several scientific procedures may be recorded for one animal. This may occur in toxicity testing, for example, when animals are catheterised so that repeated samples can be taken from the blood or digestive system, the same animal being used to test the effects of different chemicals. A separate scientific procedure would be recorded for each chemical tested (since the purpose of the procedure would be different in each case), but only one animal would be involved.

In the Netherlands, 1 131 128 animals were used in experiments

* Statistics for Northern Ireland are collected by the Northern Ireland Department of Health and Social Services and published separately as *Statistics of scientific procedures in Northern Ireland*. In 1988 (latest available statistics), there were 14 designated scientific procedures establishments in Northern Ireland. In this year, a total of 18 316 procedures were performed by 238 licensees. Nine thousand seven hundred and ninety-eight of the procedures were carried out in universities, 741 in commercial concerns, and 7777 in Northern Ireland government departments and NHS hospitals (Department of Health and Social Services, Northern Ireland 1988).

(one animal = one experiment) in 1987 (last available statistics). Estimates of animal use in biomedical research elsewhere in the world are less reliable. Estimates of animal use in the USA, for example, vary from around 17 to 22 million (Office of Technology Assessment 1986), to as many as 70 million animals per year (Rowan 1984).

In Britain, the 3.3 million scientific procedures performed in 1989 were carried out under the authority of 3276 project licences (78 per cent of all project licences held in that year). 70 per cent (2288) of the project licensees reporting procedures were based in universities and polytechnics, 13 per cent (412) were in research institutes (that is, public bodies and non-profit-making organizations) and 12 per cent (392) in commercial organizations. Figure 2.1 shows the locations of licensees working on animals in 1989.

Figure 2.2 shows that the majority of the scientific procedures (61 per cent) were performed by the licensees working in commercial organizations, whilst 23 per cent of all procedures were carried out in universities and polytechnics, and 10 per cent in research institutes.

Figure 2.3 shows that by far the majority (84 per cent) of animals used in these procedures were rodents (mice, rats, guinea-pigs,

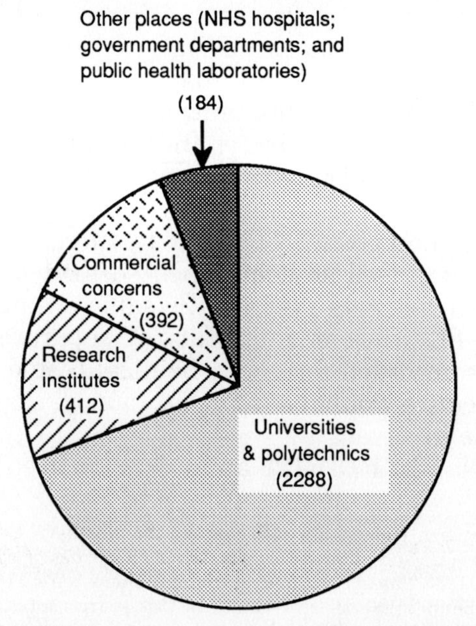

Fig. 2.1 Place of work of project licensees reporting scientific procedures on animals in Great Britain, in 1989 (number of licensees shown in brackets).

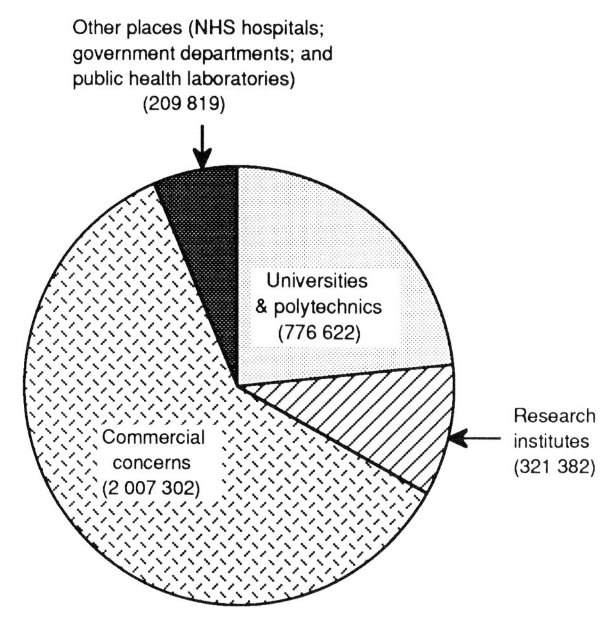

Fig. 2.2 Places where scientific procedures were performed on animals in Great Britain, in 1989 (number of procedures shown in brackets).

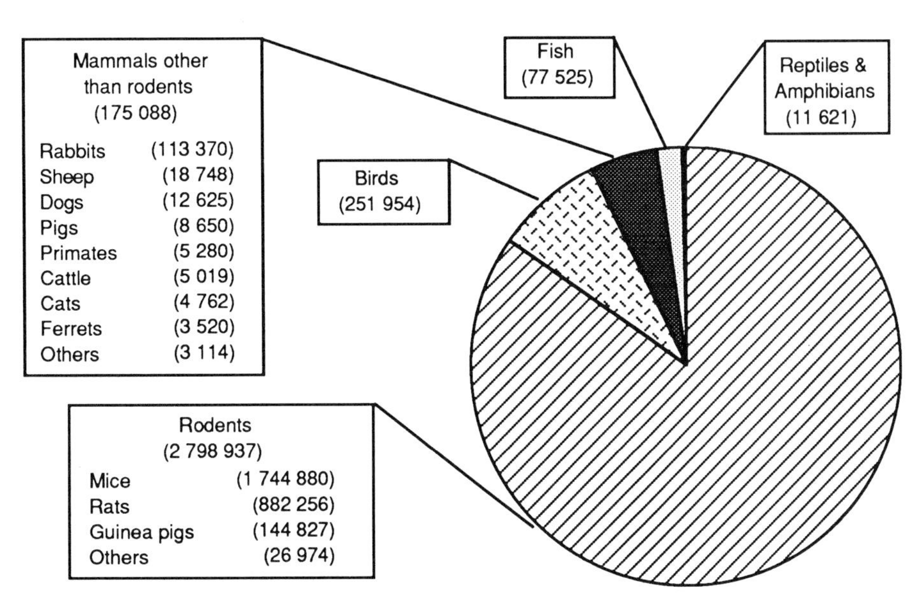

Fig. 2.3 Species of animal used in scientific procedures in Great Britain in 1989 (number of procedures shown in brackets).

hamsters, gerbils, and others). Fifty-three per cent of all procedures involved mice, 27 per cent rats, and 5 per cent other rodents. The remaining procedures were carried out on birds (8 per cent), fish (2 per cent), reptiles and amphibians (0.3 per cent), and on mammals other than rodents (5 per cent). Rabbits were the most commonly used mammal in this last category (accounting for 65 per cent of the non-rodent mammalian species used, and 3 per cent of the total species used). Sheep, dogs, pigs, cattle, non-human primates, cats, ferrets, horses, donkeys and mules, and other species, taken together, were used in less than 2 per cent of the total scientific procedures.

Worldwide, as might be expected, mice and rats are the major experimental subjects. In the USA, mice and rats account for between 80 and 90 per cent of all animals used in laboratories (see Office of Technology Assessment 1986). In the Netherlands mice and rats accounted for 81 per cent of the animals used in experiments in 1987.

The British statistics record that the majority of animals are used in studies classified as having an applied purpose. Figure 2.4 shows that, in 1989, 79 per cent of procedures were performed in the course of applied studies, with the remaining procedures being involved in research aimed at improving fundamental scientific

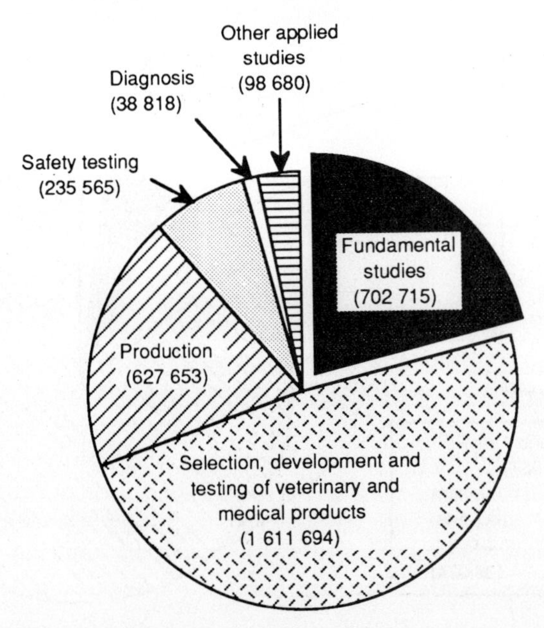

Fig. 2.4 Purpose of scientific procedures performed on animals in Great Britain in 1989 (number of procedures shown in brackets).

knowledge. Of the applied studies, the main category of purpose was the selection, development and testing of medical and veterinary products. In 1989, this broad category included 49 per cent of all procedures. Nineteen per cent of all procedures were for the production of biological products (serum, monoclonal antibodies and other blood products), and for the production and maintenance of tumours and of infectious diseases; uses which were not covered by the Cruelty to Animals Act 1876. The remaining applied purposes were the safety testing of chemical products other than medicinal products (7 per cent of all procedures), the diagnosis of disease, including the investigation of suspected food poisoning (1 per cent) and other studies (3 per cent), such as the development of surgical techniques, the development of diagnostic techniques in medicine, the use of animals in education and training, and ecological studies.

The British statistics do not record the severity banding for the procedures carried out. The statistics do, however, record the use of anaesthesia, dividing procedures into the three groups shown on p.11. In 1989, 20 per cent of all procedures were carried out wholly under anaesthesia, the animals being killed before they had recovered from the anaesthetic; 63 per cent of procedures were minor enough to be carried out without anaesthetic, or were procedures which caused more than minor pain and distress, but for which anaesthesia would have frustrated the object of the experiment; and 17 per cent of procedures involved anaesthesia followed by recovery of the animal.

Information about the likely severity of procedures performed on animals is contained, uniquely, in the Dutch statistics of animal experiments. Whether these data are also representative of the likely severity of procedures in the UK, or worldwide, is unknown, but they are the only available statistical indicators of the amount of pain and distress which might be caused to animals in biomedical research. In the Netherlands, before each experiment, researchers are required to estimate the discomfort likely to be experienced by the animals. Discomfort is defined so as to include 'impairment of the animal's health', or 'appreciable pain, injury or other grave distress caused to the animal' (Veterinary Public Health Inspectorate in the Netherlands 1989). The Dutch statistics record that 57 per cent of experiments registered in 1987 were assessed as being likely to involve minor discomfort for the animals, 21 per cent as likely to involve moderate discomfort, and the remaining 22 per cent of procedures as likely to involve severe discomfort. Table 2.2 gives examples of the procedures which might have been included in each category of discomfort (Veterinary

Table 2.2 Categories of discomfort

Minor (57 per cent)
For example:
- simple blood sampling
- rectal examination
- vaginal smear sampling
- force feeding of innocuous substances
- taking of X-rays in unanaesthetized animals
- killing without prior sedation
- terminal experiments under anaesthesia
- restraint in pens (animals can stand or lie down)
- immunization without adjuvants

Moderate (21 per cent)
For example:
- frequent blood sampling
- pyrogenicity testing
- insertion of indwelling cannulae/catheters
- use of casts (plaster of Paris)
- immobilization or restraint (e.g. in an inhalation chamber)
- skin transplantation
- caesarian section
- recovery from anaesthesia
- immunization with incomplete adjuvants

Severe (22 per cent)
For example:
- collection of ascitic fluid
- total bleeding without anaesthesia (excluding decapitation)
- production of genetic defects, e.g. muscular dystrophy, haemophilia
- prolonged deprivation of food, water, or sleep
- immobilization under muscle relaxants without sedation
- some experimental infections
- carcinogenicity research with tumour induction
- application of painful stimuli, induction of convulsions
- LD50 and LC50 tests
- immunization in footpad or with complete adjuvants

From *Animal Experimentation in the Netherlands, Statistics 1987*. Veterinary Public Health Inspectorate in the Netherlands 1989.

Public Health Inspectorate in the Netherlands 1989). This kind of ranking of procedures according to their severity is open to criticism, and the limitations of such a scheme of classification are considered in Chapter 5 (section 5.2.5).

Overall, in Britain, the number of experiments on animals reported under the Cruelty to Animals Act 1876 decreased in each

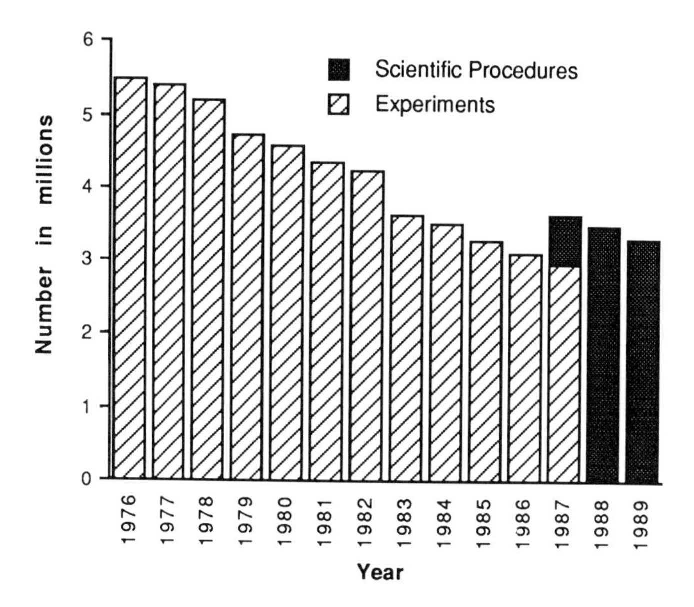

Fig. 2.5 Experiments (1976–87) and Scientific Procedures (1987–9) performed on protected animals, in Great Britain, 1976–89.

year after 1976, and this trend appears to have continued in returns under the 1986 Act (Fig. 2.5). In 1987, researchers were required to report both the procedures carried out under the 1986 Act, and the number of experiments which would have been recorded under the 1876 Act. Figure 2.5 shows that the number of scientific procedures was greater than the number of experiments in 1987. This is because, under the 1876 Act, 'experiments' referred only to those procedures whose outcome was unknown. The change in legislation, to cover all 'scientific procedures' involving animal subjects, brought in a considerable number of procedures (677 502 in 1987) whose outcome was 'known' and which would not have been regarded as experiments under the 1876 Act. These additional procedures include the use of animals in the production of antiserum, the passage of tumours, the maintenance of infectious diseases, and forensic work—none of which was covered by the 1876 Act. That the downward trend in animal use has indeed been continued is shown by the fact that the number of experiments recorded under the 1876 Act decreased from 1986 to 1987, and the number of scientific procedures recorded under the 1986 Act decreased from 1987 to 1988, and again from 1988 to 1989.

The reasons for this decline in animal use are not certain, but some speculations can be offered. In the first place, molecular biology has assumed greater prominence in biomedical research over the past ten years or so, and workers in fundamental research may have shifted from, say, physiological work using whole animals, to molecular work using *in vitro* systems, such as tissue culture. Such a change has also occurred in the pharmaceutical industry, where *in vitro* work and computer models have replaced the use of large numbers of animals in the initial screening of chemicals for pharmaceutical properties. A second factor may have been the increased costs of buying and maintaining animals. This may have had an impact on the scale of animal use in universities, polytechnics and research institutes in particular. In addition, the recent opening up of debate about animal experiments (particularly with the discussions leading up to the passing of the 1986 Act), the climate of public opinion, which now seems to be against some forms of animal use (such as the testing of cosmetics), and the work of some of the antivivisectionist groups, may have combined to make researchers think more carefully about their use of animals, so perhaps influencing the overall scale of use.

REFERENCES

Department of Health and Social Services, Northern Ireland (1988). *Statistics of scientific procedures on living animals, Northern Ireland 1988*. DHSS, Belfast.

Home Office (1988). *Statistics of scientific procedures on living animals, Great Britain 1987*. Command 515. HMSO, London.

Home Office (1989). *Code of practice for the housing and care of animals used in scientific procedures*. HMSO, London.

Home Office (1990*a*). *Guidance on the operation of the Animals (Scientific Procedures) Act, 1986*. HMSO, London.

Home Office (1990*b*). *Statistics of scientific procedures on living animals, Great Britain 1989*. Command 1152. HMSO, London.

Office of Technology Assessment, US Congress (1986). *Alternatives to animal use in research, testing and education*. US Government Printing Office, Washington DC.

Paton, W. (1984). *Man and mouse: animals in medical research*. Oxford University Press, Oxford.

Rowan, A. N. (1984). *Of mice, models, and men*. State University of New York Press, Albany.

Veterinary Public Health Inspectorate in the Netherlands (1989). *Animal experimentation in the Netherlands: statistics 1987* (English summary). PO Box 5406, 2800 HK Rijswijk, The Netherlands.

3

Benefits of biomedical research involving animal subjects

In the previous chapter, there was a tacit assumption that there may be good moral reasons for using animals in biomedical research. Whether there might, indeed, be such good moral reasons clearly depends on how the benefits of biomedical research using animals are assessed, and whether these benefits require the use of animals. If some use of animals is necessary to achieve benefits which it would be morally wrong to forego, the moral community, it may be argued, has a responsibility, in particular cases, to 'weigh' these benefits against the 'cost', in terms of animal suffering, imposed by the research.

The assumption that it is indeed possible to weigh benefits and costs is made in a central requirement of the United Kingdom's Animals (Scientific Procedures) Act 1986, namely that:

> in determining whether and on what terms to grant a project licence the Secretary of State shall weigh the likely adverse effects on the animals concerned against the benefit likely to accrue as a result of the programme to be specified in the licence (section 5.5).

A rationale for assessing and weighing benefits and costs is developed over the next few chapters.

Here, as a first step towards carrying out such a weighing, we ask what potential or likely benefits of biomedical research should be weighed against the likely cost to the animals involved, and how best these benefits might be assessed. The attempt, which follows, to answer this question begins by setting it in an historical context. It then examines two conflicting moral positions on the kind of benefits needed to justify, in principle, the use of animals, and tries to reconcile these, suggesting some practical implications. A further moral argument, about necessity, is then examined and its practical implications again drawn out. In conclusion, a scheme for the assessment of potential and likely benefits of research using animals is suggested.

3.1 HISTORICAL PERSPECTIVE

There can be no doubt that the use of animals in medical research in the past has proved worthwhile for human purposes, with consequent benefit to human and animal health. The use of animals in anatomical and physiological research is an ancient practice. As early as 500 BC

Alcmaeon of Croton was able to find the function of the optic nerve by cutting through this structure in the living animal and recording the ensuing blindess. In a similar way the author of the Hippocratic text *On the heart* (about 350 BC) cut the throat of a pig which was drinking coloured water, in order to study the act of swallowing. He also opened the chest of another living animal and described how the auricles and ventricles of the heart were beating alternately (Maehle and Tröhler 1987).

The importance of using living animals in such research (until about 1850 without the benefit of anaesthesia) is that this could provide information not available from experiments on human or animal bodies, owing to the changes which take place after death. The use of animals to test toxic substances (begun as early as the second century BC—see Maehle and Tröhler 1987), and in pharmacological research, has also provided information from which human and animal health have benefited. For much of history, indeed, the only alternative envisaged to the use of animals for these purposes would have been the use of human beings; and from time to time until the eighteenth century, it was advocated that criminals in particular should be the subjects of vivisection or toxicological tests (Maehle and Tröhler 1987). By the nineteenth century, however, even the use of dead human bodies aroused sufficient antipathy for the strict provisions of the 1832 Anatomy Act to be required; and the classic defence of experimental science by Claude Bernard and other scientists of the time was firmly based on the assumption that the use of animals was the only alternative to the now unthinkable use of human beings.

Numerous publications cite benefits which have accrued from the use of animals in biomedical research. Since 1927, the London-based Research Defence Society has published the texts of annual lectures, given in memory of the Society's founder, Stephen Paget. Many of these lectures have been devoted to consideration of the advantages of animal research. Recent examples include lectures by Sir John Vane, describing the benefits of using animals in drug research and testing (Vane 1985); and by Professor I. A. Silver, discussing the benefits which have accrued in veterinary medicine, through the use of animals in research (Silver 1987). Other relevant works include Sir William Paton's book *Man and mouse* (1984), and a report of the American Medical Association's Council on Scientific Affairs (1989). The latter publication provides a 'brief glossary', which, the authors argue, 'reveals only a small part of the crucial role animals have played in contributing to major advances in reducing morbidity and mortality for all of society'. This glossary covers medical advances through research on aging (including improved understanding of the pathology of Alzheimer's disease), AIDS, anaesthesia, autoimmune diseases, basic genetics, behaviour (including the development of neurosurgical procedures), diseases of and defects in the cardiovascular system, childhood diseases, cholera, convulsive disorders, diabetes, gastrointestinal tract surgery, hearing, haemophilia, hepatitis, infection, malaria, muscular dystrophy, nutrition, ophthalmology, organ transplantation, Parkinson's disease, treatment of pulmonary disease and injury, prevention of rabies, radiobiology, reproductive biology (including the development of the contraceptive pill), skeletal system (including orthopaedic surgery), treatment of spinal cord injuries, toxoplasmosis, trauma and shock, yellow fever and virology.

Whether the benefits alluded to above could have been achieved without the use of animals seems unlikely. This, however, is the kind of hypothetical historical question which no one is in a position to answer. It is often argued, of course, that some of the most notable improvements in the health of Western populations during the present century owe more to better nutrition, hygiene and public health measures than to biomedical research (see, for example, McKeown 1979; also Sharpe 1988). Health standards in Third World countries today similarly might be radically improved by relatively simple methods such as oral rehydration therapy and the encouragement of breast feeding, the product of epidemiological and clinical research rather than of research using animals. Such arguments, quite clearly, are important and should be taken into account when priorities for research are being determined. They arose, however, at a time when infectious diseases appeared to be diminishing and when it was both advisable and popular to issue cautionary warnings about the rhetoric of scientific medicine. With the advent of AIDS in the developed and especially the developing world, there may now be less need for such warnings, since the need for fundamental research is so evident; and while it may be appropriate to emphasize the behavioural aspects of HIV transmission, or for that matter behavioural aspects of the aetiology of heart disease and some cancers, there are formidable difficulties in persuading people to change their sexual, dietary and other customary forms of behaviour. Rather than seeing epidemiological and clinical research as a preferred alternative to research using animals, it thus seems more appropriate to see these as complementary.

Furthermore, it is difficult to see how basic scientific information could have been provided without research using animals. Without this, we would lack most of the knowledge required for any practice of human or veterinary medicine much beyond bonesetting, herbal remedies and faith healing. Human and animal life, moreover, might well be prey to a great variety of epidemics, poisons, and pains of childbirth and dying which can now be prevented or controlled. Without the use of animals in research, the surviving human population might be leaner and fitter (in evolutionary terms), but the cost to less fit individuals, or to those succumbing to serious accidents, would be the absence of most means available of prolonging their lives and alleviating their pains. Whether or not other means might have evolved had animals not been used is, of course, another hypothetical historical question which cannot now be answered. The fact simply is that these benefits were achieved in this way. As matters have developed, it is doubtful whether even the most radical champion of animal welfare could consistently abandon all that has been learned from research using animals, simply because of the suffering which this abandonment (if feasible) would entail for so many other animals, let alone human beings. In so far as fallible human knowledge can determine therefore, the moral community's commitment to the common and individual good appears to have been more adequately expressed by the use of animals than it would have been by not using them.

The undoubted benefits of animal use in the past, however, do not mean that the continued and unquestioning use of animals in biomedical research today is thereby also morally justified. The argument advanced powerfully by Sir William Paton's 'test of deletion' (Paton 1984) does indeed make it difficult, if not impossible, to draw a line across the page of history at any point in the last hundred years and to say that we willingly would have foregone the knowledge gained by animal use in research after that date. But the same argument presumably would hold had the relevant research been conducted secretly on involuntary human subjects: public exposure of such work doubtless would immediately call a halt to it, without necessarily entailing that all the benefits of such knowledge would be rejected. Acceptance of the undoubted benefits of medical progress in wartime does not necessarily entail moral approval of war; and it might even be argued that it is unethical to refuse to use the limited amount of reliable published research data deriving from unethical Nazi experiments on humans, if this information cannot now be discovered otherwise by ethical means, and particularly if it can be used (as some Nazi hypothermia research apparently can be—see Moe 1984) to save human lives. Whether or not such an argument is accepted however, it should be added, will also depend on the moral weight given to the feelings and consciences of people still alive whose friends and relatives were the subjects of the experiments.

3.2 CHANGING MORAL SENTIMENTS

In turning from the past to the present, two significant changes may be noted. First, there is the possibility of using alternatives to animals in research. This subject will be discussed in Chapter 6. For the present purpose, it may be sufficient to note two requirements for any procedure to be a real alternative to the use of animals. The first is that it must be no more morally unsatisfactory than the use of animals: the involuntary use of humans, that is, would not satisfy this requirement. The second requirement is that any alternative must be at least as satisfactory as the use of animals for the relevant scientific purposes.

The second significant change is that it is no longer normally considered morally justified in Britain and many other countries to use human beings in research without ethical committee approval and voluntary informed consent, nor to use higher animals in certain ways for research purposes. With reference to human beings, this shift in moral sentiment can be supported by powerful ethical arguments concerned with justice and respect for persons. The fact that moral sentiment (among scientists and the public) concurs with these arguments can be seen as a consequence of wider and deeper moral reflection on human action. The principle of informed consent is not totally without exceptions: a good example would be when the human subject of research is an unconscious patient who may benefit from an experimental therapy. It may also be argued that some information may be too distressing or too technical to burden patients with. The point, nevertheless, is that voluntary consent, based on adequate information, is now the normal expectation, and that exceptions have to be argued for.

The change in moral sentiment concerning the use of higher animals is less widespread and exceptions to it are more likely to be allowed. To take an example: it is not clear at what point in the progress of an epidemic such as AIDS (and with what probability of success in the development of a vaccine or an effective treatment) those who consider testing on chimpanzees to be unethical would be willing to make such an exception. On what moral grounds might such an exception be justified?

In the case of human beings, clearly, no involuntary testing of a vaccine could be morally defensible: the only justifiable course would be to call for volunteers. But human duties to chimpanzees are not sufficiently securely agreed to ensure that their use could not be justified if, say, the epidemic were sufficiently widespread and sufficient benefit was likely to accrue from the use of chimpanzees. Even in a case such as AIDS, however, the probability of success in the development of a vaccine might be insufficient to justify the serious adverse effects which the research might cause to individual chimpanzees, and the threat posed to this endangered species, by the capture of chimpanzees in the wild. It would be difficult to overcome these moral objections by revising accepted ethical standards to accommodate such a use of animals. The case, rather, would be analogous to killing in wartime, when it is still believed that it is immoral to kill people, but it is also concluded that this might be the lesser evil in these extreme circumstances.

A number of factors and considerations would be important in deciding whether chimpanzees might or might not be used in extreme circumstances, under the 'ethics of emergency'. These factors might include the extent and urgency of the threat of the epidemic to human populations; the extent and urgency of the threat to animal species of their members being used in research; and the likely severity of the adverse effects caused to humans by the epidemic, and to chimpanzees by the research procedures. The likelihood of success in developing a vaccine and the rapidity with which this could be achieved would be a further consideration which might help, exceptionally, to justify using chimpanzees. If, on the other hand, we were dealing with an epidemic involving the rapid transmission of a fatal virus by means which, unlike those in the case of HIV, were unavoidable by behavioural or public health means; and if the risk to humans of imminent death from the epidemic was likely to be as great as any adverse effects of their involvement in research, it might be possible, again exceptionally, to justify proceeding directly to research on human subjects.

At present, however, speculation of this nature remains hypothetical. The factors to be taken into consideration in such exceptional circumstances, and the relative weight to be given to them, can be determined only under the pressure of the imminent need to make a decision. In this context, it is essential to emphasize that the ethics of emergency is not the same as the ethics of normal practice. The normal must be established before the emergency is considered. Since emergency, by its very nature, is the exception, the normal obligation at present is to require the strongest possible reasons to justify all uses of higher primates in biomedical research. This again can be seen as a consequence of wider and deeper moral reflection on human action than in the past.

3.3 TWO MORAL POSITIONS ON BENEFITS

What are the benefits of biomedical research which might serve sufficiently serious purposes to be weighed against the cost to the animals used? Two moral positions in response to this question can be identified. The first of these, stated in its most extreme form, is that in order for the purposes to be sufficiently serious, it must be possible to predict some benefit accruing from the research to the prevention, diagnosis or treatment of disease, ill health or abnormality in humans or animals. An immediate objection to this position, however, is that it is not always possible to predict such a benefit with any likelihood, with the possible exception of certain forms of applied research. It is, after all, precisely the aim of the research to discover, for example in the development of a new drug, whether the compound offers the benefits which the scientist hopes for, without the harmful effects which he hopes to avoid.

It would seem necessary therefore to modify this first moral position to say no more than that a benefit accruing to the prevention, diagnosis or treatment of disease, ill health or abnormality in humans or animals must be *possible and intended.* This restatement, however, radically changes the first position. In its original form, justification turned on what (it was implied) could be measured by scientific criteria. For its modified form, scientific criteria are still necessary (to exclude the impossibility of benefit), but the focus of interest now also includes the criterion of intention; and the difficulty with intention is that a responsible adult is normally the first judge of his own intentions. It would be very difficult, on these grounds alone, to dispute the claim of any research worker that his work met the criteria. As modified therefore, this first moral position seems to require too little to prove that the purposes of research are sufficiently serious.

In another sense, however, it might be argued that the first moral position requires too much. The position interprets benefit, or doing good, solely in terms of those forms of good associated with therapeutic advances. But clearly, this is not the only form of good associated with scientific research. The advancement of scientific knowledge in itself, it is argued, is a good.

This argument is the basis of the second moral position, which is that the advancement of scientific knowledge is itself a benefit. In the most extreme form of this position, any such advance is a sufficiently serious purpose to justify the use (although not necessarily every use) of animals in research. Scientific research, it may be added, is by definition not undertaken without both the possibility and the intention of adding to the sum of human knowledge. (Science is knowledge: research is the systematic search for it.) The very fact that such work is done according to scientific criteria, according to this argument, excludes the possibility of its being worthless from a scientific point of view.

There are two immediate objections to this second position. One is that its blandness conceals great variations in the quality of review given to proposed research (between, say, work funded by the Medical Research Council, and a surgeon making some money available for a research assistant). Such a position might well make the public suspicious of peer review as a way whereby scientists collude to obfuscate the issue: the need for publicly defensible arguments in this area is emphasized further below. The second, related, objection is that once again it is very difficult, and perhaps impossible on the grounds stated alone, to dispute any research worker's claim that his research meets the criteria required, namely that its intended goal was to add to the sum of human knowledge, and that it was possible that the research would achieve this. On this basis therefore, there would seem no way of saying, at least in terms of the potential benefits, that the use of animals in fundamental research is ever unjustified.

Against this, however, it may be argued that internal criteria can be identified which make it possible to determine the relative importance of the additional good (in terms of knowledge) which a particular piece of scientific research is likely to produce. Within a particular scientific discipline, that is, the community of scientists should be able to distinguish between, on the one hand, research which is likely simply to add to knowledge of particular facts, and thus is at best of minor importance, and, on the other hand, research which is likely to be much more important, in the sense that it will develop the core of ideas in that branch of science, and thus lead to the unification of very many facts or hypotheses.

The main difficulty about this argument is that while, in some cases, awareness of a significant gap in scientific knowledge may lead some scientists to undertake a particular line of research, the results of which may unify hitherto unrelated facts or hypotheses and, as it were, fill the gap—this is far from invariably the case. Awareness of a gap in scientific knowledge may give no indication of the extent or nature of what is missing: in this respect, the metaphor of scientific knowledge as a series of oases in an uncharted and shifting desert (or islands in an unfathomed sea) of ignorance, may be more appropriate. The experience of many scientific disciplines suggests that some of the greatest advances in them have come through serendipity and informed curiosity. In fundamental research then, it may be possible to distinguish between work which has had more, or less, important results. But in either case, what might have been dismissed, even by other scientists, as 'mere curiosity' may have played a significant part; and prospectively, the community of scientists within a particular discipline cannot necessarily be relied upon to provide criteria which will reliably predict the relative worth of different research projects.

These considerations apply not only to predicting which pieces of fundamental research are likely to fill a crucial gap (or remove a hampering bottleneck) in basic scientific knowledge, but also to predicting which are likely to provide benefits which can be applied to therapeutic advance. A variety of examples of this are cited in a note at the end of this chapter, and two are given here.

1. A laboratory was conducting a study of the integration of the nervous and endocrine systems. For purely theoretical reasons, this included studying the release of growth hormone, which led to the observation that the hormone was released in a pulsatile fashion. This discovery in turn is leading to a complete change in the administration of growth hormone therapy in children, although the research workers had no prior interest in the administration of this therapy. The benefit was entirely unintended.

2. In 1957, it was discovered that, in the intestine of the guinea-pig, there appeared to be two different types of receptor to an endogenous chemical called serotonin (5-HT). At the time, most scientists thought that this was simply a curiosity, and certainly of no clinical importance. One of the research workers continuing work on serotonin in the early 1960s was even advised by a leading authority to give it up, because it was a 'waste of time'. Much later, when receptors were further subdivided, it was realized that one of the serotonin receptors (5-HT_3) was associated with emesis, or vomiting. It was recently discovered that drugs which could oppose the actions of serotonin on 5-HT_3 receptors could oppose the vomiting caused by highly emetogenic substances such as the drug cisplatin, used to treat cancer. The development of the 5-HT_3 receptor antagonists, it is hoped, will allow the majority of cancer patients to take their cisplatin without vomiting: the first of these drugs has now been approved in Europe (1989/1990). This therapeutic advance is thus the result of information built on pure knowledge gained thirty years earlier, the benefit of which was entirely unforeseen.

Benefit in terms of therapeutic advance may therefore derive from research undertaken with no intention other than the benefit of advancing scientific knowledge. Even this general aim of 'advancing scientific knowledge' may not be formulated in the mind of a particular scientist deeply curious about a problem which has arisen from his previous research. The most it may be possible to say, in this case, is that the scientist works on the assumption that his research is of a kind which, if successful, will advance scientific knowledge (if only by disproving a particular hypothesis, so making him and other scientists examine alternative possibilities); and thus will add to the fundamental knowledge of, say, biochemistry or physiology, on which future therapeutic advances may be based. There is, in short, no certain way of predicting that any properly conducted piece of fundamental research will not contribute, at some point, to the advance of scientific knowledge, nor later (if the research is in a relevant biomedical field) to therapeutic advance.

Does this mean that the potential benefit from any properly conducted piece of scientific research is sufficiently serious to justify the use (although again, not necessarily every use) of animals in research? The second moral position on this question would seem to suggest this, at least in its more extreme form. On the other hand, the scientific community (and through it, the public) has an interest in whether or not research is conducted in those ways most conducive to the good of advancing scientific knowledge. Certainly, we can never exclude the possibility that benefit might accrue from work which is not congruous with anything hitherto known or researched in a particular discipline. In such a case, however, the scientific community would have a duty (in relation to funding or licensing of the work) to say that the case for there being not only potential, but also probable, benefit requires to be argued rather than assumed.

In assessing the arguments for the potential and probable benefit of a particular research project, the scientific community might be expected to ask, for example, whether the researcher is fully aware of all relevant earlier work in that area, and whether, against this background, the question being asked can be stated logically, in terms of some definable objective which the investigator has the technical training, ability, and expertise to follow. If such requirements of scientific competence, and of the quality of the working hypothesis and design of the research can be met, and if the researcher then makes a well argued case to suggest that the intended work might establish a principle which will fit together hitherto unrelated information in a significant way, his arguments will have to be taken very seriously by peers. Even if ultimately they are based on a hunch, the fact that the above mentioned conditions have been met will shift the balance toward probable as well as potential benefit.

The need to argue for, rather than to assume potential and probable benefit, is no less important in the case of research which is congruous with work hitherto undertaken in the same area. Just as there is the possibility of prejudice against work which appears incongruous, so there is the possibility of prejudice in favour of established lines of research. Science is a human activity, so that judgements about potential benefits are fallible, and must be made on a balance of probabilities. Because of this, it is important that the scientific community should not only seek to advance the public interest in research work which it judges most conducive to the good of advancing scientific knowledge, but should be *seen to do so.* This implies that bodies considering research proposals, for example, should be responsive to public attitudes; for, whilst the public might not have enough knowledge to hold positive views on particular research priorities, its negative views might be well-founded.

The requirement that the scientific community should be seen to advance the public interest in this way, modifies the extreme form of the second moral position described above. The relevant judgement is no longer left to the individual researcher, deciding whether his project is of potential benefit. Rather it is that of the scientific community in dialogue with public opinion, and is based on probable as well as potential benefit. This modification brings the two moral positions much closer to one another. They can be brought closer still, if in the first position's requirement that benefit be intended, the intention is attributed not only to the individual scientist, but also to the scientific community. Shifting the onus of intention from that of the individual scientist to that of the scientific community may also help to satisfy the first moral position's requirement that the intention should be of benefit accruing to the prevention, diagnosis and treatment of disease, ill health, or abnormality in humans or animals. Clearly, the biomedical scientific community as a whole is concerned with this kind of benefit as well as with the benefit of adding to scientific knowledge. This formulation of the scientific community's intention that animals should be used for sufficiently serious purposes, thus may meet the requirements of both moral positions, while allowing for fallibility of judgement in any particular case.

From this discussion, it seems that the question of potential and probable benefit ultimately must depend on the good faith of the scientific community in its stewardship of the public interest and the welfare of animals. But this, again, requires not simply to be assumed, but to be shown in the intentions of the scientific community, reflected in the benefits which it values. Clearly, it is important for this purpose that the scientific community should be seen to value benefits which add to knowledge, not least because such knowledge is indispensable to the larger moral community for making judgements about what is, as well as what is conducive to, the good of the whole and the good of individuals. But the identification of these goods relies on judgements of value as well as fact about the quality of the individual and social life, the quality of the environment and the welfare of other species and of future generations. It is important therefore for the scientific community to be seen to value benefits which make a positive contribution to these also.

For all this to be shown, clearly, good communication between scientists and the public is necessary. It also is necessary that the public understands the legal framework within which research using animals is carried out, and in particular the crucial requirement of the UK Act that, in determining whether or not a licence should be granted, the 'likely adverse effects' on the animals should be 'weighed' against 'the benefit likely to accrue' from the research project.

Public confidence in such decisions is more likely if the factors and interests taken into account in making them are well known and widely agreed to. A variety of these factors are included in the Working Party's *Scheme for the assessment of potential and likely benefit*, which is described at the end of this chapter, and is set out in Chapter 7. This scheme is intended as an aid to communication between the scientific community and public opinion. Its usefulness, however, will depend upon whether not only scientists but also all who profess moral concern for animals agree that attention to the details as well as the rhetoric of relevant judgements, about both facts and values, are of critical significance for the act of moral weighing.

3.4 THE ARGUMENT FROM NECESSITY

An argument still vigorously used in defence of the use of animals in biomedical research is that their use is 'necessary'. At its simplest, the idea is that animal use is a 'necessary evil'—something not necessarily blameworthy, but at least morally disquieting, for which a defence is required. Such a defence is provided by the argument from necessity, but only if certain qualifications are carefully observed.

The defence of necessity can be seen in terms of the Common Law doctrine that an act which is normally unlawful may be lawful if it is done as the only way of achieving a greater and lawful benefit. One example of this might be when a highjacker is killed in order to save other passengers. Or more routinely, when surgeons operate, their defence against the offence of cutting a patient's body with a knife is that the lesser evil of doing this is necessary, in order to avert the greater evil from the threat of disease to the patient's life. Necessity is therefore the principal justification; the consent of the patient, although required, is not by itself a sufficient justification.

In ethics, as in law, however, the defence of necessity requires to be examined carefully and the necessity proved. It has to be shown:

(1) that the evil prevented is greater than that done; and
(2) that there is no less drastic method of achieving the stated aim.

Thus, if a scientist claims that it is necessary to use animals in a particular project in order to achieve some goal, he is required morally to demonstrate at least four things:

(1) that the goal is worthwhile;
(2) that it has a high moral claim to be achieved;
(3) that there is no less drastic method of achieving it; and
(4) that there actually is some reasonable possibility of the project achieving the goal.

If the goal is, say, a cure for some major crippling or life-threatening disease, the first and second of these conditions would be fulfilled. But the scientist would also need to show the third and fourth; and with special reference to the third, he would need to show that his goal could not be achieved by using fewer animals or different species, or by procedures less harmful to the animals, or indeed by more morally and no less scientifically acceptable alternatives.

As these conditions suggest, justification by necessity often relies on crucial judgements of a scientific rather than a moral nature. This means, importantly, that any judgement that the use of animals is necessary, is normally an interim judgement. That is, it may change over time and with scientific advance, so that the necessity of animal use may diminish. An example of this is that when polio vaccine was first developed, the virus was grown in cells freshly removed from the kidneys of monkeys; and in order to extract their kidneys, up to half a million monkeys were killed for this purpose in a single year. At that time, it might have been argued, this was necessary in order to prevent a common, often paralysing, disease which had long proved unamenable to research. It was necessary also, because earlier attempts to find a vaccine had proceeded to human trials prematurely, in some cases with fatal results (see Paul 1971). Later, however, when it became possible to grow the virus in human cells in continuous culture, the use of monkeys could no longer be defended as necessary, despite usage in some remaining countries.

Justification of the use of animals in biomedical research by the defence of a diminishing necessity (proved by strict argument against the presumption of giving the maximum protection to the animals), may be seen as a further way of finding some common ground between those who are concerned to reduce the number of animals used (and ultimately to replace animal use by alternative methods), and those who are concerned that scientific advance and its benefits to human and animal health should not be impeded by arbitrary restrictions on animal use.

The defence of necessity, it has been noted, may be invoked in the case of research with a worthwhile goal which has a high moral claim to be achieved. In most instances, the assessment of this goal and claim is best made by judgements which take into account all the known morally relevant factors and interests, such as those set out in the Working Party's assessment scheme. In this context, however, it might be argued that research into some forms of disease, ill health or abnormality has, in principle, a lower moral claim than research into other forms. This might be because the former conditions were less serious, in terms of the suffering and distress inflicted by them, or because they affected only very small numbers of people. This argument, however, is open to both moral and scientific objections. The fact that very few people suffer from a serious condition, or that very few suffer seriously from a widespread minor condition, may make research specifically undertaken into these conditions seem to have a lower moral claim, if that claim is considered simply on crude quantitative or qualitative utilitarian grounds. But this assessment can be objected to on non-utilitarian moral grounds; and also because in practice, if research into either a rare or serious disease or a common and minor condition is undertaken, it is likely to involve and depend upon fundamental research of the kind already seen to be a worthwhile good in itself. The possibility of unforeseen therapeutic benefits deriving from fundamental research and respect for the autonomy and feelings of those suffering from conditions of the kind described, thus may mean that research into these conditions has a much higher moral claim than may appear at first sight.

Questions about the benefits and necessity of using animals in testing chemical products such as cosmetics and pesticides commonly are raised with the intention of suggesting that such work also has a low moral claim. Here again matters are much more complex than they may appear at first sight, and these questions are discussed in greater detail in Chapter 8.

A different area which again may raise doubts about the moral claim of research is that of work done on animals to benefit other animals, an example of which might be research on parvovirus in dogs. The reason for undertaking such research, it is sometimes suggested, was that people would pay for a vaccine which gave them more 'use' out of their companion animals. Similar arguments might be advanced to suggest that research on the diseases of farm animals is motivated by the usefulness of healthy farm animals to humans. Against these claims, however, it can be argued that the development, for example, of antibiotics to treat bovine mastitis actually relieved the animal of a very painful condition. This is a very different matter, morally, from research intended primarily to make an animal yield more meat or milk, particularly if this induces painful conditions in the animal. Further counter-examples include behavioural research on animals designed to improve the conservation of animals in the wild. Perhaps the most important moral consideration here is that the existence of self-interested human motives alongside the human intention to benefit animals does not invalidate the latter, nor necessarily lower the moral claim of research designed to benefit animals.

3.5 ASSESSMENT OF BENEFITS

In the absence of any scientifically and morally acceptable alternative then, some use of animals in biomedical research can be justified (albeit by different moral reasons for different people) as necessary to safeguard and improve the health, and alleviate the suffering, of human beings and animals. These benefits, in turn, depend on the advancement of fundamental scientific knowledge. But even when no therapeutic or other practical benefit can yet be derived from it, any significant advance in scientific knowledge is an inherent good, and *may* serve as a justification for using animals to that end.

Not every addition to scientific knowledge, however, is sufficiently significant to justify the use (let alone every use) of animals. Some uses may have adverse effects too serious to justify them at all. In other cases the adverse effects may be considered disproportionately serious in relation to the significance of the results gained. In the latter case especially, both the potential benefits of a particular research project and the likelihood of the project achieving those benefits need to be assessed carefully before they, in turn, are weighed against the likely adverse effects to the animals.

Precisely because the benefits concerned are potential and likely, their assessment is not an exact science. Their assessment does involve the most exacting scientific scrutiny of the particular project's aims, approach and procedures, and of the project workers' ability to achieve their goals. But when this scrutiny has been completed, the overall assessment of how likely it is to achieve its potential benefits is a matter of judgement. For that judgement to gain the confidence of both the scientific community and of informed public opinion, it is essential that the questions which have been asked before arriving at it, are seen to have been the most relevant, exhaustive and exacting.

The scheme outlined in Table 3.1, and given in detail in Chapter 7, sets out a series of features, which the Working Party believes ought to be taken into account when assessing the benefits of research involving animal subjects. It is not proposed as a definitive or exhaustive list, but as a first attempt, arising from a dialogue between scientists and informed public opinion, to spell out the kinds of question which need to be asked. It is hoped that this list will commend itself sufficiently for others to use, adapt and refine, and in this way for such assessment to become more relevant, exhaustive and exacting over time.

The scheme is intended for use in a variety of contexts. Initially, it might be used by research workers considering undertaking a particular piece of work. A critical and careful rating of the features in parts 1 and 2, concerning the potential benefits and proposed approach, would provide a basis for the necessarily more impressionistic overall assessment of the likely benefits, in part 3. If the latter was low, the research workers might well decide to re-examine their goals, approach or procedures with a view to undertaking work with a greater potential and likelihood of benefit. If, on the other hand, they decided that the potential or likely benefit, although low, was still worthwhile, they would have to take this into account when deciding about their use of animals.

The scheme might most usefully be employed if the research workers' own rating were subject to peer review by colleagues and if it was used for this purpose. If an independent assessment agreed

Table 3.1 Summary of the features which should be taken into account in assessing the potential and likely benefits of a research project involving animal subjects

---

1. **The *potential* benefits of the project,** in terms of:
   the project's potential social, scientific, economic, educational, and/or other value;
   its originality, timeliness, pervasiveness, and applicability.
2. **The quality of the proposed approach**, in terms of
   the scientific merit of the proposed general approach, including:
   (a) its relevance to the potential benefits;
   (b) the quality of the working hypothesis and the experimental design; and
   (c) appreciation of any relevant background literature and of other work in progress;
   the necessity and validity of the proposed scientific procedures, including:
   (a) their applicability to the proposed approach;
   (b) the necessity to use animals at all, or of the species proposed;
   (c) the need to use procedures of the proposed severity;
   (d) the need to use the proposed number of animals; and
   (e) the maximization of information to be obtained from each animal, consistent with welfare constraints;
   the quality of the project workers and facilities.
3. **The *likelihood* that the potential benefits will be realized, given the proposed approach.**

---

with the research workers' own, this would greatly strengthen the case for proceeding when the rating was high and for reconsidering the project if it was low. The probability that many projects may be rated as medium increases the advisability of independent assessment. Assessment of the project, using the scheme, might also be undertaken by senior colleagues in a research institution, or by a local research review (or 'ethics') committee.

The scheme, it must be emphasized, has an educational rather than a legalistic aim. It may be useful to those officially assessing projects, whether for funding or under provisions such as the UK Animals (Scientific Procedures) Act; and any project whose proposers have used a scheme of this kind seems more likely to merit funding or licensing. But the point of the scheme will be lost if it is linked too closely with funding or licensing procedures. It is designed rather to assist research workers, in consultation with others, to improve the scientific quality of their work. As such, it embodies an important philosophical presupposition. It presupposes that fair-minded scientists are going to look at their own prospective work and judge it by these criteria. The same questions are then going to be asked independently of their peers. The two will then be compared, and the original scientists—fair minded, just, and open—will consider the results and decide whether or not to proceed. The scheme thus appeals to professional self-regulation, which rests on a view of human beings as by nature rational and benevolent. In the present climate of public opinion, with a considerable amount of opposition to any use of animals in research, it is also a matter of professional self-interest that the public understands why and what animal use is necessary. This understanding will be assisted if the public is aware that scientists examine critically the considerations raised by a scheme of this kind.

The use of such a scheme then, is in the interest both of biomedical science and of animal welfare. Its explicit and detailed character means that critics (save those who oppose all use of animals) will lack good reasons for opposing projects which can be shown, convincingly, to have a high rating. But where the rating is low, this will provide a structured rationale for limiting animal use. Knowledge that a scheme of this kind is used, is likely, moreover, to improve communication between scientists and lay people. The status of good science and the welfare of animals alike can only benefit from this.

Note

Unintended or unforeseen benefits of fundamental research

It is rewarding to trace the historical origins of some important groups of drugs. The first example is provided by anti-inflammatory drugs which are of such widespread use in the treatment of arthritis and other conditions. In 1930, Kurzrok and Lieb observed that the addition of semen caused isolated strips of human uterus either to contract or to relax. It was a matter of scientific curiosity to reveal and, if possible, identify the active substance in semen. Later, an active principle was isolated from the semen and the prostate gland, which appeared chemically to be a lipid-soluble acid and was termed 'prostaglandin'. Further analysis revealed that this was to be regarded as a generic term for a complex group of inter-related substances. Their origin was traced to arachidonic acid, a constituent of phospholipid in cell membranes. The first step in biosynthesis is the formation of an intermediate by the action of an enzyme, cyclo-oxygenase. In recent years, one of the most productive discoveries in the field of pharmacology is that aspirin and other non-steroidal anti-inflammatory drugs act by inhibiting this enzyme. Moreover, the steroidal anti-inflammatory drugs prevent the release of arachidonic acid from phospholipids. In addition to providing an explanation for the mode of action of such important and widely used drugs, this research has revealed that prostaglandins are ubiquitous, naturally occurring substances, which have widespread actions as mediators or modulators of a variety of physiological functions. One action of therapeutic importance, recalling the original observation of Kurzrok and Lieb, is that on the uterus. Prostaglandins can be used to initiate labour.

A similar example is provided by the discovery of the renin–angiotensin system, which we can again trace back to a rather curious observation. The idea had grown that the kidneys were in some way connected with high blood pressure or hypertension, and in 1898 Tigerstedt and Bergman found that the intravenous injection of a saline extract of kidneys into an experimental animal produced a rise of blood pressure. They coined the term 'renin' to describe the unidentified active principle producing this effect. Further investigation showed that renin was actually an enzyme, produced by the kidney, which acted on a precursor in the plasma proteins to produce the substance angiotensin I: this in turn acts as a precursor for the final active agent, angiotensin II. Angiotensin I is converted into angiotensin II by the action of an enzyme 'angiotensin converting enzyme', derived from the lungs. The discovery of an enzyme of this sort is a challenge to the pharmacologist to find something that will inhibit or block it. As a result, 'angiotensin converting enzyme inhibitors' were discovered and these are now predominant in the management of both hypertension and heart failure. There is a further exciting development. It was discovered that angiotensin was the principal regulator of the secretion from the adrenal cortex of aldosterone, a hormone which is of vital importance in the maintenance of fluid and electrolyte balance in the body.

Another example of the value of scientific knowledge for its own sake is provided by the intellectual challenge of defining receptors or subtypes of receptors for active substances in pharmacology. One example has been given in the foregoing chapter, quoting work on serotonin receptors. Another is that the recognition and classification of receptors for chemotransmitters of nerve impulses in the central nervous system, has led directly to an understanding of the pathology, and the introduction of clinically useful drugs for the treatment of mental diseases such as anxiety, depression, schizophrenia and Parkinson's Disease. It is worth noting also that the search for a naturally occurring substance which might conceivably act on the morphine receptor, led to the discovery of the enkephalins and endorphins. A further important point, mentioned in Chapter 2, is that many clinical procedures, such as blood transfusions, renal dialysis and transplants, rely on fundamental basic knowledge in the fields of physiology and immunology. It is of course obvious too, that the development of new drugs in pharmacology rests on a foundation of basic physiology and biochemistry.

REFERENCES

American Medical Association: Council on Scientific Affairs (1989). Council report on Animals in Research. *J. Am. Med. Assoc.* **261,** 3602–6.

Kurzrok, R. and Lieb, C. C. (1930). Biochemical studies of human semen II. The action of semen on the human uterus. *Proc. Soc. Exp. Biol. Med.* **28,** 268–72.

Maehle, A-H. and Tröhler, U. (1987). Animal experimentation from antiquity to the end of the eighteenth century: attitudes and arguments. In *Vivisection in historical perspective* (ed. N. A. Rupke), pp. 14–47. Croom Helm, London.

McKeown, T. (1979). *The role of medicine.* Blackwell, Oxford.

Moe, K. (1984). Should the Nazi research data be cited? *Hastings Center Report* **14,** 5–7.

Paton, W. (1984). *Man and mouse.* Oxford University Press, Oxford.

Paul, J. R. (1971). *A history of poliomyelitis.* Yale University Press, New Haven, Connecticut.

Sharpe, R. (1988). *The cruel deception: the use of animals in medical research.* Thorsons, Wellingborough.

Silver, I. A. (1987). Animals and medicine (the fifty-fifth Stephen Paget Memorial Lecture, 11 November 1986). *Conquest* **176,** 1–9.

Tigerstedt, R. and Bergman, P. G. (1898). Niere und Kreislauf. *Skand. Arch. Physiol.* **8,** 223–71.

Vane, J. (1985). How animals discover drugs (the fifty-third Stephen Paget Memorial Lecture, 14 November 1984). *Conquest* **174,** 1–12.

4
Pain, stress, and anxiety in animals

4.1 INTRODUCTION

How does a human know that an animal is in pain? Widespread agreement between people might be reached about the suffering of a chimpanzee. But what about an earthworm or an octopus? What criteria should be used when assessing whether or not animals with such distinct nervous systems and behavioural repertoires as these feel pain? A sceptic might be forgiven for supposing that answers to these questions are totally dependent on the intuitions, beliefs and prejudices of the person who cares to reply. This chapter examines the problem of assessing pain, stress, and anxiety in other animals since its resolution is of such importance to the rest of this report. An attempt is made to produce reasonably explicit criteria that might be widely accepted and applied.

The grounds for supposing that animals have a subjective life somewhat similar to that of a human are not nearly as self-evident as is sometimes supposed. Some argue that thoughts and feelings are essentially subjective and private and, therefore, that no human can ever know exactly what it is like to be another human being, let alone a different kind of animal. Whatever might be said about this argument, nobody has direct access to the thoughts and feelings of other animals, nor can a mouse, say, describe to humans what it is thinking or how it is feeling when it is undergoing a particular experimental procedure. Therefore, fundamental uncertainty exists when it comes to assessing the thoughts and feelings of other species. Their subjective experiences, if they have any, may be totally different from those of humans, reflecting their different ways of life and the different ways in which their bodies work. Furthermore, human interpretation of what is observed in other animals can, at present, only be based on projections from humans together with a good knowledge of the natural history and behaviour of the animals in question.

While the task might seem hopeless at first, everybody (or almost everybody) makes judgements about other humans' feelings. If it is possible to specify the types of information that encourage such judgements, then the exercise of projecting into other animals may become more transparent. In adopting a 'top down' approach in order to investigate an animal's capacity for experiencing pain, distress, anxiety and so on, it is convenient to apply two types of test:

1. Has the animal anatomical, physiological and biochemical *mechanisms* similar to those which in a human are known to be correlated with such experiences?
2. Does the animal *perform* in similar ways to humans who are believed to be suffering?

Before applying the tests of mechanism and performance to pain, stress and anxiety specifically, the general issues of brain development, cognitive ability and awareness of self are discussed, since these considerations lie behind assumptions that another animal has similar subjective experiences to humans. How elaborate is the computing equipment of the animal? How is the equipment used in dealing with complicated problems? Does the animal show signs of self awareness? Having considered these general points, lists of criteria are drawn up which might be used when a judgement must be made about whether an animal is capable of experiencing the adverse states of pain, distress and anxiety. In each case, the criteria are applied to some real animals. Other discussions of this general approach can be found in Rose and Adams (1989), and Bateson (in press).

In assessing the criteria that might be used to decide whether an animal is capable of experiencing those states, such as pain and distress, which humans find unpleasant or, in extremes, intolerable, it is important to appreciate that they are likely to be interrelated. Rather like fitting together the pieces of a jigsaw puzzle, the evidence needs to be considered as a whole in order to build up a useful picture of the animal's capabilities. For that reason, it is important to emphasize that no single criterion provides an all-or-none 'litmus' test for the existence of subjective experience.

4.2 BRAIN DEVELOPMENT

The capacity to suffer is plausibly related to the complexity of an animal's behaviour and its capacity to process information. In broad terms, complexity of behaviour and information-processing capacity are undoubtedly associated with the size and structure of an animal's brain. To be more accurate, they are linked to the size of the brain relative to the size of the body since, even when complexity of behaviour and information processing are equal, brain size is strongly correlated with body size.

Starting with a simple case, earthworms have a central nervous system which consists of a cord of neurones running parallel to the long axis of the body. Bundles of nerve cells are found at intervals along the nerve cord. These bundles are called 'ganglia'. In the third body segment, just above the pharynx (a muscular passage leading from the mouth to the gut), several such ganglia are fused together, forming a 'cerebral ganglion'—a primitive kind of brain (see Edwards and Lofty 1977). If it is removed movement can occur normally, but the worm loses its ability to respond to external environmental conditions (see Barnes 1980). In other respects this 'brain' is, however, extremely simple.

The insect central nervous system also consists of a nerve cord and ganglia. At the anterior end of the nerve cord a cerebral ganglion or 'brain' is composed of several fused ganglia. Many of the insect's activities are controlled by the posterior ganglia, which control activities such as walking and respiration in the absence of any input from the brain (decapitated insects can live for months—Wigglesworth 1980). However, the brain is required for the control of feeding in blow flies (Dethier 1976) and for learning in honey bees (Marler and Terrace 1984). Octopuses, along with squids and cuttlefishes, belong to the zoological group called cephalopoda. According to Russell-Hunter (1979) 'by invertebrate standards, cephalopods have enormous brains'. The brain–body weight ratio of modern cephalopods also exceeds that of most fish and reptiles (Packard 1972). Only the most advanced of the bony fishes (teleosts) have brain weights similar to *Octopus vulgaris*. Furthermore, much of an octopus's basic movement is controlled by the ganglionated nerve cords of the arms which contain almost three times as many neurones as the brain (Wells 1978). The brain weight therefore 'represents only the more specialised sensory integrative, higher movement control and learning parts of a rather diffuse nervous system' and it is clear that 'one is dealing with an animal that might well be expected to possess a central nervous capability approaching or exceeding that of many birds and mammals' (Wells 1978). The cephalopod brain shows a hierarchical organization as in vertebrates (see, for example, Boycott 1961). The lower centres provide inputs to the more peripheral ganglia (such as those in the arm nerve cords) and produce a few of the animal's basic movements, such as the regular respiratory rhythm. The higher brain centres are concerned with sensory analysis, memory, learning, and 'decision-making'. These latter parts contain vast numbers of interconnected neurones and function as association areas, storing sensory inputs from past experience. Together, they *might* be considered to be analogous to the cerebrum of higher vertebrates (Russell-Hunter 1979).

Vertebrate brains may be divided into three parts: the hind-brain, midbrain and forebrain. The forebrain is the 'highest' part of the brain, and is divided into two parts, the diencephalon (consisting of the thalamus and hypothalamus) and the telencephalon, or cerebrum. The cerebrum, and in particular its outer layer, the cerebral cortex, is thought to be the seat of conscious experience in man. All of the different groups of vertebrates possess a cerebrum, consisting of two cerebral hemispheres, and all mammals possess a cerebral cortex—a thin cladding of grey matter (the cell bodies of neurones) covering the cerebral hemispheres. A similar cortex is found in reptiles and birds. In reptiles, a distinct lamina (sheet) of grey matter is found on the dorsal surfaces of the cerebral hemispheres, whilst in birds similar neurones are found compressed into a bulge, the Wulst, found on the top of the cerebral hemispheres. Only small patches of similar nerve cells have been found on the surfaces of the cerebrum in fish and amphibians (see Macphail 1982; Walker 1983). Some authors suggest that these patches are similar to, or evolutionary forerunners of, the mammalian cerebral cortex (see Macphail 1982). Unfortunately, although these differences in cerebral structure are interesting, it is difficult to assess their significance in determining the mental capacities of members of the different vertebrate groups. A general point is that structural differences do not always imply differences in performance.

Figure 4.1 suggests that vertebrate cerebral hemispheres show a trend of increasing size from fish to amphibia to reptiles to birds to mammals, and from the 'lower' to the 'higher' mammals (that is, 'up the evolutionary scale'). In addition, in some mammals, but not in other vertebrates, the surface area of the cerebral hemispheres is greatly increased by its being organized into folds or sulci. It is important, however, that in interpreting Fig. 4.1 allowance is made for the body weights of the different animals.

Figure 4.2 shows a plot of overall brain weight against body weight for representatives of four of the vertebrate groups—fish, reptiles, birds and mammals (Jerison 1969). The figure shows that 'higher' vertebrates (birds and mammals) have significantly greater brain weights than 'lower' vertebrates (fish and reptiles) of similar body weights. No significant difference exists between the brain weights of birds and mammals of equivalent body weight; nor do the fish and reptiles differ.

The relative brain size of humans is far greater than that of any other animal including our closest relatives, the chimpanzees. Even so, other vertebrates do possess structures that are akin to the human cerebrum.

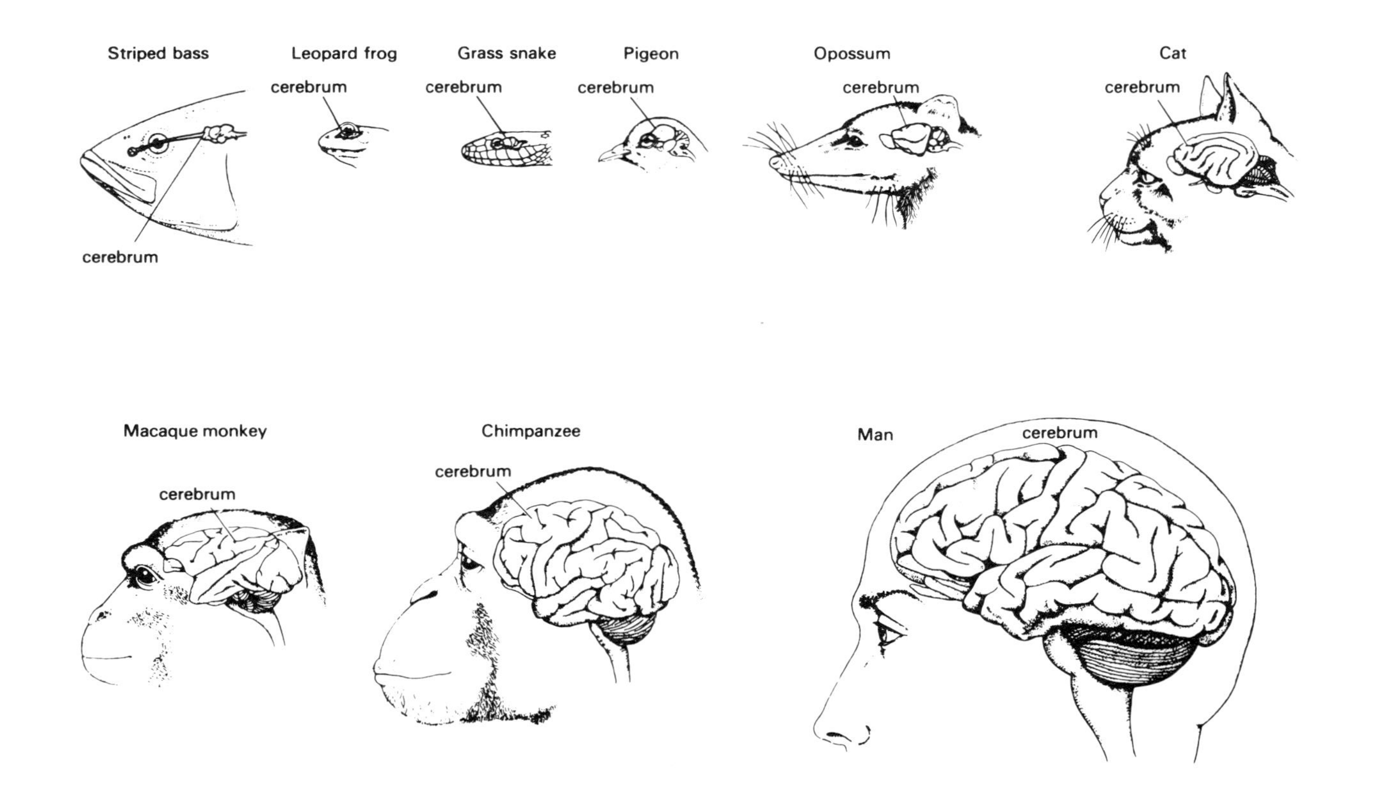

Fig. 4.1 Cerebral hemispheres in vertebrate brains. Drawn to the same scale (from Hubel 1979).

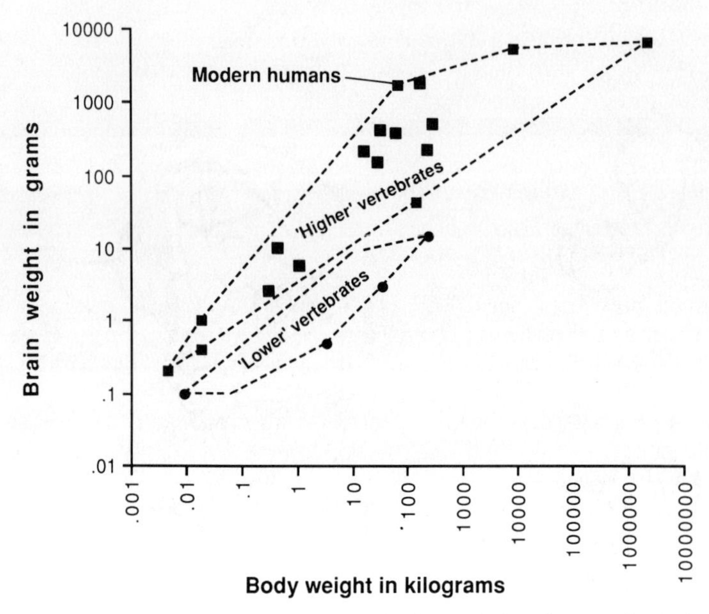

Fig. 4.2 Plot of brain weight against body weight (log scales) for representatives of four vertebrate groups (adapted from Jerison 1969, © University of Chicago Press). '"Higher" vertebrates' includes birds and mammals; '"lower" vertebrates' includes fish and reptiles.

4.3 GENERAL COGNITIVE ABILITIES

In addition to asking questions about relative brain size in an attempt to assess the capacity for suffering, knowledge of how the animal deals with complicated problems is helpful. As with relative brain size, the reasoning is indirect and stems from the top-down approach which has been adopted here. Humans suffer and are also extremely intelligent. If an animal is found to solve problems in ways that demonstrate high cognitive abilities then it may also be equipped with the capacity for suffering. However, the association between cognition and suffering is not required and such evidence is merely part of the jigsaw that has to be assembled when attempting to infer whether or not animals suffer.

Cognition or intelligence in animals is not easily defined (see Blakemore and Greenfield 1987; and Pearce 1987). Seemingly complex behaviour may be generated by very simple rules. Therefore, even when behavioural complexity can be defined in anything other than vague terms, it is not altogether satisfactory as an index.

Many animals perform highly elaborate and appropriate activities the first time they are called upon to do so and without apparently having to learn how to do so. The elaborate nest-building by wasps is one example of these 'instincts' (Smith 1978). Such instances give pause when complex forms of behaviour that were thought to be peculiarly human have subsequently been found in other animals. Jane Goodall (1971), for example, describes how two chimpanzees, David Greybeard and Goliath, used grass stems as tools, to 'fish' for termites....

> I observed how they scratched open the sealed-over passage entrances with a thumb or fore-finger. I watched how they bit the ends off their tools when they became bent, or used the other end, or discarded them in favour of new ones. Goliath once moved at least fifteen yards from the heap to select a firm-looking piece of vine, and both males often picked three or four stems, whilst they were collecting tools, and put the spares beside them on the ground until they wanted them.... Most exciting of all, on several occasions they picked little leafy twigs and prepared them for use by stripping off the leaves. This was the first recorded example of wild animals not merely *using* an object as a tool, but actually modifying an object and thus showing the crude beginnings of tool-*making* (p. 48).

Amongst the mammals, tool use is not restricted to primates. An impressive example comes from sea otters, which use small stones to detach molluscs (shellfish) from rocks and then to open the shells (Griffin 1984, citing work by Kenyon 1969). Birds, too, may fashion and use tools. Griffin (1984, citing work by Lack 1947; Bowman 1961; Millikan and Bowman 1967) describes behaviour in birds very similar to that observed in chimpanzees by Goodall (1971). Some of the species of finch found in the Galapagos Islands use tools to obtain insects, for food, from crevices. Such reports seem to provide compelling evidence that some animals, some birds and mammals at least, do exhibit intentional behaviour. Indeed, several authors, including Hubbard (1975) and Griffin (1984), suggest that the use by animals of objects as tools to solve certain problems provides evidence of anticipation of a goal. However, similar behaviour has been observed in animals possessing much simpler nervous systems, such as insects. Griffin (1984) again has an example. Assassin bugs use certain 'tricks' to capture the workers of termite colonies (McMahan 1982). Assassin bugs, it seems, use termite exoskeletons to 'fish' for termites, just as chimps and birds use sticks to probe for such insect food. Evidence about brain size and function in insects might lead us to suppose that the assassin bugs are not, in fact, behaving intentionally, but using simple rules. If this view is adopted, then what of the 'higher' animals, such as birds and mammals? Their behaviour might be explained in a similar way.

When inferring that the ability to perform complex behaviour also implies awareness, is an element of learning crucial? Here again, the answer is not straightforward. Rats that have been repeatedly removed from their cages for injection may bite when taken out of a cage—anticipating, so it would seem, an injection. A dog, bringing a lead to its owner, seems to be hoping for or looking forward to a walk. The difficulty, in these and other cases, lies in deciding how such behaviour can be explained. Are these animals thinking about the future and behaving intentionally (‘*if I bite this hand then . . . I might not be injected*’; ‘*if I bring this lead then . . . I might be taken for a walk*’)? Or are these examples of learning by simple rules (‘*hand = injection, then bite*’; ‘*lead taken to owner = walk, then take lead*’)? The rules for learning are sometimes so automatic that the mere ability to learn is not in itself the mark of a complicated brain. Capacities to perform simple associations between neutral and biologically significant stimuli are known in many rather primitive invertebrates (see Marler and Terrace, 1984). In the context of a discussion of suffering, it is worth noting, for instance, that insects seem to be capable of learning associations between neutral and noxious stimuli. Fruit flies (*Drosophila* species) can learn to associate certain odours with impending electric shock, and behave so as to avoid being shocked (Marler and Terrace 1984).

As McFarland (1989) noted, the important criterion may be how the animal learns. Dickinson (1980) made a helpful distinction between ‘procedural’ and ‘declarative’ representations of acquired knowledge. When a rat has learned to approach a feeder on hearing a tone, it might simply follow the mechanical procedure that, on hearing a tone previously associated with the arrival of the food, it moves towards the feeder. That would involve a procedural representation of the association between the tone and food. Alternatively, it may have learned in a rather more general way that the tone reminds it of food, and, on being so reminded, it starts to forage in ways that were previously successful. The second, declarative form of knowledge is more complicated, but an experiment by Holland and Straub (1979) suggests that it may be attributed plausibly to rats. Rats were made ill with an injection of lithium chloride after eating the type of food which had previously been associated with a tone. Later these rats were found to approach the feeder much less readily on hearing the tone than rats that had been made ill without any association with the food. It was as though the tone reminded them of the food which then reminded them of feeling ill.

Somewhat similar to declarative knowledge is the ability to

interpret symbolic relationships between stimuli, such as *same as* and *different from*. However, care must be taken when interpreting the evidence, since it is not easy to show that an animal has actually formed such a concept. Premack (1983), for example, argues that it is likely that all vertebrate species (and at least higher invertebrates such as *Octopus*—see Wells 1978) can answer (by their behaviour) such questions as 'Do these items look [or, equally, feel, taste, smell, sound] alike?' and 'Have I seen [felt, tasted, etc.] them before?' These same animals, however, may not be able to judge that the relationship between the items is, for example, *the same*: this requires an animal presented with, say, five apples, three cotton reels and two peanuts, to be able to judge that the relationship between the members of each of the three groups of objects is that they are the same and, further, that all three relationships are equivalent—that is, the common feature uniting the three groups is that the members of the groups are the same. Premack (1983) has suggested that the ability to judge such relations is limited to primates.

Recent work by Pepperberg (1987), however, suggests that 'symbolic comprehension of the concepts *same/different* can no longer be considered the exclusive domain of primates' (p. 429). Pepperberg first taught a talking African Grey parrot, Alex, to use vocal English words to describe over 80 objects and to categorize these according to colour, shape, and material (for example, wood, hide). She then trained and tested the parrot on the relationships *same* and *different*. Alex could respond with the correct category ('colour', 'shape', or 'mah-mah' [matter]) when asked 'What's same?' or 'What's different?' about pairs of objects (such as, for 'What's same?', a red wooden triangle and a green hide triangle; and, for 'What's different?', a red wooden square and a blue wooden square). Pepperberg's work suggests that Alex can understand the vocal concepts of *same* and *different* although it is not yet clear if he can learn to respond directly to 'relations between relations', as described by Premack (1983).

Another aspect of complex processing often suggested as an important component of awareness is the ability to respond selectively to first one and then another feature of the environment. In a broad sense, such ability seems to be common to most kinds of animal. Evidence about selective attention in animals has come from rather complicated psychological experiments, which are reviewed (for vertebrates) by Macphail (1982). The controversy regarding the possibility of selective attention in fish is lively. Macphail's survey provides 'good evidence' that mammals (such as rats) and birds (such as pigeons) possess mechanisms of selective attention or (in Macphail's words) 'at least for the proposal that phenomena taken by some theorists to demonstrate selective attention are found in pigeons as well as rats'. Wells (1978) reviews experiments similar to those described by Macphail, carried out using *Octopus*. This evidence suggests that *Octopus* may be capable of selective attention and, further that 'individuals may choose to solve a discriminatory problem in different ways, depending on the context and their own previous experience' (p. 206).

The survey in this section shows how difficult the evidence is to interpret. Anticipation of the future, often thought to be a mark of reflective consciousness and great cognitive ability, may be generated by rather simple learning rules. Apparent anticipation of the future can occur in the absence of a cerebrum. It is not easy to establish that an animal's behaviour reveals cognitive abilities of the type that humans would term intentional when fellow humans are seen behaving in similar ways. However, of these various aspects of cognitive ability, evidence for declarative representation of knowledge ('*I know that I was hurt in that place*') is perhaps the most relevant when considering criteria for suffering (McFarland 1989). It should be noted that difficulties operate in both directions and the behaviour of an animal is all too readily devalued. Comparison of the intelligence of different species is fraught with difficulty because a test that may measure accurately the ability of one animal, may be totally inappropriate for another adapted to solving different but, in abstract terms, equally difficult problems. Consider, for instance, a well-trained dog's ability to use scents in finding drugs or a bomb (for which it receives a food reward). This acquired skill exceeds vastly anything that humans can do using their noses. The capacities of animals with very different bodies or natural histories from humans are often underestimated. In general, as experimenters have become more adept in tapping the abilities of other animals, more and more power has been attributed to species that were thought to be incapable of any learning.

4.4 AWARENESS OF 'SELF'

Suffering is not the same as awareness of self. Nevertheless, if an animal could be shown to be aware of itself then the grounds for supposing that it could suffer under certain conditions would be greatly strengthened—even if no other evidence were available. It does not, of course, follow that if no evidence for self-awareness can be obtained, the animal never suffers. The philosophical dimension of self-awareness is considered in Chapter 11. Here the concern is with aspects of self-consciousness that might be accessible from the study of an individual's behaviour.

Humans are able to think about (or reflect on) what they are doing and feeling at any particular time, what they have been and what they might become. Every adult has a concept of self, of what it means to be 'me' as distinct from somebody else. Humans do not seem to be born with such self-consciousness and the time of its emergence is not clear-cut. A variety of psychological abilities, taken together, allow an adult human to 'look in' on his or her own mind. These abilities do not emerge all at once, but at different times during a child's early life. Some of these psychological abilities, together with their approximate time of appearance (as described by Kagan 1981) are shown in Table 4.1. Most child psychologists believe that children gain a full 'awareness of self' some time during their second year of life (see, for example, Lewis and Brooks-Gunn 1979; Kagan 1981). It seems that a child under about two years of age, whilst being conscious (so that it is, for example, aware of being hurt), does not have full 'reflective' consciousness—it is not fully aware of itself as an entity distinct from others. Interestingly, the stepwise development of self-awareness, as illustrated in Table 4.1, seems to be correlated with the maturation of the child's brain (Kagan 1981). Kagan notes that the length and degree of neuronal branching and maturation of the layers of the cerebral cortex only approach adult levels at around two years of

Table 4.1 Some psychological components of reflective consciousness and their approximate time of appearance during child development

Approximate age of child	Child's emerging psychological attributes
Less than 8 months	Ability to recognize events which have occurred previously
Between 8 and 16 months	Ability to relate past events to the present and to make inferences
After 16–17 months	Awareness of ability to select and control actions: the child knows what it can do
Later in second year	Awareness of self as an entity with particular attributes: the child knows how it feels and how it is similar to and different from others

Drawn from an account by Kagan 1981.

age, that is, at about the same time as full reflective consciousness is thought to appear.

Kagan describes the first four psychological attributes listed in Table 4.1 as 'necessary . . . but not sufficient conditions' for the appearance of full self-awareness at around two years of age. These clearly overlap with the cognitive abilities considered in the previous section. Young children, as well as mentally retarded or senile adults, may possess some, but not necessarily all of these attributes. The gradation is important when considering the behaviour of another animal, since clear-cut, all-or-nothing evidence for or against the animal's self-awareness is not necessarily expected.

Two other attributes have been considered in the development of reflective consciousness in humans and its evolution in other species: the ability to recognize oneself in, for example, a mirror and the ability to deceive others. Lewis and Brooks-Gunn (1979) have studied young children's responses to mirrors. Although children as young as nine months showed behaviour indicative of self-recognition, behaviour directed at marks put onto their bodies was not exhibited by infants under 15 months old. The proportion of infants showing self- and mark-directed behaviour, when presented with mirrors, increased with age, changing most markedly at around 18–21 months. This correlates nicely with Kagan's (1981) observation of the emergence of 'an awareness of self as an entity with particular attributes' towards the end of the child's second year (see Table 4.1).

Gallup (1970, 1977) has described how, after a few days experience with a mirror, chimpanzees use the mirror to engage in 'self-directed' behaviour: they may use the mirror, for example, to groom the parts of their body which they cannot see directly, and to pick food from between their teeth; they may also blow bubbles, make faces and so on into the mirror. Gallup has also described specific mark-directed behaviour in chimpanzees exposed to mirrors. Having first removed the mirrors, he anaesthetized the chimpanzees and marked their faces with non-irritant red dye. He then observed the number of times the animals touched their faces, in the absence of a mirror and then following reintroduction of the mirrors. Gallup found that in the absence of mirrors the chimpanzees seldom, if ever, touched the marks on their faces. Following reintroduction of the mirrors a great deal of mark-directed behaviour (touching, pointing at and examining the mark) was observed. This suggests that the animals had recognized their reflections and realized that the marks seen in the mirror were on their own faces. Similar self- and mark-directed patterns of behaviour have been observed in orangutans but not in other primates (such as monkeys, macaques, baboons and gibbons). Of all the animals investigated, the evidence at present suggests that humans and Great Apes (such as chimpanzees and orangutans) alone have the ability to recognize themselves in a mirror. Other animals respond as if their reflection was another member of the same species and may engage in aggressive or social responses towards the reflection. It must be appreciated, however, that lack of self recognition in a mirror may indicate an inability to understand how mirrors work, rather than lack of self awareness.

So-called 'tactical deception' seems to provide compelling evidence for animals understanding enough of their own intentions to manipulate those of others. According to Whiten and Byrne (1988), tactical deception occurs 'when an individual is able to use an "honest" act from his normal repertoire in a different context to mislead familiar individuals' (p. 233). These authors have collected many reports of acts of apparently deliberate deception by certain primates, including chimpanzees, baboons and gorillas. Here is one example:

> Paul, a young juvenile baboon, approached an adult female, Mel, who was digging in hard dry earth for a large rhizome to eat. Paul looked round: in the undulating grassland habitat no baboons were in sight, although they could not be far off. Then he screamed loudly, which baboons do not normally do unless threatened. Within seconds Paul's mother, who was dominant to Mel, rushed to the scene and chased her, both going right out of sight. Paul walked forward and began to eat the rhizome.

Whiten and Byrne (1988) suggest that, in performing such acts of deception, these primates are acting as 'natural psychologists' (see also Humphrey 1986). That is, in order for such deceptions to be effective, an animal must understand the psychology of the individuals it intends to deceive. Such an understanding must (Humphrey would argue) be based on the animal's own self awareness, on its understanding of its own thoughts and feelings.

Whether apparent deceptions by animals represent examples of truly intentional behaviour is the subject of much debate. The 1988 Whiten and Byrne paper, published in *Behaviour and Brain Sciences*, is followed by commentaries written by a variety of workers interested in this area of research. Many of these commentaries doubt that Whiten and Byrne's collection of anecdotal reports demonstrate intentionality. Some scepticism is justified, because people can be easily fooled—a robot using well-recognized learning rules could do many of the things reported as intentionality. Nevertheless, for a given animal, the cumulative impact of well described instances of apparently intentional behaviour, combined with a brain structure which would suggest the capacity for reflective consiousness, might suggest that such an animal is conscious of what it is about. The present anecdotal evidence raises the possibility, at least, that some birds and mammals may exhibit truly intentional behaviour. Unfortunately, evidence is either lacking, or difficult to interpret in other kinds of animal. Perhaps, as Whiten and Byrne suggest, more systematic experimental work might shed a stronger, clearer light in this area.

A variety of abilities are likely to be involved in self-awareness. They do not all need to be present in equal measure for some glimmerings of reflective consciousness to exist. As Dawkins (1980) has put it: 'Different animals might possess some or all of these attributes to different extents, so that it may not be possible to say that an animal is either conscious (possessing all elements) or not (possessing none)' (p. 25). If she is right, a choice is not forced between, on the one hand, regarding the animal as an 'automaton' and, on the other hand, endowing it with full-blown, human-style reflective consciousness. This view is developed in a thoughtful discussion by Crook (1983) who lists five levels of conscious awareness, beginning with those systems that monitor the interdependence of different systems within the body. These arguments suggest that degrees of reflective consciousness will be found in different species just as they are in human infants of different ages.

4.5 PAIN

Colloquially, the term 'pain' is used to describe any unpleasant experience, physical or emotional. Thus humans may speak of the pain of a long separation or a bereavement. Here, however, the concern is with the physical kind of pain which, in general, helps humans to avoid damage, or prevent further damage, to their bodies. This kind of pain may take many forms (see Melzack and Wall 1982 for a full review). It may, for example, vary in intensity from the tolerable to the excruciatingly intolerable. It may be hard to localize, a dull diffuse ache, as in deep pain which originates in the viscera and in bones, muscles, tendons and so on, or it may be easier to pin-point, as in most cutaneous pain. It may be relatively short lasting, with a well-defined cause, such as an injury, or it may be more persistent and its cause may be more difficult to identify. To complicate matters further, similar events may evoke quite different responses in different individuals, or even in the same individual in different circumstances.

It is clear that discussions of pain in humans deal with a subjective state that has many facets. As might be expected, the neural mechanisms responsible for the production of pain sensations appear to be complex, and the way in which the body produces feelings of pain, in all its subtle dimensions, has yet to be fully understood. Nevertheless, some simplifying generalizations might be of use when considering other animals.

Most human pain responses have two components:

1. Nociception: the *physiological* perception of and response to painful stimuli which can be an automatic, reflex process not involving the highest parts of the brain.
2. Pain proper: the conscious *emotional* experience of pain, which involves nerve pathways in the highest part of the brain, the cerebrum (see Fig. 4.3).

Figure 4.3 provides a much simplified representation of the nervous pathways involved in both nociception and the experience of pain (for more detail see, for example, Melzack and Wall 1982, and Kitchell and Erickson 1983). In nociception, potentially harmful (noxious) stimuli may evoke electrical activity in special nerve endings (sensory receptors called 'nociceptors'), found in the skin and in other organs and tissues. These nociceptors are sensitive to potentially damaging extremes of temperature, to mechanical forces, such as those which occur when a muscle is pulled, and to chemical substances which are released when body tissues are damaged. Above a certain threshold level, electrical activity in the nociceptors evokes nerve impulses which are transmitted along nerve fibres to the spinal cord, and thence along spinal nerve tracts to the lower parts of the brain, the brain stem and thalamus.

As Iggo (1984) notes, nervous output from these lower parts of the brain is capable of initiating complex, coordinated, muscular and physiological responses to 'painful' stimuli, in the absence of any input to the higher centres of the brain—that is, in the absence of any conscious sensation of pain. In humans, however, impulses do pass from these lower to the higher brain centres, in particular the cerebral cortex, where the emotional experience of pain arises. Surgery carried out on humans has shown that removal of parts of the frontal cerebral cortex may remove this emotional aspect of pain—patients perceive the painful stimulus, but it does not bother them (Freeman and Watts 1946; Foltz and White 1962).

Figure 4.3 also indicates some of the inbuilt inhibitory nervous pathways, descending from the brain or found within the spinal cord, which may modify nervous input in the ascending pathways and so modify, sometimes ameliorating, the perception and experi-

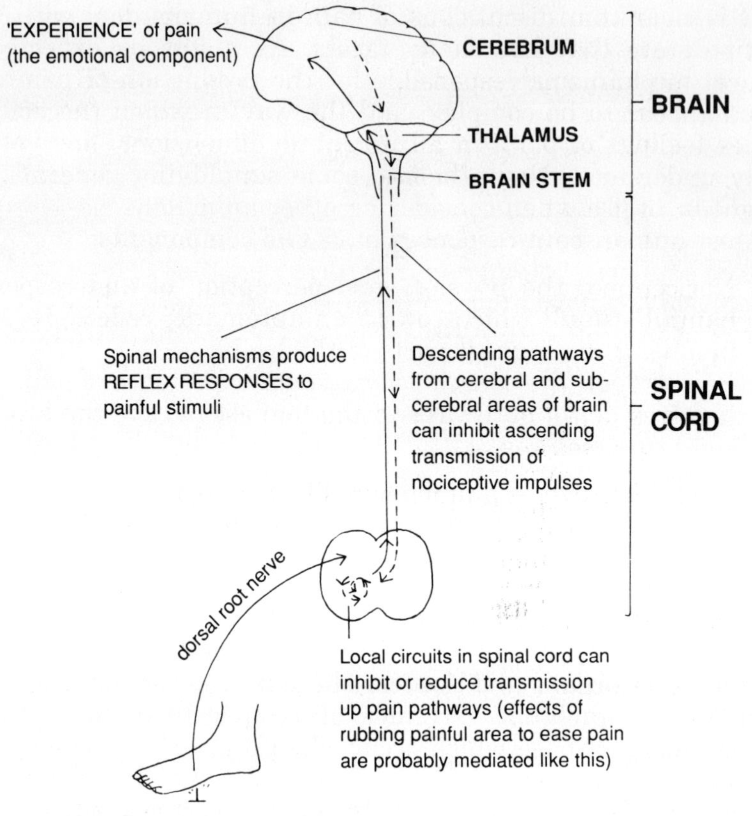

Fig. 4.3 Simplified representation of human pain 'pathways': ⟶, excitatory (nociceptive) pathways; – – →, inhibitory (anti-nociceptive) pathways.

ence of pain. Several chemical substances, required for transmission from one neuron to the next, are involved in the pain process. For example, opioid substances, which are released within the central nervous system (the brain and spinal cord), are able to modify pain perception. When combined with special receptor sites, they may stimulate the inhibitory pathways and so modify the transmission of nociceptive impulses. These chemicals (the enkephalins and endorphins) and their receptors are found especially in areas of the brain known to modify pain perception.

Externally administered drugs (such as morphine) act as analgesics, reducing the response to painful stimuli.

While physiological changes—such as increased heart and breathing rates, or constriction of the surface blood vessels—are associated with noxious stimulation and pain, such responses may all be produced without the involvement of the cerebral cortex. They occur without the person being aware of experiencing pain. In the human frontal lobotomy studies, removal of the higher centres changes the emotional experience of pain.

Recently developed techniques allow examination of patterns of energy utilization within the brains of both humans and animals. Since brain activity requires energy, these patterns show which parts of the brain are active at a given time. In one technique, a special form of glucose (2-deoxyglucose) is injected into the bloodstream. When it reaches the brain, it is not broken down like normal glucose so as to release the energy required by active neurones. It stays trapped and accumulates within the neuron. When such glucose is radioactively labelled, the accumulation of the trapped end products can be detected by radioactivity measurements, using a totally non-invasive technique, called positron emission tomography (Buchsbaum *et al.* 1983). Such techniques have shown that in response to painful stimulation of, for example, the hand, activity increases in the cerebral cortex (the outer layer of the cerebrum), particularly the frontal parts, of the brain. Studies of blood flow show also that the blood supply to the frontal parts of the brain increases during mildly painful stimulation, another indication that these parts of the brain are engaged in the pain response (see Iggo 1984).

After they have been hurt, humans' behavioural responses may persist so that, for example, they 'protect' the affected part of their body (perhaps by becoming less active, tending the part, limping and so on). They rapidly learn to act promptly in circumstances that were associated with painful experiences. In this way they serve further to protect themselves from potentially damaging events. Humans can be 'questioned' behaviourally by means of choice tests to find out whether one potentially painful experience is less unpleasant than another. They can also be asked how much they must be paid to endure a potentially painful experience—or how much they will pay to avoid it. Marian Dawkins (1980, 1990) has been at the forefront in applying these behavioural techniques in order to discover how much an animal may be deemed to suffer under particular conditions.

It is already plain that a reasonably long list of mechanisms and types of behaviour associated with the experience of pain might be

Table 4.2 Possible criteria for the experience of pain in animals

1. Possession of receptors sensitive to noxious stimuli, located in functionally useful positions on or in the body, and connected by nervous pathways to the lower parts of a central nervous system.
2. Possession of higher brain centres (especially a structure analogous to the human cerebral cortex).
3. Possession of nervous pathways connecting the nociceptive system to the higher brain centres.
4. Receptors for opioid substances found in the central nervous system, especially the brain.
5. Analgesics modify the animal's response to stimuli which would be painful for a human; the animal chooses to take analgesics in 'painful' situations.
6. The animal's response to stimuli that would be painful for a human is functionally similar to the human response (that is, the animal responds so as to avoid or minimize damage to its body).
7. The animal's behavioural response persists and it shows an unwillingness to resubmit to a painful procedure; the animal can learn to associate apparently non-painful with apparently painful events.

looked for in an animal. Table 4.2 summarizes these potential criteria for the experience of pain. It must be emphasized that such criteria should be considered *together* in order to draw any useful conclusions about the animal's experience of pain. On its own, the behavioural evidence is ambiguous, since persistence of a response after the stimulus has ended, memory, anticipation and learning, are readily explained in terms of simple neural mechanisms which do not require the postulation of any form of subjective experience on the part of the animal. Even so, such observations are all pieces in the overall jigsaw. Other ways of examining the animal may not have been yet thought of and may suggest themselves as a more thorough examination of the animal's complete way of life is made.

The animal must be looked at as a whole, interacting with its environment, since the evidence must be set in an appropriate context: is its response *functionally* similar to that expected of a human? Life-style is important since a horse, for instance, grazes normally after surgery in circumstances that might be expected to incapacitate a human for several weeks. Given a co-operative and caring social environment, a human's dependence on other people during recovery makes sense. Similarly, in the absence of individuals that will feed a stricken animal, a horse's apparently normal feeding after substantial injury also maximizes its chances of survival.

Table 4.3 shows the results of application of the criteria for

Table 4.3 Available evidence in the animal kingdom for the criteria used in the identification of pain

	Invertebrates			**Vertebrates**				
	Earthworm	Insect	Octopus	Fish	Amphibian	Reptile	Bird	Mammal
Nociceptive system	?	–	?	?	?	?	+	+
Brain structure analogous to human cerebral cortex	–	–	?	+	+	+	+	+
Nociceptors connected to higher brain structures	–	–	?	?	?	?	?	+
Opioid-type receptors	+	+	?	+	+	+	+	+
Responses modified by analgesics	?	?	?	?	?	?	+	+
Response to noxious stimuli persists	–	–	+/–	?	?	?	+	+
Associates neutral with noxious stimuli	–	+	+	+	+	+	+	+

identifying pain to some selected invertebrates (earthworms, insects and octopuses) and vertebrates (fish, amphibia, reptiles, birds and mammals). Much remains to be discovered, but taking the evidence as whole, the most obvious divide is between the vertebrates and the invertebrates. Earthworms show a characteristic withdrawal reflex response when subjected to a sharp poke. Muscles running lengthwise along the animal's body contract rapidly, so pulling the worm back into its burrow or away from the stimulus. However, if stimulation is repeated the earthworm may cease to respond at all (it is said to have habituated to the stimulus). Opioid substances and their receptors have been found in insects (El-Salhy *et al.* 1983; Stefano and Scharrer 1981). While many insects do not respond to potentially painful stimuli in a way that would help them to avoid or minimize damage to their bodies, some seem capable of learning associations between neutral and noxious stimuli (Marler and Terrace 1984). Wigglesworth (1980) suggests that insects may be particularly responsive to certain stimuli. He describes how the bug *Rhodnius* reacts violently, attempting to escape, when a heated needle is brought close to its antennae. Eisemann and others (Eisemann *et al.* 1984) refer to the 'writhing of insecticide poisoned insects ... and the production of sounds, repellent secretions and alarm pheromones by insects under attack'. However, Eisemann *et al.* (1984) also describe how insects will continue to feed whilst being eaten by predators, parasitoids or even, in the case of the male praying mantis, their sexual partners. Wigglesworth (1980) reports that:

> For the most part insects behave as though their integument is insensitive to pain. They show no manifestation of pain on cutting the cuticle: they cannot cry out but they do not flinch or run. Whereas a nip with forceps is very painful to us, a caterpillar treated in this way shows no sustained signs of agitation (p. 8)

Eisemann *et al.* (1984) report that no examples are known of insects showing protective behaviour towards injured body parts, such as limping after a leg injury, or becoming less active after abdominal injuries.

One invertebrate group stands out as being different from the others—the cephalopods. Both Young's (1965) and Wells' (1978) models of learning in the octopus include a 'pain' pathway leading into the vertical lobe of the brain, but Wells states that, although such a pathway is 'generally assumed to signal "pain" ... there is no proof of this and it might well carry "pleasure" or any other signal'. Octopuses have distinct and easy-to-recognize responses to noxious stimuli. They will withdraw, either the arm or the whole body, may produce a cloud of ink from the ink sac (normally used as a 'smoke-screen' which confuses potential predators), and will usually change colour.

The nervous pathways involved in nociception in mammals are as shown in Fig. 4.3. Similarly, in all other vertebrate groups nerve fibres transmit sensory information from receptors at the surface of the body to the spinal cord, and thence along spinal nerve tracts to the lower parts of the brain. Opioid-like substances and their receptors have been found in all of the vertebrate groups (see, for example, Pert *et al.* 1974; Simantov *et al.* 1976). All of the vertebrates show avoidance and escape responses to stimuli which would be painful for humans (see, for example, Kavaliers 1988). Only in mammals have physiological studies been used to show that the cerebral cortex is involved in responses to potentially painful stimuli. Anatomical studies have, however, demonstrated the existence of nervous pathways from the thalamus to the cerebral cortex in amphibia, reptiles and birds, as well as mammals. Similar pathways have been found in shark brains but, as yet, there have been few studies on the forebrains of bony fish (see Macphail 1982). Sensory information, therefore, may reach the higher centres of the brain in all vertebrates. Do these higher centres provide for the capacity for experiencing pain proper?

Until recently, the forebrain of fishes, amphibia and reptiles was thought to be almost entirely concerned with a sense of smell, so giving these animals little capacity for higher brain functions. Anatomical evidence has, however, shown that in these vertebrates the olfactory nerve pathways project into only small parts of the cerebral hemispheres (see Walker 1983). A variety of sensory input (nerve impulses from the sense organs and receptors) is integrated and motor output (nerve impulses going to the muscles and internal organs) is organized in the forebrains of fish, amphibia and reptiles. Macphail (1982) has suggested that: '. . . there is no reason to suppose that the fish telencephalon [cerebrum] is wholly taken up with sensory and motor analysis, that, in other words, there may be ample 'space' for higher level processes' (p. 56). Amphibians and reptiles are not likely to be any different.

Consideration of relative brain sizes and, in particular, the size of the cerebral hemispheres (see Figs 4.2 and 4.3), suggests that, if these 'lower' vertebrates have complex processing abilities (including perhaps the capacity for experiencing pain), these might not be as highly developed as in mammals. Walker (1983), for example, supports:

> . . . the idea that the functions of the different divisions of the brain are more or less similar in all vertebrates, and that therefore we should

expect . . . higher psychological functions to be present roughly in proportion to how well the forebrain seems to be developed. A fish, frog, or lizard does not have much by way of cerebral hemispheres, and should therefore have very little cognition.

Mammals, including humans, all have a similar brain structure and so may all be able to experience pain in some way. Iggo (1984) has suggested that the nature and extent of this experience may, however, be related to the size of the cerebral hemispheres, so that mammals such as rats and rabbits may have different kinds of experience from, say, dogs and cats which have larger cerebrums. Apes, whose cerebral hemispheres approach human dimensions, may have experiences which are more closely similar to those of humans.

Drugs which act as analgesics in humans may also reduce or remove certain behavioural signs associated with pain in mammals (for example, Taylor and Houlton 1984). If behavioural signs of pain are reduced, this might suggest that the animal is experiencing pain in a similar way to a human. The main problem here, though, lies in the lack of information on the availability and specificity of analgesics for use in most species of animal (see Flecknell 1986). When this is the case, an argument based on the pain-reducing use of a substance runs the risk of becoming circular: 'analgesics' are those substances that eliminate 'signs of pain' which are those signs eliminated by 'analgesics' (Bateson in press). Mammals may also choose to take analgesics for themselves: Colpaert and co-workers (1980), have shown that rats with chronic inflammation of the joints will select drinking water containing analgesic in preference to a highly palatable sweet solution, whereas normal rats prefer the sweet solution. Such self-administration of pain relief may perhaps be taken as evidence that rats do experience pain in some way.

In conclusion, while most invertebrates may not be able to experience pain, cephalopods like the octopus, have much more complex nervous systems and behaviour and give cause for concern. While the character of their bodies is very different from that of vertebrates, the general level of organization suggests that these animals might well be able to experience pain.

The evidence presented here suggests that all vertebrates may have the ability to experience pain in some way. It is possible, however, that this experience varies both qualitatively and quantitatively between the different vertebrate groups, owing to variation in development of the frontal lobes of the brain in the different groups. Iggo (1984) suggests that the difference in pain experience between humans and other animals may be one of degree, in that higher animals may 'have the capacity to experience, in an unspecified but lesser degree, some of those aspects of pain that cause humanity to execrate it'.

4.6 STRESS

'Stress' is used colloquially to describe the state associated with difficulties in coping. The demands, or stressors, placed upon people may take many forms. A change in environment (such as a move from the country to a town), a changed pattern or level of activity (such as an interrupted night's sleep, or an unusually demanding day at the office), a change in one's own body (perhaps some persistent pain), or an emotional change (such as the ending of a long standing relationship) may all be examples of possible stressors. When faced with such stressors, humans may engage, knowingly or unknowingly, in a variety of behavioural, psychological or physiological changes which help to adapt to the challenges superimposed on the usual way of life. Things become problematic when, for example, the stressor is severe or prolonged, so that the body systems are stretched to their limits or, in extremes, simply cannot deal with the challenge. When humans are aware that such problems arise, they commonly describe themselves as 'distressed'.

When faced with a short-lasting stressor, such as a sudden and/or novel event, pain, frustration, or high intensity stimulation (such as a loud noise, or a bright light) an alarm reaction may occur (see Selye 1973, for example). Part of the nervous system, the sympathetic part of the autonomic nervous system, is activated and the hormones adrenalin and noradrenalin are secreted from the inner part of the adrenal gland (the adrenal medulla). This nervous and hormonal activity produces a variety of physiological changes which serve to get us ready for action—perhaps to run away from the stressor. Such physiological changes include an increased cardiac output (increased rate and power of heart beat), increased blood flow to the muscles and, associated with this, constriction of the surface blood vessels (giving the skin an ashen coloration), as well as mobilization of the body's energy stores and dilation of the pupils. These responses are relatively short term, helping to adjust quickly to the potential threat posed by the stressor.

With more prolonged stressors, or perhaps with repeated activation of the alarm response, a complex series of hormonal changes may occur. These changes may influence just about every physiological system in the body (see Moberg 1985). Although the pattern of hormonal change varies with different stressors, secretion of corticosteroid hormones from the outer part of the adrenal gland (the adrenal cortex) appears to be a common feature in many, but not all, such responses (Archer 1979). The corticosteroids help the body to mobilize its resources by, for example, promoting release of glucose from storage into the bloodstream, increasing heart rate and, in the longer term, enhancing blood cell production. Such chronic stress responses help us to adapt to persistently challenging situations. A delicate balance is maintained, however, between the adaptive effects of these responses and their potential harmful effects on the body: if a stressor is prolonged or severe, or several stressors act at the same time or in succession, long-term mobilization of the body's coping responses may result in harmful (pathological) changes. These may include arteriosclerosis and associated high blood pressure (which may lead to heart failure), stomach ulcers, reduction in resistance to infection, depression or abnormal behaviour (Archer 1979) as well as alteration of the body's metabolism (usually resulting in weight loss or a slower rate of weight gain in growing children) and fertility or sexual problems (Moberg 1985).

The stress responses are controlled and integrated by the brain. Centres below the level of the cerebral cortex control all of the hormonal, nervous and behavioural responses described, so that all of the responses can take place without our being aware of them. Indeed, much of the time humans *are* unaware of the complex and subtle coping responses occurring everyday within their bodies. However, many human stress responses have psychological components, particularly when difficulties are experienced in coping with more long-term stressors. These are discussed below under anxiety. Humans are aware of being stressed and find it unpleasant or, if intense or prolonged, unbearable.

Turning again to an animal, as was done for pain, the criteria for identifying stress are set out in Table 4.4. It is necessary to look at how the animal responds to potentially stressful situations and to examine the animal's normal way of life in order to make an informed guess about the kinds of situation that might be stressful for it. If, for example, the animal lives in wide open spaces, confinement in a small space, even briefly, might be a potent stressor; conversely, if the animal normally lives in confined spaces, placing it in an open arena might be stressful. Novel or suddenly-appearing stimuli might be used as a general test for the alarm reaction type of response. Since human stress responses, like pain responses, vary both within and between individuals and according

Table 4.4 Possible criteria for the experience of stress

1. Possession of a sympathetic nervous system and an associated hormonal system.
2. Production of hormones similar to adrenalin, noradrenalin, and/or corticosteroids.
3. Possession of higher brain centres (especially a structure analogous to the human cerebral cortex).
4. The animal shows behavioural, and/or physiological changes when faced with a potential stressor.
5. These responses may involve the sympathetic nervous system and adrenal hormones.
6. Behavioural studies suggest that higher brain centres are involved in the animal's responses to stressors (for example, the animal remembers and anticipates stressful situations and can learn to associate apparently non-stressful with apparently stressful events).
7. Physiological studies (for example, brain ablation and metabolism studies) suggest involvement of higher brain centres in the animal's responses to stressors.

to the nature of the stressor, the pattern of response is not easily characterized. Almost any physiological change might be an indicator of stress. Activation of the sympathetic nervous system, secretion of adrenalin and noradrenalin, and/or raised levels of corticosteroids in the blood are, however, components of many human stress responses. Similar responses in an animal might suggest that it is stressed, physiologically, in a similar way to a human, although such activation and secretion may occur in the course of many other unstressed states.

It will be much less easy to decide whether there is a psychological aspect to this physiological stress response. Whether the animal is aware that it is stressed and whether it experiences the worrying, emotional aspects of stress in the way that a human might, will depend (as for the experience of pain) on the involvement of the highest centres of the brain, in particular the cerebral cortex. Does the animal remember previously stressful experiences, showing unwillingness to resubmit to similar procedures, and learn to associate neutral and stressful events? Are the higher brain centres involved in the animal's response in stressful situations?

Most work on stress responses has been carried out using mammals (see Archer 1979). Other mammals share with humans 'a largely identical set of behavioural and physiological responses to [stressful] situations' (Archer 1979, p. 5). Mammals, according to Archer, share with humans some psychological mechanisms for activating stress responses. Such mechanisms detect potential danger by comparing what is perceived with what is expected: when the features of the environment cannot be interpreted in terms of prior knowledge of the environment, the discrepancy can evoke emotions, such as anxiety, anger and fear, and the same sort of physiological reactions, such as increased adrenocortical function, in both humans and other, non-human, mammals. Non-human mammals may, however, differ from humans in that they do not have the ability to activate stress reactions 'as a result of circumstances interpreted in complex cognitive terms' (see Archer 1979 pp. 4–5).

While the evidence for stress is less complete and satisfactory than for pain, it seems likely that all vertebrates may be stressed in a similar way to humans. The way in which the different groups of vertebrates may 'experience' stress may be related to brain development in a way similar to that described for pain.

4.7 ANXIETY

Human emotional states such as anxiety and fear may arise as correlates of pain or stress, or may arise quite independently of any specific physiological changes. Once again, such states are difficult to describe or define with any precision. Nevertheless, humans know what it is like to feel anxious or fearful and may be able to describe their feelings to fellow human beings. Such states are usually perceived as unpleasant, sometimes very unpleasant, and, if prolonged, may compromise our psychological and physiological well-being. These states do, however, have 'survival value' since they serve to increase our alertness or vigilance, getting us ready to face a perceived threat, so helping us to avoid or escape from that threat.

Several authors have attempted to discriminate between anxiety and fear, but none of these distinctions is entirely satisfactory. Kitchen *et al.* (1987), for example, have described anxiety as 'a generalised, unfocused response to the unknown' and fear as 'a focused response to a known object or previous experience'. Humans, however, often talk of 'fear of the unknown': are they really fearful, or are they anxious? It is probably best not to be too concerned with niceties of meaning here. In this discussion anxiety is used as a blanket term to cover all of the different human states of apprehension, which may include fear.

Rowan (1988) has described several behavioural and physiological signs which may accompany anxiety in humans:

(1) motor tension: shakiness and jumpiness;
(2) hyperactivity of the sympathetic nervous system: producing, for example, sweating, increased heart and breathing rates (a pounding heart, quick shallow breathing), frequent urination, diarrhoea;
(3) apprehensive expectation: anticipation of, and perhaps attempts to avoid, what lies ahead;
(4) hyperattentiveness: increased vigilance (wariness) and scanning (constantly looking about).

The most effective anti-anxiety drugs in humans are the benzodiazepines. These bind to specific receptors, found in parts of the central nervous system which are associated with the expression of emotion.

Possible criteria for identifying anxiety are given in Table 4.5. As Rowan (1988) has pointed out, lack of evidence does not necessarily mean that the animal is unable to experience anxiety. The animal might have quite different physiological or behavioural systems capable of giving rise to anxiety. Even in humans the mode of action of several anxiolytics does not appear to involve binding to benzodiazepine receptors (File 1987).

Benzodiazepine receptors have yet to be found in invertebrate species: Nielsen, Baestrup and Squires (1978) failed to find such receptors in earthworms, locusts or squid, amongst other invertebrates (octopuses have yet to be investigated). However, several writers suggest that anxious states *might* exist in *Octopus*. Mather (1989) describes how: 'when I recently had two octopuses in one aquarium and one of the animals changed from its known daytime

Table 4.5 Possible criteria for the experience of anxiety

1. Possession of higher brain centres (especially a structure analogous to the human cerebral cortex).
2. Possession of benzodiazepine receptors in the central nervous system.
3. Specific behavioural and physiological signs (for example, motor tension; hyperactivity of the sympathetic nervous system; apprehensive expectation; hyperattentiveness) are observed in conditions which might be expected to cause anxiety in the animal.
4. Human anti-anxiety drugs reduce or remove these signs.
5. Higher brain centres appear to be involved in the production of the animal's anxious state.

activity in a hidden location, and took on a coloration previously associated with aversive stimuli, I had surmised before I saw the other animal attack it that there was a problem'. Although rejecting man's ability to empathize with the octopus, Wells (1978) also seems able to detect fear or sickness in his animals. Packard and Sanders (1969) have described a particular type of colour pattern (the 'broad, or conflict, mottle') which is worn when an octopus is undecided whether to attack or not. This has been shown to happen when the animal receives a shock from an object which it previously approached with confidence—perhaps this coloration indicates that the animal is now anxious about what will happen when it next approaches the object?

Mammals are used routinely in testing new human anti-anxiety drugs. In the main, two kinds of anxious situation may be used:

(1) the animal is given electric shocks when carrying out its normal activities such as feeding or drinking. Thus anxiety is generated since the animal is placed in a conflict between its need to feed and drink and its need to avoid damage;
(2) the animal is placed in a novel situation such as a brightly lit open space.

Mammals put into such situations show the characteristic signs of anxiety (motor tension, inhibition of activity, autonomic hyperactivity, apprehensive expectation and vigilance and scanning) mentioned previously. The efficacy of an anti-anxiety drug is then measured in terms of its ability to reduce or abolish such signs in the animals. Gray (1982) has argued that the neurophysiological and behavioural results from such drug studies strongly indicate that an emotional state, analogous to human anxiety, occurs in animals (see Rowan 1988). Gray has not, however, suggested that these emotional states are identical in humans and animals. As Iggo (1984) has done for pain, Gray argues that the frontal lobes of the brain play an important role in mediating anxiety and that human and non-human mammalian anxiety differ since the frontal lobes are more highly developed in humans than in other mammals. Similarly, anxious states might be expected to be different in different kinds of non-human mammal, since the size of the cerebral cortex varies between the different mammalian orders (see Fig. 4.1).

To conclude, while most invertebrates are unlikely to experience anxiety, the evidence suggests that octopuses *may* do so. Mammals seem to experience anxiety. The evidence from non-mammalian vertebrates is inconclusive, but suggests that these animals may rather than may not experience anxiety.

4.8 GENERAL CONCLUSIONS

Although pain, stress, and anxiety are all defined as subjective experiences, our assessment that another individual experiences these things is not simply based on whim. Such capacities or states are inferred from a collection of independent pieces of evidence. The more that the evidence points to one conclusion rather than another, the more plausible the inference. Of course, people differ in their readiness to accept a plausible inference. Some will only trust the evidence of their own sensations. Others are only prepared to generalize from their own feelings to other human beings. The point that is stressed in this chapter is that if it is rational to do that, it is no less rational to extend the generalization to other species.

Many mechanisms and behavioural signs are associated with pain in humans, although no sign can itself be regarded as diagnostic of pain and may also occur in conditions in which pain is unlikely to be involved. When generalizing to other animals, particularly those with nervous systems different from humans, good observational data on their normal behaviour, requirements and ecological conditions are needed. None the less, the fuzziness of the boundary between pain and non-pain asserts itself as, one by one, the criteria for recognizing pain cease to apply as less and less complicated animals are considered. The same is true for stress and anxiety. And it has become obvious that continuities between mere sentience and full reflective consciousness must be accepted.

It is no surprise that the criteria used to recognize subjective states should be exceedingly fuzzy at the edges. Even though the law (in the United Kingdom at least) insists on precision when dealing with animals, a cut off point which is anything other than arbitrary cannot be provided. However, criteria that are based on modern methods of measuring and investigating behaviour and on analyzing the character of the nervous system can be used. If the animal stops activities which it habitually performs in conditions that might be supposed to produce pain, if it learns to avoid those conditions, and if it has parts of its nervous system dedicated to damage avoidance, then some grounds for worrying that it might feel something exist. On this basis, *Octopus* is a cause for concern. But clearly, those animals showing least evidence of suffering pose far fewer ethical problems for research than those that are very similar to humans.

It is a matter of taste where projection and empathy stop. The law governing the use of animals in research in the United Kingdom has clearly recognized that fact by giving special protection to dogs, cats and equines, the animals with which people identify most. Equally clearly, tastes change. Twenty years ago it would have been hard to find an insect physiologist wondering whether or not insects felt pain. Nowadays, a number do (for example Eisemann *et al.* 1984; Wigglesworth 1980).

Although the emphasis in this chapter has been on the top down approach, knowledge of the ecology and behaviour of an animal very different from humans may change the way that humans think about that animal. When environmental conditions cause changes in the animal which suggest that it is defending itself from a challenge, its welfare may be in jeopardy. Scientists who work with intact animals are increasingly willing to give the animal the benefit of the doubt, which certainly seems to be the appropriate ethical stance. Is it not better to treat an insentient animal as though it were sentient, rather than the other way round?

REFERENCES

Archer, J. (1979). *Animals under stress.* Edward Arnold, London.

Barnes, R. D. (1980). *Invertebrate zoology.* Saunders College, Philadelphia, Pennsylvania.

Bateson, P. (In press). The assessment of pain in animals. *Animal Behaviour.*

Blakemore, C. and Greenfield, S. (1987). *Mindwaves.* Blackwell, Oxford.

Boycott, B. B. (1961). The functional organization of the brain of the cuttlefish *Sepia officinalis. Proc. R. Soc.* **B153,** 503–34.

Bowman, R. I. (1961). Morphological differentiation and adaptation in the Galapagos finches. *Univ. Calif. Pub. Zool.* **58,** 1–326.

Buchsbaum, M. S., Holcomb, H. H., Johnson, J., King, A. C. and Kessler, R. (1983). Cerebral metabolic consequences of electrical cutaneous stimulation in normal individuals. *Human Neurobiol.* **2,** 35–8.

Byrne, R. and Whiten, A. (1987). The thinking primate's guide to deception. *New Scientist* **116** (1589), 54–7.

Colpaert, F. C., de Witte, P. C., Maroli, A. N., Awouters, F., Niemgeens, C. A. and Janssen, P. A. J. (1980). Chronic pain. *Life Sci.* **27,** 921–8.

Crook, J. H. (1983). On attributing consciousness to animals. *Nature* **303,** 11–14.

Dawkins, M. S. (1980). *Animal suffering: the science of animal welfare.* Chapman and Hall, London.

Dawkins, M. S. (1990). From an animal's point of view: motivation, fitness and animal welfare. *Behav. Brain Sci.* **13,** 1–61.

Dethier, V. G. (1976). *The hungry fly.* Harvard University Press, Cambridge, Massachusetts.

Dickinson, A. (1980). *Contemporary animal learning theory.* Cambridge University Press, Cambridge.
Edwards, C. A. and Lofty, J. R. (1977). *Biology of earthworms* (2nd edn). Chapman and Hall, London.
Eisemann, C. H., Jorgensen, W. K., Merritt, D. J., Rice, M. J., Cribb, B. W., Webb, P. D. and Zalucki, M. P. (1984). Do insects feel pain?—A biological view. *Experientia* **40,** 164–7.
El-Salhy, M., Falkmer, S. Kramer, K. J. and Spiers, R. D. (1983). Immunohistochemical investigations of neuropeptides in the brain, corpora cardiaca, and corpora allata of an adult lepidopteran insect, *Manduca sexta* (L). *Cell Tissue Res.* **232,** 295–317.
File, S. E. (1987). The search for novel anxiolytics. *Trends Neurosci.* **10,** 461–3.
Flecknell, P. A. (1986). Recognition and alleviation of pain in animals. In *Advances in animal welfare science 1985* (ed. M. W. Fox and L. D. Mickley), pp. 61–78. Martinus Nijhoff, Dordrecht.
Foltz, E. L. and White, L. E. (1962). Pain 'relief' by frontal cingulumotomy. *J. Neurosurg.* **19,** 89–100.
Freeman, W. and Watts, J. W. (1946). Pain of organic disease relieved by frontal lobotomy. *Lancet* **i** (245), 953–5.
Gallup, G. C. (1970). Chimpanzees: self-recognition. *Science NY* **167,** 86–7.
Gallup, G. C. (1977). Self recognition in primates—a comparative approach to the bidirectional properties of consciousness. *Amer. Psychol.* **32,** 329–38.
Goodall, J. (1971). *In the shadow of man.* Collins, London.
Gray, J. A. (1982). *The neuropsychology of anxiety.* Oxford University Press, Oxford.
Griffin, D. R. (1984). *Animal thinking.* Harvard University Press, Cambridge, Massachusetts.
Holland P. C. and Straub, J. J. (1979). Differential effects of two ways of devaluing the unconditional stimulus after Pavlovian appetitive conditioning. *J. Exp. Psychol. Anim. Behav. Proc.* **5,** 65–78.
Hubbard, J. I. (1975). *The biological basis of mental activity.* Addison-Wesley, Reading, Massachusetts.
Hubel, D. H. (1979). The brain. *Sci. Amer.* **241** (3), 38–47.
Humphrey, N. K. (1986). *The inner eye.* Faber and Faber, London.
Iggo, A. (1984). *Pain in animals.* Third Hume Memorial Lecture, 15 November 1984. Universities Federation for Animal Welfare, Potters Bar.
Jerison, H. J. (1969). Brain evolution and dinosaur brains. *Amer. Natur.* **103,** 575–88.
Kavaliers, M. (1988). Evolutionary and comparative aspects of nociception. *Brain Res. Bull.* **21,** 923–31.
Kagan, J. (1981). *The second year: the emergence of self-awareness.* Harvard University Press, Cambridge, Massachusetts.
Kenyon, K. W. (1969). *The sea otter in the eastern Pacific Ocean.* North American Fauna, no. 68. US Bureau of Sport Fisheries and Wildlife, Washington DC.

Kitchell, R. L. and Erickson, H. H. (1983). eds. *Animal pain: perception and alleviation*. American Physiological Society, Bethesda, Maryland.
Kitchen, H., Aronson, A. L., Bittle, J. L., Mcpherson, C. W., Morton, D. B., Pakes, S. P., *et al.* (1987). Panel report on the colloquium on recognition and alleviation of animal pain and distress. *J. Amer. Vet. Med. Assoc.* **191,** 1186–91.
Lack, D. (1947). *Darwin's finches*. Cambridge University Press, New York.
Lewis, M. and Brooks-Gunn, J. (1979). *Social cognition and the acquisition of self*. Plenum Press, New York.
McFarland, D. (1989). *Problems of animal behaviour*. Longman, Harlow, Essex.
McMahan, E. A. (1982). Bait-and-capture strategy of termite-eating assassin bug. *Insectes Sociaux* **29,** 346–51.
Macphail, E. M. (1982). *Brain and intelligence in vertebrates*. Clarendon Press, Oxford.
Marler, P. and Terrace, H. S. (1984). *The biology of learning*. Springer-Verlag, Berlin.
Mather, J. A. (1989). Ethical treatment of invertebrates: how do we define an animal? In *Animal care and use in behavioral research: Regulations, issues and applications* (ed. J. W. Driscoll), pp. 52–9. Animal Welfare Information Center, National Agricultural Library, USA.
Melzack, R. and Wall, P. D. (1982). *The challenge of pain*. Penguin Books, Harmondsworth.
Millikan, G. C. and Bowman, R. I. (1967). Observations on Galapagos tool-making finches in captivity. *Living Bird* **6,** 23–41.
Moberg, G. P. (1985). Biological response to stress: Key to the assessment of animal well-being? in *Animal stress* (ed. G. P. Moberg), pp. 27–49. American Physiological Society, Bethedsa, Maryland.
Nielson, M., Baestrup, C., and Squires, R. F. (1978). Evidence for a late evolutionary appearance of a brain-specific benzodiazepine receptor. *Brain Res.* **141,** 342–6.
Packard, A. (1972). Cephalopods and fish: The limits of convergence. *Biol. Rev.* **47,** 241–307.
Packard, A. and Sanders, G. D. (1969). What the octopus shows to the world. *Endeavour* **28,** 92–9.
Pearce, J. M. (1987). *An introduction to animal cognition*. Erlbaum, Hillsdale, New Jersey.
Pepperberg, I. M. (1987). Acquisition of the same/different concept by an African Grey parrot (*Psittacus erithacus*): Learning with respect to categories of color, shape, and material. *Anim. Learn. Behav.* **15,** 423–32.
Pert, C. D., Aposhian, D., and Snyder, S. H. (1974). Phylogenetic distribution of opiate receptor binding. *Brain Res.* **75,** 356–61.
Premack, D. (1983). The codes of man and beasts. *Behav. Brain Sci.* **6,** 125–67.
Rose, M. and Adams, D. (1989). Evidence for pain and suffering in other animals. In *Animal Experimentation* (ed. G. Langley), pp. 42–71. Macmillan, London.

Rowan, A. N. (1988). Animal anxiety and animal suffering. *Appl. Anim. Behav. Sci.* **20,** 135–42.

Russell-Hunter, W. D. (1979). *A life of invertebrates.* Macmillan, London.

Selye, H. (1973). The evolution of the stress concept. *Amer. Sci.* **61,** 692–9.

Simantov, R., Goodman, R., Aposhian, D. and Snyder, S. H. (1976). Phylogenetic distribution of a morphine-like peptide 'enkephalin'. *Brain Res.* **111,** 204–11.

Smith, A. P. (1978). An investigation of the mechanisms underlying nest construction in the mud wasp, *Paralastor* sp. (Hymenoptera: Eumenidae). *Anim. Behav.* **26,** 232–40.

Stefano, G. B. and Scharrer, B. (1981). High affinity binding of an enkephalin analog in the cerebral ganglion of the insect *Leucophaea maderae* (Blattaria). *Brain Res.* **225,** 107–14.

Taylor, P. M. and Houlton, J. E. F. (1984). Post-operative analgesia in the dog: A comparison of morphine, buprenorphine and pentazocine. *J. Small Anim. Pract.* **25,** 437–51.

Walker, S. F. (1983). *Animal thought.* Routledge and Kegan Paul, London.

Wells, M. J. (1978). *Octopus.* Chapman and Hall, London.

Whiten, A. and Byrne, R. W. (1988). Tactical deception in primates. *Behav. Brian Sci.* **11,** 233–73.

Wigglesworth, V. B. (1980). Do insects feel pain? *Antenna* **4,** 8–9.

Young, J. Z. (1965). The organisation of a memory system. *Proc. R. Soc.* **B162,** 47–79.

5

Animals as experimental subjects

Strategies for recognizing, assessing, and reducing the costs imposed on animals in biomedical research

The previous discussion asked whether various kinds of animals are capable of experiencing adverse states such as pain, distress, and anxiety. Although the nature of such experiences may vary both qualitatively and quantitatively between the different animal species, it seems that the vertebrates and 'higher' invertebrates (such as *Octopus*), at least, may experience those states which humans find unpleasant or, in extremes, intolerable. Hence, experiments can impose 'costs', in terms of suffering, on these animals. If we wish to examine the moral justification for imposing such costs on animals used in biomedical research, we need to find ways of assessing the severity of the costs involved, and to identify strategies for reducing these costs wherever possible. The degree of cost imposed in a particular piece of research will, in part, be related to the kind of animal used in the experiments, since different animals have different capacities for experiencing adverse states. Costs will also vary according to the type of experiment performed on the animals, since different scientific procedures and husbandry conditions (and indeed even who carries out the procedures and in what facilities) are likely to cause different kinds and amounts of harm to the animals. However, in order to be in a position to assess the likely costs imposed on these animals, we must first find ways of recognizing adverse effects, such as pain, distress and anxiety, caused to the animals by the experiments.

5.1 RECOGNITION OF ADVERSE EFFECTS IN ANIMALS

Researchers using animals, and others involved in the care of experimental animals, frequently have statutory as well as moral duties to predict, recognize, assess, and so far as possible reduce and relieve any pain, or other adverse effect, suffered by the animals. This is so in the European Community (EC), for example, since, according to Directive 86/609/EEC, on the protection of animals used for experimental or other scientific purposes:

'all experiments shall be designed to avoid distress and unnecessary pain and suffering to the experimental animals'; 'in a choice between experiments those which . . . cause the least pain, suffering, distress or lasting harm shall be selected'; 'the well-being and state of health of the animals shall be observed by a competent person to prevent pain or avoidable suffering, distress or lasting harm'; and 'arrangements shall be made to ensure that any defect or suffering discovered is eliminated as quickly as possible.'

In the UK, licensees under the 1876 Cruelty to Animals Act were required to determine whether an experiment was 'calculated to give pain' and to 'prevent the animals feeling pain', as well as to be able to recognize when animals were in 'severe pain . . . likely to endure' and 'suffering considerable pain'. Now, under the Animals (Scientific Procedures) Act 1986, applicants for project licences are required to assess the likely 'severity' of the scientific procedures which are to be undertaken, so that this can be balanced against the potential benefits of the work (Home Office Guidance Notes 1990, para. 4.6, p. 9). Severity encompasses not only pain but also suffering, distress, or lasting harm. Such terms are taken to include 'death, disease, injury, physiological or psychological stress, significant discomfort or any disturbance to normal health, whether in the short-term or long term' (Home Office 1990, para. 1.10, p. 2). Project licences, when granted, contain a condition which states a maximum degree of severity authorized for that particular project. Both personal and project licensees are required to ensure that this condition is not breached; that any pain, suffering or distress caused to the animals is minimized and that the maximum severity is approached 'only when absolutely necessary' (Home Office 1990, para. 4.17, p. 11). In addition, the two Named Persons under the 1986 Act (the Named Veterinary Surgeon and the Named Person in Day-to-Day Care) must be able to recognize, and should inform the relevant personal and project licensees, when an animal's condition gives cause for concern.

It is clear that, in order to fulfil these statutory requirements (which themselves reflect moral duties), researchers and those responsible for the care of the animals need to find ways of recognizing and assessing signs of pain, discomfort and distress, in their daily practice.

5.1.1 Ways of recognizing adverse effects in animals

Those who spend time working with animals are able to recognize when 'something is wrong' with those animals. They seem to be able to recognize when an animal is suffering adverse effects, and may even be able to quantify those effects.

By observing particular animals under normal conditions (that is, when they are not involved in an experiment), norms of behaviour and physiology for species, strains and individuals can be established, so permitting recognition of deviations from these norms. As was noted in Chapter 4, such observations, interpreted with reference to the species' general biology and likely human experience under similar conditions, can lead to plausible inferences about the pain, distress or anxiety experienced by the animals.

Perhaps the most important maxim here is 'know your animal', an ideal realized by the 'good stockman' (Oldham 1988). This is the approach offered by Morton and Griffiths (1985). On the basis of the experience of 'animal technicians, animal nurses, research workers and veterinary surgeons who have worked with animals for many years', Morton and Griffiths have drawn up a list of behavioural, physiological and clinical criteria which they believe can be used in the recognition of pain, distress, and discomfort in mammals. They also describe how deviations from the norm in appearance; behaviour, including 'provoked behaviour' (the animals' reaction to external stimuli); food and water intake; and clinical signs relating to cardiovascular, digestive, and nervous criteria, may be related to the severity of the adverse effects experienced by the animals (Tables 5.1 and 5.2).

Some of the criteria listed in Table 5.1, such as body weight and pulse rate, are objective and measurable. Others, such as appearance and behaviour, are more subjective, requiring interpretation by an observer who has training in the recognition of such signs and a knowledge of the norms for the particular animals involved. For particular experiments it may be possible to draw up lists of 'cardinal' signs and provide a scoring scheme which indicates the action to be taken when certain combinations of signs are observed. Table 5.3 shows the signs established by Morton (1987) for studies involving corneal grafting and orthopaedic fracture work.

It is important that a variety of criteria be assessed, since an animal may be normal in some respects, whilst showing strong deviations from the norm in other ways. This is also emphasized in a report by a Working Party of the Association of Veterinary Teachers and Research Workers (Sanford *et al.* 1989), on the assessment of pain in animals. Sanford *et al.* have put forward a scheme for the systematic assessment of pain, which includes examination of the animals for signs of disease, assessment of

Table 5.1 Signs and their severity in the assessment of pain, distress, and discomfort in mammals

	Normal (0)	Mild (1)	Moderate (2)	Severe (3/4)
Appearance		Coat loses sheen, hair loss, starey—harsh		
		Failure to groom, soiled perineum		
		Discharge from eyes and nose		
	Eyelids partly closed			
			Eyes sunken and glazed	
		Hunched up look		
		Respiration laboured, abnormal panting		
				Grunting before expiration; grating teeth
Food/water intake	Reduced			Zero (prolonged)
	Faecal/urine output reduced			Zero
Behaviour	Away from cage mates, isolated;			Unaware of extraneous activities or bullying from mates
		Self mutilation		
		Restlessness, reluctant to move, recumbent		
		Change in temperament		
		Squealing, howling, etc, especially when provoked		
Clinical signs Cardiovascular	Strong pulse			Weak pulse
		Cardiac rate increased or decreased		
	Abnormal peripheral circulation			
		Pneumonia, pleurisy		
Digestive		Altered faecal volume, colour, consistency		
		Abnormal salivation		
		Vomiting (high frequency)		
				Boarded abdomen as in peritonitis
Nervous (musculoskeletal)			Lameness and arthritis	
		Twitching		Convulsions

From Morton and Griffiths (1985).

Table 5.2 Possible interpretation of scores from an overall assessment of an animal

Total score	Overall assessment
0 to 4	Normal
5 to 9	Monitor carefully, should consider the use of analgesics and sedatives
10 to 14	Ample evidence of suffering, some form of relief must be seriously considered; should be under regular observation; seek expert advice; consider termination
15 to 20	Relief should be given, unless the animal is comatose. Is it a worthwhile experimental animal, because physiologically it is likely to be abnormal? There is ample evidence of severe pain. If likely to endure, terminate the experiment

From Morton and Griffiths (1985).

possible physiological correlates of pain (dilation of pupils, changes in blood pressure and in heart and breathing rates, sweating, erection of hair, etc), indications of abnormal activity, examination of the animal's gait, posture, facial expression, resistance to handling, patterns of vocalization and response to analgesics as well as assessment of the presumed mental state of the animal (for example, whether the animal is 'bright' or 'dull', 'alert' or 'depressed', 'aware' or 'unaware', 'aggressive', 'timid', 'excitable', and so on). Another report, by a Working Party of the Laboratory Animal Science Association (Wallace *et al.* 1990), adopts a similar approach. The report concentrates on six common laboratory species; rats, mice, guinea-pigs, hamsters, gerbils and rabbits. For each species, descriptions of signs which are believed to indicate pain and distress are given under a series of headings, including general appearance, respiratory and ocular characteristics, changes in defaecation/urination, behaviour, abnormal activity, posture, locomotion, vocalization and other signs. A list of 'key signs' for each species is abstracted.

5.1.2 Comments on schemes for the recognition of adverse effects

The schemes described above offer a useful starting point in the recognition of adverse effects in animals. There are, however, several limitations to be borne in mind when considering their application.

Table 5.3 Important clinical signs to assess the severity of adverse effects and determine further action in studies involving (a) corneal grafting and (b) tibial fractures in the rabbit. A total score of ⩾ 10 would require some predetermined action, e.g. call licensee.

	Clinical signs	Rating
(a)	Contact lens come out	10
	Scratching or rubbing eye	10
	Cells in anterior chamber of eye	10
	Eye discharge: heavy to moderate	10–5
	Eyelids: tightly to half closed	10–5
	Temperature (°C): >40 to 39+	10–5
	Behaviour: depressed, lethargic	5–3
	Response to light touch near eye: violent avoidance or trembling	5–3
	Eating and drinking: nothing to some; Body weight lost: 200 to 100g since start	10–5
(b)	Plaster removed or come off	10
	Chewing at plaster/self-mutilation	7
	Temperature (°C): >40 to 39+	10–5
	Behaviour: depressed, lethargic or abnormal	5–3
	Response to light touch on limb: violent avoidance to trembling to vocalization	5-3
	Eating and drinking: nothing to some; Body weight lost: 200 to 100g since start	10–5
	Swelling and bruising around fracture site: extensive to moderate	7–4

From Morton (1987).

Non-mammals

The schemes refer mainly to mammals, although Sanford *et al.* list some signs which may indicate pain in birds and reptiles. As expected, these authors have found that it becomes more difficult to recognize pain, distress, and so on, the further one moves away from humans. Thus there are no published guidelines for the recognition of pain and distress in the 'lower' vertebrates (fish and amphibia), or in invertebrates, such as *Octopus*, which might be able to experience pain and distress. Experienced researchers and animal keepers may, however, be able to recognize adverse effects in these animals. By careful observation of the norms for such animals, workers may at least recognize when 'something is wrong' with the animals, even if they are unable to say exactly what that 'something' is in terms of pain or distress.

Subjective assessment

Many of the criteria in the assessment schemes are open to interpretation by the observer. Hence, in order to decide what constitutes a departure from the norm in appearance (coat condition, gait and posture, etc), behaviour (level of activity, temperament, etc) or mental state (dull or alert, etc) of an animal, an observer must become familiar with the norms for that particular animal and be trained in the recognition and interpretation of such signs. In this context, Beynen and co-workers (Beynen *et al.* 1987; Beynen *et al.* 1988) have suggested that there is a difficulty in applying such schemes, since there may be considerable variation in interpretation of signs by different observers, even when these observers are appropriately trained. Beynen and his colleagues attempted to assess discomfort in gallstone-bearing mice (1987) and in rats with enlarged livers (1988), using a scoring scheme based on that proposed by Morton and Griffiths (1985). Several veterinarians (three in the case of mice, and six for the rats), working independently, assessed the animals according to nine separate criteria. Scores of zero to three were assigned to each criterion, with a score of zero indicating no abnormality and scores of one to three indicating increasing departure from normality. In both studies, there was considerable between-observer variation in the scores assigned to some of the criteria. This led Beynen *et al.* to conclude that it is important for such assessment to be performed by more than one observer. It should be noted, however, that in this work, the overall scores indicated only mild severity. The observers, therefore, were having to look for and score quite subtle effects, perhaps giving more scope for between-observer variation than would be the case in studies involving more obvious changes in the animals. It would, therefore, be interesting to examine the extent of between-observer variation in studies involving adverse effects of more substantial severity.

Recently, in an attempt to avoid problems of subjective assessment, Barclay *et al.* (1988) have developed a technique for assessing the 'disturbance' caused to laboratory rats and mice by a variety of procedures, including restraint and different kinds of injections. In this technique, the normal exploratory activity of each animal is first assessed by placing the animal in a novel environment (say, a new cage) and recording its movements using an electronic monitor. The animal's activity is then assessed after a particular procedure has been carried out and a statistic, the 'disturbance index' (DI), is calculated. The DI indicates the extent of any change in normal activity shown by the animal, and is said to show the degree of upset caused to the animal by the procedure.* As Barclay *et al.* point out, however, a changed pattern of activity may be only one of a combination of signs that an animal is suffering pain or distress. Hence, if a procedure yields a non-significant DI it cannot be assumed that the animal is not experiencing pain or distress and, by the same token, if a significant DI is recorded, this does not necessarily indicate that the animal is in pain or is distressed. The technique might be useful in pilot studies to help select the least disturbing scientific procedure for use in a particular experiment. It is, however, important that other criteria are examined before any firm conclusions are drawn.

Stress and distress

The schemes described above involve the use of overt behavioural, physiological and clinical signs in order to assess when something is wrong with an animal. Some of these signs (shown in Table 5.1), such as squealing or howling when an area of the body which might be expected to be causing pain is touched, limping, guarding certain areas of the body and so on, may be associated specifically with pain. Others of these signs may also indicate pain but, equally, may indicate that the animal is distressed but not experiencing pain. Generally, consideration of the nature of the procedure(s) performed on an animal and observation of its response to analgesics can help sort out what the animal is likely to be experiencing (whether pain and associated distress, or distress without pain), and determine appropriate methods of alleviation.

A difficulty arises since animals showing no such overt signs may still be experiencing stress, distress and even pain. Some primates may not show signs of pain (Sanford *et al.* 1989), perhaps because, if they did, they would run a risk of losing their place in the social hierarchy. Animals kept in restricted spaces may look and behave normally, that is, showing none of the abnormal signs shown in Table 5.1. In spite of their normal appearance such animals may still be physiologically and psychologically stressed. As described in Chapter 4, measurements of hormone and blood glucose levels, and of blood composition, can be used to assess whether animals are stressed by certain procedures or husbandry

* Note that the 'activity' criterion in Barclay *et al.* corresponds with the 'provoked behaviour' criterion of Morton and Griffiths.

conditions. Such stress may simply be part of the animal's normal adaptive, or coping response, and hence may not be perceived by the animal as being unpleasant. If prolonged or more severe, such stress responses may start to threaten the animal's well-being and may eventually produce some of the more overt signs listed in Table 5.1, as well as internal pathological changes, such as stomach ulcers and arteriosclerosis and associated high blood pressure. According to Moberg (1985), in between these coping and pathological stages animals may go through a stage of 'vulnerability to pathological changes', a 'pre-pathological state'. In this stage the animal's well-being is compromised but there may be no overt signs that anything is wrong. Moberg (1985, 1987) has proposed that assessment of the prepathological state can be used as an early indicator of adverse effects of stress. He suggests, for example, looking for signs of immunosuppression, which might enhance the possibility of disease, or using measurements of protein metabolism as an early indicator of a reduction in growth rate. Although these physiological measurements are time consuming and expensive to perform using large numbers of laboratory animals, they may, like the activity monitoring, be of use in pilot studies to help to choose the least distressing procedures or husbandry techniques for use in particular experiments.

5.2 EXAMINING THE COSTS IMPOSED ON ANIMALS BY EXPERIMENTS

Although not without its difficulties, the experience of those involved in the care and use of laboratory animals suggests that it is possible for humans to recognize and roughly quantify adverse effects in at least some kinds of animal. The evidence presented in Chapter 4 supports this claim and also reveals its limitations. Publications such as those of Morton and Griffiths, Sanford *et al.* and Wallace *et al.* may provide useful starting points, to assist in recognizing possible signs of pain and distress in animals. There is a need for those involved in the care and use of laboratory animals to be trained in the recognition of adverse effects and to take time to observe and get to know the particular characteristics of the animals they propose to work with.

Against this background, it should also be possible to provide estimates of the costs, in terms of adverse effects, likely to be imposed on animals when they are used in biomedical research. A variety of features of any given experimental situation will contribute towards the overall harm or suffering caused to animals, and each of these features should be considered in arriving at an overall assessment of the cost of a particular piece of research. For each feature, it is also possible to identify strategies for reducing costs. Such strategies should be employed wherever possible.

5.2.1 Species of animal used

It is important that costs be assessed with respect to the particular species of animal involved. The costs imposed by particular techniques will vary according to:

1. The species' normal way of life; so that, for example, experiments requiring animals to be kept in isolation are likely to impose fewer adverse effects when performed using normally solitary animals than when using normally social, gregarious animals.
2. The species' level of sensitivity and awareness. It has already been suggested (Chapter 4), that different species of animals have different capacities for experiencing pain, distress, etc, and the costs of experiments can be minimized by using species with the least possible capacity. It should be remembered that variation within species can also affect overall costs. A particular procedure, carried out on animals of a particular species, may cause different amounts of suffering depending on the age and strain of the animals used, or even the individual animals involved.

Species differences are recognized, in a very broad sense, by most laboratory animal protection laws. Worldwide, almost all such laws refer exclusively to vertebrates, presumably because it is thought that these animals are able to suffer as invertebrates cannot. The 1986 EC Directive covers 'any live non-human vertebrate, including any free-living larval and/or reproducing larval forms, but excluding fetal or embryonic forms'. The UK Animals (Scientific Procedures) Act goes further, extending its protection to cover fetuses from halfway through incubation or gestation. In neither case is concern extended to invertebrates, though the UK Act gives the Home Secretary power to 'extend the definition of protected animal to include invertebrates of any description' if, in the future, this is considered appropriate. Our examination, in Chapter 4, of the sensitivity and level of awareness of different kinds of animal suggested that there is the possibility that the invertebrate cephalopods (which include the octopuses, cuttlefish, and squids) might be able to suffer in the vertebrate sense. At least, there is no evidence to suggest that they are incapable of such suffering. There is therefore a need for researchers to be alert to the possibility of suffering in cephalopods, and perhaps in other 'higher' invertebrates, and, where appropriate, to give these species the benefit of the doubt in assessing the costs of involving them in scientific procedures. Since there is no reason to assume that cephalopods are any less able to suffer than vertebrate fish and amphibia, there may indeed be a case for extending legal protection to cover these invertebrate animals.

International legislation also makes distinctions between different vertebrate species. Article 7.3 of the 1986 EC Directive states that researchers must use animals (defined as vertebrates) with 'the lowest degree of neurophysiological sensitivity' possible. UK law gives special protection to non-human primates and also to dogs, cats, and equidae. Project licences authorizing the use of such animals should be granted only after it has been shown that 'no other species are suitable' or that 'it is not practicable to obtain animals of any other species that are suitable'. Similar laws exist in some other European countries: under Norwegian law, for example, monkeys, dogs and cats are 'better protected' (Office of Technology Assessment 1986).

Non-human primates are given special protection because, in general, they show a highly developed capacity for experiencing pain, distress and so on (see Chapter 4), and have complex behavioural and psychological needs (see section 5.2.3). Some species of primate are only available in the wild and their supply for research purposes may impose severe, additional, costs (see section 5.2.2). Taken together, these considerations provide strong reasons for avoiding the use of primates in research. However, because of their close evolutionary relationship with man, primates may sometimes provide the best, or only suitable, animal models for human beings in certain kinds of research. Thus, there may be great moral tension between these reasons, both for using and for not using, certain primates in research. The justification for any proposed use of primates must, therefore, be considered very carefully indeed. These points were emphasized in a document submitted to the Home Secretary by the UK Fund for the Replacement of Animals in Medical Experiments and the Committee for the Reform of Animal Experimentation (1987), which set out several important recommendations on the use of non-human primates as laboratory animals. Some of these recommendations, which should be considered when assessing and attempting to reduce the costs of animal experiments, are listed in Table 5.4. All of the recommendations listed in the Table, with the exception of a complete ban on the use of endangered species, were accepted in principle by the Home Secretary (see Anon 1987). These principles are taken into account when the Home Office inspectors decide whether or not to license such work under the Animals (Scientific Procedures) Act 1986.

Table 5.4 Some recommendations concerning the use of non-human primates as laboratory animals

1. That the use of primates should not be permitted when the objectives of the research could be achieved with other species. (This is a requirement of the UK Animals (Scientific Procedures) Act 1986.)
2. That members of endangered species should not be used in laboratories, and that no species (e.g. the baboon) should be deemed acceptable as a laboratory animal merely because it is plentiful in its country of origin or is considered a pest there.
3. That, wherever practicable, when primates have to be used, animals bred in captivity as opposed to wild-caught animals should be used.
4. That particularly careful consideration should be given to the choice of species, and that lower order primates (such as *Callitrichidae*) should wherever possible be used in preference to higher order primates (such as *Cercopithecoidea*).
5. That life time records should be kept for all primates bred by suppliers or user institutions and, from the date of importation, for all wild-caught animals. (This is a requirement of the UK Animals (Scientific Procedures) Act 1986.)
6. That the use of primates should be restricted to institutions which provide facilities of a very high standard for their husbandry and welfare, including adequate provision for general health care, social contact, exercise, recreation and privacy. (The UK Home Office *Code of practice for the housing and care of animals used in scientific procedures* 1989, addresses these points.)
7. That primates should not be kept in isolation (that is, without aural and visual contact with other members of the same species), except in special circumstances (that is, in quarantine) and then only for specified short periods; and that long-term individual caging should be regarded as acceptable only for special reasons and that, wherever practicable, animals maintained singly should be allowed regular periods in communal recreation areas and/or given suitable in-cage inclusions to provide them with entertainment. (These points are also addressed in the Home Office *Code of practice* 1989.)
8. That, on a scale of severity from mild to moderate to substantial, all possible steps should be taken to restrict primate use to procedures which are mild or toward the mild end of the moderate category. (In the UK, all Project Licence applications proposing procedures of substantial severity involving non-human primates are referred to the Animal Procedures Committee.)
9. That institutions approved for primate use should be encouraged to have research advisory committees to provide local advice for licensees on whether proposed procedures were essential, were sufficiently well designed and were acceptable when scientific merit was considered in relation to the overall severity of the proposed procedures.
10. That special training should be undertaken by licensees wishing to work with primates, and by animal technicians and veterinary surgeons having responsibilities at institutions where primates are used.

Source: FRAME/CRAE (1987).

Dogs and cats (and, in the UK, equidae), are afforded special protection, presumably because of the status of these species as human 'companion' animals. The law in several countries puts special restrictions on the supply of dogs and cats, and the reasons for this are discussed in section 5.2.2. Whilst, from our analysis in Chapter 4, there is no particular reason to suppose that dogs and cats are able to suffer any more than say, rodents, or pigs, when involved in scientific procedures, it might be argued that dogs and cats have a special, in-built, need for human companionship, which is less easy to provide in a laboratory, than in the home. Furthermore, there may be thought to be a special relationship of 'trust' between humans and such companion animals, which is broken when these animals are used in research. For these reasons, the use of dogs and cats in research might be seen as imposing greater costs, in terms of animal suffering, than the use of other mammalian species, such as rats, mice or pigs. Whether this greater cost is real is open to debate. (Indeed, section 5.2.4 suggests that a relationship of trust between the animal and its handler can actually help to reduce the costs of certain scientific procedures.) Nevertheless, many members of the public (of which scientists form a part) *perceive* the cost as being greater. Such perceptions are important in our moral analysis, and may be as legitimate a basis for giving these animals special protection as concerns based strictly on scientific evidence.

Researchers have legal, as well as moral, duties to choose the species, strain and age of animal that is most suitable for the experimental purpose, yet has the least possible capacity for experiencing adverse effects, consistent with the aims of the experiment. Choosing a suitable species will involve a number of considerations, related to the type of experiment proposed. Table 5.5 lists some of the points which should be taken into account. The Table is based on an original by Martin and Bateson (1986), which listed points to think about when choosing a species for use in a study of behaviour, and which has been adapted so as to be relevant to any biomedical research project. As noted in the Table, some of these considerations have important implications in the assessment of costs, and separate sections of this chapter are devoted to their examination.

5.2.2 Animal supply

It might be expected that the costs of particular experiments will vary according to whether the animals used have been purpose-bred, have been taken from the wild, or have been obtained from

Table 5.5 Choosing a species for use in a biomedical research project. A list of points to think about in cases where a choice of species is possible (only some points may be relevant in any one case)

- What is the species' capacity for appreciating what is being done to it during an experiment? Would a species with a reduced capacity be equally suitable? (*this section*)
- Is the species readily available for study in captive conditions, or is it easily seen in its natural habitat? If it has to be imported what are the conditions for collecting it in its country of origin and what are the quarantine requirements? Does it breed well in captivity? Is anything known about its genetics? Is a controlled programme of breeding possible? (*see section 5.2.2*)
- Is it tolerant of human presence? Does it handle well if it is to be kept in captivity or is hand-reared? Would a more tolerant species be equally suitable? (*see sections 5.2.2 & .4*)
- What are its special housing or husbandry requirements? Is providing optimum husbandry and housing likely to be practically and financially feasible? Will the experimental conditions require any special housing or husbandry (perhaps isolation, or a special diet)? If so, is the species likely to fare well under such conditions? Would another species be more suitable? (*see section 5.2.3*)
- What are its life history characteristics, such as gestation period, age of independence and age of sexual maturity? Is its life span long enough to make repeated measurements possible and/or short enough to make studies of development possible?
- Is much else known about its natural history, anatomy and physiology? Is there an extensive biological and behavioural literature available for this species?
- Is its physiology and/or behaviour suitable for the particular problem to be investigated? If a human problem, or a general problem relevant to many species, is to be studied, is this particular species a suitable model?

Adapted from Martin and Bateson (1986).

other sources, such as stray animals. Article 21 of the 1986 EC Directive states that all mice, rats, guinea-pigs, golden hamsters, rabbits, non-human primates, dogs, cats and quails used in research should be purpose-bred, unless the individual EC country chooses to make a special or general exemption to this rule. Establishments supplying animals may not supply feral or stray animals for use in research (although, again, some countries may make exemptions to this rule). In the UK, unless a special exception is made, the law requires that mice, rats, guinea-pigs, hamsters, rabbits and primates used in experiments are 'bred at a designated breeding establishment' (that is, purpose-bred) or 'obtained from a designated supplying establishment'; (the majority of animals obtained from such establishments will be purpose-bred but some, notably some primates, may be obtained from the wild). The supply of dogs and cats is further restricted, in that all such animals must be purpose-bred. This provision aims to ensure that stray animals are not used in work covered by the legislation. Vertebrates other than those listed above may be obtained from any legally acceptable source.

Purpose-bred and wild-caught animals

Although it is permissible to use wild-caught animals in biomedical research in the UK and elsewhere, purpose-bred animals are usually the animals of choice. Purpose-bred animals make better scientific models since their genetic constitution and early environment (including any diseases to which they are exposed) are known and can be controlled. Wild-caught animals may be used when the species of choice is not available from breeding establishments (perhaps because it is difficult or impossible to breed the species successfully in captivity, or because demand for such animals is very low), or when the object of the research is to study the biology of a particular wild animal *per se*.

The use of wild animals may impose certain costs which are not imposed when purpose-bred animals are used. Wild animals must usually be subjected to the stresses of trapping and, in the case of animals imported from abroad, to a period of quarantine. Capture of non-human primates, in particular, may impose costs not only on those animals which are trapped, but on other members of the population: the capture of young animals may involve restraining or even killing their parents and other members of the family group. Attention must also be paid to the conservation status of the animals used. If an animal is endangered, the costs of using it in research may extend to the whole of the population from which it is taken or even to the whole species, since removing animals from the wild may reduce the size of the breeding pool and so increase the likelihood that that particular population or species will become extinct.

Particular scientific procedures and husbandry systems may cause more severe adverse effects in wild than in purpose-bred animals. Animals bred specifically for research will, to a certain extent, have been tamed (or 'domesticated') by artificial selection and by their early experiences, so that they are adapted to human contact and to laboratory-type housing conditions. Since wild animals have not been so adapted, either by their early experience or by selective breeding, they may be more stressed by scientific procedures and laboratory conditions than purpose-bred animals. These extra costs of using wild animals are recognized by Article 7.3 of the EC Directive, which states that 'experiments on animals taken from the wild may not be carried out unless experiments on other animals would not suffice for the aims of the experiment'.

Animals with congenital abnormalities

A large number of purpose-bred strains are available for use in biomedical research. Each strain has its own distinctive characteristics, and researchers are able to choose the strain with characteristics best suited to the problem they are studying. Some strains have congenital abnormalities or pathologies which may make them especially useful models in certain areas of research. Athymic nude mice and rats, for example, are widely used in studies on the mechanism of the immune response. The animals are hairless and, because they lack a thymus gland, have depressed immune responses and are particularly susceptible to disease. In conventional animal housing their lifespan is short, but when maintained in isolators (disease-free, barriered conditions), the animals have near normal life expectancies (Harlan Olac Ltd 1990). The characteristics of a number of other strains of mice, bearing harmful genetic defects, are shown in Table 5.6 (Harlan Olac Ltd 1990). As can be seen from the table, such congenital abnormalities can, in themselves, give rise to animal suffering. STR/Ort mice, for example, are expected to suffer pain from their congenital arthritis. Such animals may also require special care. Diabetic mice need regular injections of insulin in order to prevent them from lapsing into diabetic coma.

Transgenic animals

At present, most of the laboratory animals available from commercial suppliers are the products of many generations of controlled, selective breeding. Recently, a new method of altering the genetic make-up of animals has become available: the direct transfer of genetic material (DNA) from one animal to another. This is often achieved by microinjection of the new genetic material into fertilized one-cell embryos, followed by implantation of these altered embryos into a surrogate mother. The stages in the production of 'transgenic' mice by microinjection of DNA are shown in Figure 5.1. Two other methods are available for the transfer of the genetic material: infection of fertilized embryos with retroviruses carrying the foreign DNA, or incorporation of embryonic stem cells, infected

Table 5.6 Some strains of mice bearing congenital abnormalities

Strain	Characteristics
Inbred mice	
AKR/Ola (Albino)	A high leukaemia strain. Approximately 90 per cent leukaemia in both sexes by 6–8 months of age. Also show a low concentration of lipids in the adrenal glands.
NZB/Ola (Black)	Develop auto-immune haemolytic anaemia (the animal's antibodies destroy the red blood cells) at around 4–10 months old.
STR/Ort/Ola (Brown, white spots)	Develop a severe degenerative non-inflammatory joint disease, particularly of the knee joint. By about one year of age 60 per cent of males and 30 per cent of females have osteoarthritis as assessed radiologically. There is also a high incidence of spontaneous hepatomas (tumours of the liver).
Mutant mice	
C57BL/Ks–db/db/Ola (Diabetic)	Develop severe diabetes. At about 4 months of age the mice stop gaining weight, and often die before they reach 10 months.
C57BL/6–ob/ob/Ola (Obese)	Become extremely obese and have a defect in body temperature regulation. The animals overeat and show increased insulin secretion and high levels of blood glucose. There may also be defects in brown adipose tissue.
MF1–hr/hr/Ola (Hairless)	Have normal coat until about 10 days of age at which time hair loss commences. Complete hair is lost from the follicle. Hairless mice are fertile but most females do not nurse their young. Hairless mice also have a mild immune deficiency.

From Harlan Olac Ltd (1990).

with the new DNA, into blastocyst-stage embryos (see reviews by Jaenisch 1988; Wilmut *et al.* 1988; and Gordon 1989).

Transgenic technology can be used to transfer at most only one or two extra genes into a single embryo, whereas in selective breeding the genetic make-up of animals can be altered by hundreds or even thousands of genes. Nevertheless, the new, transgenic, technology is a very powerful tool, since it can be used to alter particular characteristics both rapidly and precisely. Rather than working towards a particular desired characteristic

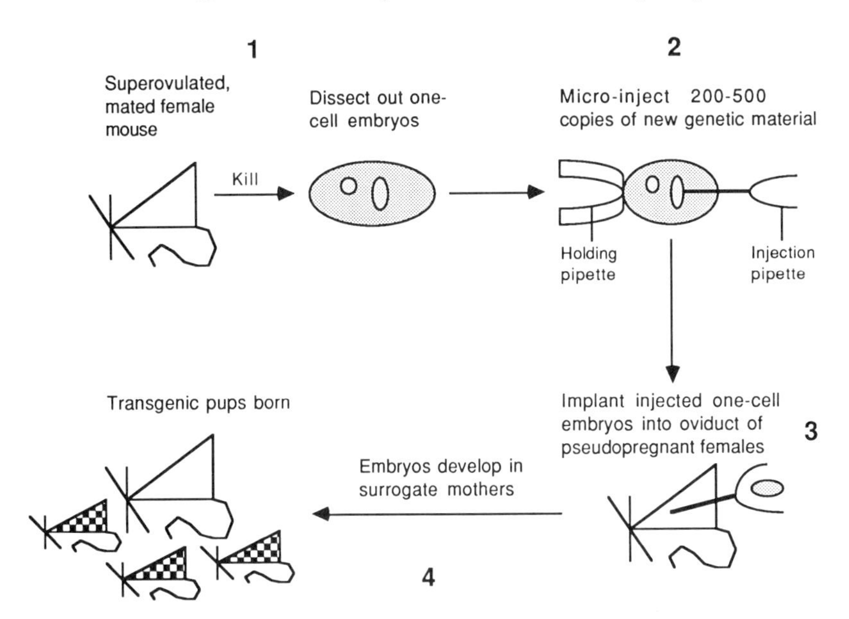

Fig. 5.1 Production of transgenic mice by microinjection of DNA. **1.** Hormones are used to induce 'super-ovulation' in female mice (30–40 eggs, rather than the usual ten, are released). The mice are mated and then killed. The one-cell fertilized embryos are then dissected out. **2.** Under a microscope, 200–500 copies of the new genetic material are injected into the 'pro-nuclei' of each one-cell embryo (initially, a fertilized egg has two pro-nuclei which soon fuse to form a single nucleus). Around 30 per cent of injected embryos degenerate within a few hours. **3.** The remaining injected embryos are implanted into the oviducts of 'pseudo-pregnant' female mice (females that have been mated with sterile males). **4.** The eggs are allowed to develop normally inside the surrogate mother mice, and pups are born about 19 days later. The expectation is that 2–5 per cent of injected embryos will survive to become transgenic mice.

through many generations of selective breeding, gene transfer offers the possibility of altering a specified characteristic in just one or two generations. There are two main technical difficulties in the production of transgenic animals: regulating the expression of the foreign DNA, and identifying and cloning genes for particular characteristics. The technical problems involved in making sure that the foreign DNA is expressed are gradually being overcome. The second problem, however, remains. The genetic basis of 99 per cent of all traits is still unknown. With around 100 000 genes being expressed in a mammal, and most traits being controlled by several genes, the task of matching genes to traits is formidable (see Bulfield 1990).

Broadly speaking, the uses of transgenic animals can be divided into two categories: those involving *expression* of the inserted genetic material, and those in which the inserted DNA causes disruption (*mutation*) of the host animal's own DNA.

In the first case, an inserted gene can be expressed in the transgenic animal (that is, it can direct a cell to manufacture a specific protein) if an appropriate regulatory sequence of DNA is inserted together with the gene itself. If a new gene is inserted together with its own regulatory sequence (that is, its own 'promoter'—the region of DNA responsible for 'switching the gene on'), the gene may be expressed in the appropriate tissue. If, on the other hand, the new gene is inserted together with a promoter from another gene, which is normally expressed in a different tissue, the new gene can be induced to express in this novel tissue. This was first shown by Brinster *et al.* (1981), who found that the herpes virus gene for the protein 'thymidine kinase' (TK) could be induced to express in the livers of transgenic mice if it was linked to the promoter region for mouse 'metallothionein-1' (MT-1), a gene which is normally expressed in the liver (see Gordon 1989). The same technique was used to produce the famous 'supermice', which show rapid and increased growth when compared with normal mice. Palmiter *et al.* (1982) inserted the gene for rat growth hormone, fused with the promoter for MT-1, into the pronuclei of fertilised mouse embryos. Some of the mice that developed from these embryos were found to carry the new growth hormone gene, which was expressed in the liver. These transgenic mice had body weights which were almost twice those of their normal siblings. Since these pioneering expriments, the technology for inducing foreign gene expression in transgenic animals has been applied in a variety of different ways in biomedical research, including the following.

1. The study of gene regulation (that is, how genes are 'switched on and off').
2. Cancer studies, in which genes suspected of causing cancer can be expressed in transgenic mice and the role that these genes play in tumour formation can be studied. One such study, which involved inserting a human oncogene linked to a viral promoter, has produced mice which are highly susceptible to the development of cancer: so-called 'Oncomice' (Stewart *et al.* 1984). These animals are likely to prove particularly useful in studying the development of cancer, in developing drugs and other potential therapies for cancer, and in testing substances suspected of causing cancer.

3. The development of gene therapy, in which inbred mice showing specific genetic defects can be used to test systems for gene therapy.
4. The study of human genetic diseases, in which gene transfer is used to induce genetic diseases in animals and the transgenic animals are used as experimental models to study the diseases and, sometimes, to investigate the possibility for developing gene therapy.
5. The study of human viral diseases, in which transgenic insertion of viral genetic material is used to produce animal models for human viral diseases, such as hepatitis B and AIDS.
6. The synthesis of important proteins, in which the technology is used to make transgenic animals which produce proteins needed for the treatment of human disease. The best developed example of such work concerns the production of the human blood clotting factors, factor VII and factor IX, in sheep's milk.
7. The production of commercially 'superior' farm animals, in which genes for commercially 'desirable' traits (such as enhanced growth or fertility) are inserted into the animals.

The second broad category of use of transgenic technology is in the disruption (or mutation) of the host animal's own DNA. Inserted foreign genetic material can end up within a host gene, so inactivating that gene. Such insertional mutations provide powerful tools for studying the function of genes involved in development and for studying human developmental abnormalities. The main advantage of the technique over other methods of causing mutations (that is, use of radiation or mutagenic chemicals) lies in the fact that the inserted DNA can be located and isolated, so also locating and isolating the host gene which has been damaged, using *in vitro* cloning techniques (see Gordon 1989).

Like selectively bred animals with harmful defects, transgenic animals may suffer from a range of adverse effects, which should be taken into account when assessing the overall costs of research in which they are involved. Transgenic technology has already been used to produce animals, useful in research, having a number of genetic diseases or developmental abnormalities. Mice which are highly susceptible to the development of cancer, which develop symptoms of human hepatitis B and AIDS, which suffer from defects of collagen formation (causing death around the time of birth), which develop insulin-dependent diabetes, and which show developmental defects, such as severely deformed limbs, have all been produced by gene transfer techniques (see review by Cuthbertson and Klintworth 1988). The introduction of foreign genes can also produce unpredictable, frequently harmful or deleterious, side effects. The insertion of human growth hormone genes, in an attempt to increase growth in pigs, has led to one of the most frequently quoted examples of this kind of problem. Although human growth hormone gene may be expressed in pigs, this does not lead to increased growth. Such pigs are found to be severely arthritic, and to have a number of other abnormalities including a lack of co-ordination of the rear legs, susceptibility to stress, and reduced fertility and libido (see Wilmut *et al.* 1988). Recently, however, Vize *et al.* (1988) have shown that transfer of a porcine growth hormone gene, fused with the promoter from the human metallothionein-IIA gene, can produce transgenic pigs showing enhanced growth, without any such detrimental side effects.

In addition to the question of costs in terms of animal suffering, which arises in both selective breeding and transgenesis, the use of transgenic technology gives rise to several special moral concerns. A particular concern lies in the risk that the new technology might intensify an old problem, which has to do with human attitudes towards animals. Transgenesis offers enormous potential for the modification and subsequent exploitation of animals (for the benefit of humans and other animals). This, it is argued, could lead (or has already led) to transgenic animals being seen in purely commercial terms. It has, for example, been suggested that the recent granting of a patent for the 'Oncomice', in the USA, 'debases animals to the level of inanimate objects' (D'Silva 1989). The tone of the language which is sometimes used to describe certain transgenic animals may lend weight to such concern. A recent report in *New Scientist*, for instance, notes that: 'Animals genetically engineered to become *living factories* that produce useful drugs or proteins in their milk, may soon become the *tools of a new industry* .... Scientists ... say they now aim to scale up the techniques they have used on mice, and apply them to rabbits to develop them as "*commercial bioreactors*"' (Watts 1990; our italics). Whether the use of such language reflects a lack of thought for the animals is open to debate. Clearly, however, there is a need for all concerned to guard against the possibility that commercial exploitation might take precedence over respect and concern for the welfare of the animals themselves.

Further moral questions, relating to transgenic technology, but which are beyond the scope of the present Working Party, concern the ethical implications of:

(1) the transfer of genes between species and, in particular, the possibility that the technology might be used to transfer genes between humans and animals;

(2) the genetic manipulation of germ line cells (eggs and sperm) and, in particular, the fear that this might lead to the practice of human eugenics; and
(3) the risks to the environment of the release of animals carrying manipulated genetic material, and, in particular, the risk that manipulated genes might be transferred to other laboratory, wild, agricultural, or domesticated animals capable of breeding with the transgenic animal.

The question of human genetic manipulation is the subject of an Advisory Committee on Gene Therapy, set up by the UK Department of Health, as well as a major Report from the European Medical Research Councils; and, in the UK, the release of genetically manipulated animals is controlled by the Advisory Committee on Genetic Manipulation, established by statute under the Health and Safety Executive.

Surgically altered animals

Many laboratory animal breeders also supply animals which have been subjected to surgical procedures, such as organ removal (for example, castration, hysterectomy, adrenalectomy), nerve transections, duct ligations (for example, ligation of one ureter), injection of specified material and blood vessel cannulation (see, for example, Charles River UK Ltd 1990). Such surgery may, of course, impose costs, in terms of adverse effects, on the animals involved. The surgery, however, is of high quality, and is performed in specialized facilities, by personnel trained to a high standard in the particular techniques involved. In some cases, it may therefore be better, in terms of overall costs to the animals, for researchers to buy in surgically altered animals, rather than to carry out the operations themselves.

Stray animals

Most countries in Europe, including the UK, forbid the use of stray dogs and cats in research. In the UK, for example, the legal requirement that dogs and cats are purpose-bred implies that the use of stray or pet animals is prohibited. However, this is not so across the world. In the USA, several States have 'pound seizure' laws, which require animal shelters (known as animal pounds) to hand over dogs and cats for research when requested. Some States have laws which allow, but do not require, the release of pound animals for research, and the majority of States have laws which neither expressly require nor forbid release. It is only in the last ten years or so that laws specifically prohibiting the release of stray animals to research institutions have been passed in the USA, and at least nine States now have such laws (Office of Technology Assessment 1986).

Two main arguments are used to support the use of stray animals in research. The first has to do with economics, that such animals are in cheap and plentiful supply when compared with purpose-bred animals; and the second is a kind of ethical argument, which says that it is better to use animals which would be killed anyway, than to breed animals specially for research. The second argument is used, in particular, in favour of the use of stray animals in experiments carried out under terminal anaesthesia, including the use of animals for the purposes of education and training (see Chapter 9). It might be argued that there is little difference between humanely killing a stray animal in an animal shelter, and anaesthetizing the same animal in a laboratory, performing an experiment, and then killing the animal (by anaesthetic overdose) without allowing it to recover.

There are, however, several arguments against the use of stray animals, which have to do with the increased costs (in terms of animal suffering) imposed by the use of such animals when compared with the use of purpose-bred animals. First, it is argued that, in recovery experiments at least, there is a direct increase in costs, since stray animals are likely to suffer more under laboratory conditions than their purpose-bred counterparts: stray animals are similar to wild animals, in that they are not adapted to the laboratory environment. A related argument, is that pet animals trust humans and that to use them in research would be a betrayal of this trust. These concerns, and also the worry that allowing the use of dogs and cats which have not been purpose-bred would encourage people to steal pets and sell them for research, have combined to produce strong public feeling that stray animals should not be used in laboratories. This public perception could also lead to further, indirect, costs, since the fear that an animal might end up in the laboratory could discourage people from handing in strays to animal shelters, or taking their own animals to be humanely killed. There is also a scientific argument against the use of strays. As scientific models, such animals have the same disadvantages as wild animals, in that their genetic constitution, early environment and disease status are not controlled and may not even be known. Both the costs and scientific arguments against the use of strays are powerful, and would seem to provide sufficient reason for avoiding their use. The provison of UK and other legislation which prohibits the use of such animals in research is to be welcomed. (See Rowan 1984, for further discussion of the US pound animal issue.)

5.2.3 Animal housing and husbandry

The way in which laboratory animals are housed and cared for is of utmost importance in determining the overall quality of their lives. Laboratory animals are *not* involved in experiments for most of the time, and for many it has been suggested that 'the most stressful part of the laboratory may be the cage itself' (Barnard 1988). The fact that the majority of animals used in laboratories have been specially bred for adaptation to the laboratory environment, does not mean that they are somehow degenerate versions of their wild counterparts, necessarily content in impoverished conditions. There is good evidence that laboratory animals, when reared in conditions more like those found in the wild, will exhibit behaviours not usually seen in the laboratory and more typical of their wild counterparts. In one interesting experiment (Boice 1977) ten albino laboratory rats were put into a semi-natural outdoor environment, consisting of a large outdoor pen (for protection from predators), with five nesting boxes sunk into the ground to which access could be gained via short pieces of drainpipe. Food was distributed between the nest boxes twice weekly and nesting material was provided once a month. Boice provided the nest boxes because he did not know whether the rats could or would make their own burrows. He found, however, that all ten of the 'pioneer' rats had dug and were residing in their own burrows after only two days in the outdoor pen. The rats bred successfully, reaching a population of about 50 rats after five generations and maintaining a stable population structure over a two-year period; they were hardy throughout the climatic extremes (which included temperatures as low as −30°C in the first winter), and showed better health and lower mortality than a control population of laboratory rats kept in standard laboratory cages. In addition to digging burrows, which were identical to burrows dug by wild Norway rats, the albino rats showed a variety of wild type behaviours: they were as strongly bound to following particular pathways (which they appeared to mark with urine) as wild rats, and their faeces were approximately twice as likely to be found above ground as in their burrows, consistent with observations on wild rats.

There is a need to consider very carefully the conditions under which laboratory animals are housed and maintained, and it is important to take housing and husbandry conditions into account when assessing the overall costs of a particular piece of research. In fact, keeping animals in the best and most appropriate conditions possible is important not only for the welfare of the animals themselves but also for the science in which they are used. In order to obtain consistent results, researchers need to use animals which are subject to the least possible stress, disease, etc., arising from housing and husbandry conditions. These considerations are recognized in much laboratory animal legislation, and special guidelines and codes of practice have been issued. In Europe, an appendix to the 1985 European Convention and the 1986 EC Directive sets out guidelines for the accommodation and care of laboratory animals. Further afield, the Canadian Council on Animal Care (which administers a voluntary national system of self regulation of laboratory animal use) publishes a two volume *Guide to the care and use of laboratory animals* (1981, 1984), which includes detailed guidance on the housing and husbandry of common laboratory species. In the USA, the Public Health Service and the National Institutes of Health have issued guidelines on the care of laboratory animals. However, recent proposals to implement more stringent regulations, under the US federal Animal Welfare Act, have been strongly opposed by the scientific community, and the most controversial new regulations (concerning standards for cage sizes, excercise for dogs and creating environments to promote the 'psychological well-being' of non-human primates) have already been drastically revised (McGourty 1989). Early in 1990 the US Office of Management and Budget rejected the amended regulations because of White House opposition (Anderson 1990*a*). The regulations are being rewritten in 'performance' rather than 'engineering'-based language, so as to allow researchers to design caging which does not conform exactly to the guidelines, provided they can show that the cages are conducive to good animal welfare. Revised regulations on the care of guinea-pigs, hamsters and rabbits were published in July 1990, but, at the time of writing, rules for the care of primates and dogs are unfinished (Anderson, 1990*b*).

In the UK, under the Animals (Scientific Procedures) Act 1986, the Home Office has issued a *Code of practice for the housing and care of animals used in scientific procedures* (1989). This Code makes recommendations regarding the design of laboratory animal facilities, maintenance of environmental conditions for the animals (suitable temperatures, relative humidities, ventilation, lighting, and levels of noise), transport of animals to the laboratory, animal accommodation, maintenance of animal health and humane killing of animals. For animal accommodation it is recommended that the 'size, shape and fittings of pens and cages should be designed to meet the physiological and behavioural needs of the animals' (para. 3.21, p. 14) and minimum cage sizes are laid down for different kinds of animals. Other general recommendations include that 'bedding and nesting material should be provided unless it is clearly inappropriate' (para. 3.28, p. 15), 'all animals must be allowed to exercise' and 'for larger species special arrangements will usually be required for social contact as well as exercise' (para. 3.38, p. 16). Whilst some generalizations are useful, precise husbandry requirements vary according to the species, age, sex and rearing of the animals to be used. The Code of Practice recognizes this, and makes special recommendations regarding certain species of animal. Some of these recommendations are summarized in Table 5.7.

The Code of Practice itself emphasizes that its recommendations are not fixed for all time. It stresses that 'animal science is a developing discipline' (para. 1.14, p. 3) and 'as understanding of how best to care for animals evolves, the recommendations . . . may need to be updated' (para. 1.17, p. 3). Much of the opposition to the proposed US regulations pointed to a perceived lack of regard for the results of research on the husbandry requirements of laboratory animals. It was argued, for instance, that proposed cage sizes were arbitrary and that research had shown that socialisation (with humans or other animals) was more important than exercise for the well-being of dogs (Holden 1989).

Clearly, research on animal housing and husbandry is important in improving the welfare of laboratory animals. Most such studies in some way use knowledge of the animals' behaviour under more natural conditions in order to suggest improved ways of keeping them in captivity. Ideas for improving animal accommodation and care must be practically and financially feasible. They should take account of the manner in which animals will be used, so that, for example, it should be possible to catch animals which are handled during experiments with the minimum of stress to the animal; and also the human requirements for caging systems, such as ease of cleaning, reasonable cost, etc. Human requirements, however, should not be used merely as an excuse for keeping animals in impoverished conditions. It should be possible to maintain high standards of laboratory animal welfare without compromising the animals' usefulness to researchers, or imposing impractical husbandry requirements; as Wallace (1984) has described in the design of the 'Cambridge' mouse cage.

Any suggested change in method of maintaining animals must indeed represent an improvement for the animals concerned. There is a need to evaluate the effects of such changes from the animal's point of view (Duncan 1978), and there are several ways in which this might be done. One approach is to use physiological and behavioural measurements to show that the animal is less

Table 5.7 Some recommendations for the housing and husbandry of some species of mammal used under the Animals (Scientific Procedures) Act 1986

Non-human primates

- Housing should provide adequate space, complexity (e.g. varied diets, cage furniture) and opportunities for social interaction. Food items such as nuts, grain, etc. may with benefit be added to bedding to encourage foraging behaviour and reduce boredom.
- Cage height should allow for a vertical flight reaction and should permit the animals to stand erect, jump and climb, and to sit on a perch without head or tail touching the cage . . . Cages should have adequate floor space for the more terrestrial species.
- Most species are highly sociable . . . and should be so housed that they have opportunity for social interaction. This can be achieved by careful design of single housing, paired or gang-caging systems with access to exercise or play areas whenever possible. Single housing should be avoided wherever possible. . . Intermittent social contact is better than none at all.

Rats and mice

- Should be group-housed unless a particular experiment requires otherwise.

Rabbits

- Wherever practicable rabbits should be group-housed.

Ferrets

- Ferrets do well when group-housed in escape-proof floor pens provided care is taken to avoid draughts. A small inner compartment to provide animals with darkness and security may be appropriate.

Cats

- Wherever practicable, cats should be housed in social groups. Where they must be housed singly, they should be let out for exercise at least once a day where this does not interfere with the procedure . . . There should be ample shelf room for resting as well as objects suitable for climbing and claw trimming.

Dogs

- Compatible dogs may be kept in pairs. Where they cannot be kept in pairs, their pens should be placed so that they can see one another, but there may be a requirement to be able to prevent this for procedural reasons.
- There is always the need for dogs to have regular human contact. Dogs should have access to adequate exercise areas and be able to exercise with other dogs. Staff time should be allocated to encouraging activity during such periods.

Farm animals and equidae

- Where practicable, these animals should be housed within sight of each other.

Extracted from Home Office Code of Practice (1989).

stressed under the changed conditions. Some of these measurements were described in Chapter 4. Such investigations must be carefully designed, since measurements indicative of stress may be influenced not only by the caging or husbandry system under evaluation, but by many other factors, including species, strain, sex, and age of the animals, environmental variables such as temperature, humidity and light (where these are not the subject of investigation), the length of time the animals are maintained under the new conditions, the source of the experimental animals, the composition of any groups of animals, and the effects of performing the measurements (for example, taking blood) on the animals (for further discussion see, for example, Brain and Benton 1979). A wide range of measurements should be used, since individual animals are likely to differ in the physiological and behavioural expedients used to cope with difficult or less than optimal conditions; and there is also likely to be significant normal background variation in many of the parameters which might be examined (see Broom 1988).

A recent study by Lawlor (1987), employed a variety of different behavioural and physiological measures to assess the effects of cage size and stocking density on growth and well-being of laboratory rats. For each cage of rats, Lawlor measured in-cage activity and total food consumption; for each individual rat, she measured weekly weight gain, and (in order to assess the rat's timidity or boldness), the behavioural responses to handling by a stranger, being placed in a relatively large arena (an 'open field'), and being confronted by a strange rat or a novel object; for sample groups of rats, Lawlor examined the microbiological profile of each animal, the weight of the adrenal glands, the body length/body weight ratio, the histological condition of the gut wall, and resting plasma corticosterone levels. It was assumed that, taken together, these measurements provided useful indicators of the animals' welfare. The experiments and their results were complex, and, accordingly, sometimes difficult to interpret clearly. However, it seemed that overall, rats caged in groups of three from weaning, with plenty of space (Lawlor recommended a minimum of 600 cm^2 per male rat weighing 250–450 g), appeared least stressed. The animals grew well, were quiet to handle, friendly to other rats and to human beings, had a healthy gut and, in the males, an unusually low resting corticosterone level. From her study, Lawlor recommended that the number of rats per cage should not exceed 12, and, ideally, should be reduced to six or three when the animals are involved in experimental procedures; and that the animals should be given more space than the current minima set out in the UK guidelines.

In fact, Lawlor's recommendations would amount to an approximate doubling of the area per animal at each weight level.

Another recent study, by Chamove (1989*a*), examined the effects on mice of a more natural cage design. Since wild mice live in burrows, and house mice also prefer burrows, Chamove suggested that cages with a burrow-like structure could cause laboratory mice less stress and make them less reactive (or 'emotional'). Vertical and horizontal dividers were added to standard mouse boxes, in order to create a burrow-like environment. The effects of rearing mice in cages with no dividers, with five vertical dividers, with nine vertical dividers, and with nine vertical dividers and one horizontal platform, were tested. Chamove found that mice reared in the more complex cages grew more rapidly, were more active and had lower adrenal gland weights than the mice in less complex environments. In behavioural tests, the mice reared in the more complex cages emerged from a novel box more quickly, and appeared more confident in an open field (walking more, defaecating less, and grooming less) than the mice reared in the less complex cages. In assessing the significance of these results, Chamove pointed out that while low adrenal weight and high growth rate can be considered to be 'good', it is more difficult to assess the behavioural characteristics. Nevertheless, he concluded, the results suggest that the mice were under less stress and were less emotional when reared in more complex cages. A more natural housing environment, Chamove suggested, would therefore lead to healthier animals. However, such a new cage design (incorporating a burrow system), would also have to fulfil the practical requirement for ease of cleaning, if it were to be used routinely in laboratory animal houses. In this case, it might be difficult to incorporate this human requirement into the cage design.

A physiological and behavioural measurement approach has also been used in studies of the effects of environmental variables on laboratory animals. Yamauchi *et al.* (1981, 1983) have investigated the effects of a range of room temperatures (12–32 °C, at steps of 2 °C) on reproduction, body weight gain, food and water intake, *post mortem* organ weights (liver, spleen, kidneys, lung, heart, adrenals, testes and ovaries) and haematological values, including red and white blood cell counts, packed cell volume, total blood protein and a variety of serum biochemical values, in rats and mice. The work suggested that the optimum temperature range was 20–26 °C, there being no significant difference in any of the measures over this range, in rooms containing rats and mice; but that a temperature range of 18–28 °C would be permissible for the breeding and rearing of these species. Temperatures for rat and mice rooms recommended in current guidelines fall within this optimum range. Guidelines under the 1986 EC Directive recommend 20–24 °C; the 1989 UK Home Office Code of Practice, 19–23 °C, although the work of Yamauchi *et al.* suggests that a slightly wider range might be employed, without affecting the well-being of the animals. Studies of the behaviour of the animals (for example, their activity patterns) at the different temperatures would provide further evidence about the suitability of different temperatures for rats and mice.

Interesting recent work by Sales *et al.* (1988, 1989) has identified a feature of the environment which has been neglected in the design of animal accommodation: 'ultrasound'. Sales *et al.* (1988) noted that many items of equipment commonly found in laboratories and animal houses (including cleaning equipment, running taps and visual display units) produce sounds above the human hearing range, that is ultrasound. Although inaudible to humans, such sound may be detected by laboratory animals and there is increasing evidence that this may cause stressful effects. In particular, Sales *et al.* (1989) have shown that ambient ultrasound can have diverse effects on the behaviour of laboratory rats, suggesting that the sound might influence both the animal's welfare and the results of experiments in which they are involved. These authors recommend that more detailed studies be carried out, involving an array of physiological and behavioural measures, in order to evaluate the potential harmful or distressing effects of ultrasound and so make recommendations on acceptable levels in laboratories and animal houses. It is also possible that further researches might reveal similar differences between animals in other sensory modalities.

A second approach to assessing welfare under different husbandry conditions involves observing the behaviour of the animals, in order to find out whether the changed environment allows the animals to perform more of their normal behavioural repertoire, and/or reduces the amount of any stereotyped behaviour (such as pacing, rocking, etc) or abnormal behaviour (such as self-biting) shown by the animals. Lawlor (1984) has applied criteria from the known behavioural repertoire of rats, in order to determine the suitability of different caging systems. Lawlor used information from the classic work of Barnett (1963) to establish a rat 'ethogram' (a list of behaviours which forms the characteristic repertoire of the species). She then examined the behaviour of rats in two different cages and listed those behaviours which it was possible for a rat to perform in each type of cage. Figure 5.2 shows the

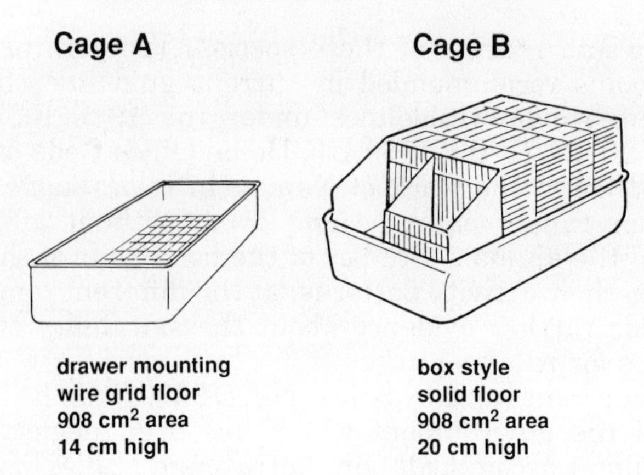

Fig. 5.2 Rat cages examined in Lawlor's (1984) study.

two types of cage considered. Cage A is of the minimum height laid down in the guidelines under the 1986 EC Directive and the 1985 Council of Europe Convention. Cage B is of the minimum height recommended in the 1989 UK Home Office Code of Practice, for rats of 250 g and above. (Various floor areas are recommended in all of these guidelines, according to the number and weight of animals maintained in the cage, and, in the UK, a cage height of 18 cm is recommended for rats weighing less than 250 g.)

Lawlor found that the added height and other features of Cage B allowed the rats to perform a wider range of behaviours than did Cage A. In particular, Cage B, but not Cage A, allowed the animals to stand on their hind legs, climb, and look out of the cage top (that is, sit in the 'normal orienting stance'). However, as Lawlor has pointed out, neither cage allowed the animals to run, jump, burrow or explore. From this analysis, Lawlor concluded that a cage height of 14 cm is not sufficient for the comfort and well-being of laboratory rats, and that a minimum cage height of 20 cm is more appropriate. This work suggests a need for revision of existing Council of Europe and EC guidelines.

Several groups of researchers are studying ways of enriching the environment of laboratory primates. The goal of this enrichment is to improve the well-being of the animals, by increasing desirable behaviour (elements of the animals' natural behavioural repertoire, such as exploration, play and foraging), decreasing undesirable behaviour (such as regurgitation, hair-pulling, self-injury or stereotyped movements) and decreasing elements of normal behaviour, such as aggression and reproductive behaviour, which have become undesirable because they occur much more frequently than is normal (Chamove 1989*b*). Enrichment devices include the addition to cages of seats or perches, bars or ropes for swinging, or play objects, provision of materials which allow the animals to forage for food, and provision of facilities for visual or physical contact with other non-human primates, or for interaction with humans. Bryant, Rupniak and Iversen (1988) observed six adult cynomolgus monkeys, which had been individually caged for most of their adult lives (they were all about eleven years old). The researchers found that in standard, individual cages (each containing a perch), the animals spent approximately 10 per cent of their time engaged in stereotyped behaviour (repetitive swaying, pacing, circling, bouncing and rocking), and occasionally engaged in non-injurious self directed aggression (chasing and biting). Each individual was transferred to an 'enriched environment', consisting of a larger cage, with a perch, containing a perspex viewing panel, through which it was possible for the animals to see into adjacent cages, a swing, a nylon rope hanging from ceiling to floor, a nylon ball, a telephone directory, a surgical glove, and a tray placed below the grid floor of the cage containing deep woodchip litter scattered with sunflower seeds and peanuts, for foraging. In the enriched conditions, stereotyped behaviour and self-directed aggression were virtually absent, but appeared on return to the home cage. Natural locomotory and exploratory activity were considerably lower in the home cage than in the enriched environment. Bryant *et al.* concluded that the larger cage size, and some of the enrichment devices, were beneficial to the animals. In particular, the monkey's attention was engaged by the foraging material, the viewing panel and the telephone directory. Such devices might be used to improve the well-being of the animal in their home cages. Other enrichment devices (the swing, hanging rope, ball, and rubber glove), however, did not prove to be of interest to the animals, showing the need to assess the animals' preferences for different activities, in order to avoid cluttering up the cage with items likely to have little beneficial effect. Housing animals in groups, or in pairs, has also been shown to improve welfare. Work by Reinhardt (1987), for example, has shown that it is possible to house rhesus monkeys in compatible, but unrelated, weaned infant–adult pairs, and also in female adult–adult pairs. This pairing seems to reduce the stereotyped behaviour shown by the adults, and allows the animals to huddle together and groom one another.

A third approach to finding the best ways of maintaining laboratory animals is to let the animals themselves choose between different sorts of environments; that is, to allow them to express a preference by 'voting with their feet' (Dawkins 1980). Interpretation of the results of such choice experiments is not always straightforward, and the following points, as noted by Duncan (1978), should be borne in mind. First, the choices given to the animals can only be relative: the animals can only be given choices between those aspects of their accommodation or care which the experimenter thinks are important for the animals. Second, the animals' preferences may be based only on short-term predictions and they may not weigh up the long-term consequences of their decisions. Third, any choice is a positive choice for the particular animal concerned: a choice which occurs on only 10 per cent of occasions may be as important for the animal's welfare at that time as a choice which occurs on 90 per cent of occasions. It may, nevertheless, be possible to assess the importance of a particular choice for an animal by asking how much work the animal is prepared to do in order to achieve the choice: will it, for example, push open a door in order to get from one, apparently less preferred, cage, into another, apparently more preferred, and if so, how much weight is it willing to push in order to gain access to the preferred cage? (see Dawkins 1990).

So far, most research using choice experiments has focused on farm animals, especially hens (Dawkins 1976, 1977). A few studies have looked at the preferences of laboratory animals. Weiss and Taylor (1984) tested adult male rats' preferences for four different cage types. They set up a central 'discrimination box' which provided access, via a short tunnel on each wall of the box, to four test cages. Preferences were assessed by measuring the time spent by each animal in each type of cage and preferences for cage shape, height and brightness were tested in separate experiments. Weiss and Taylor found that, of the different cages presented to the rats, each of 360 cm^2 inside floor space, the rats exhibited a clear preference for a more elongated, rectangular as opposed to square cage, slightly darkened (with a black rear wall), and 18 (rather than 10, 14 or 22) cm high. Chamove (1989*a*), in his work on more natural mouse caging, also recorded the preferences of the mice. As might have been expected from the physiological and behavioural measurements described earlier, the mice exhibited a clear preference for the more complex cages (those with the most 'burrows').

5.2.4 Scientific procedures performed on the animals

Many different kinds of scientific procedure are used on animals in biomedical research. It is therefore very difficult to make general statements concerning the costs, in terms of pain, distress and anxiety, caused to animals by such procedures. There have been some attempts, by experienced groups of workers, to rank common scientific procedures according to the severity of their likely effects on the animals used. (See Orlans 1987; Canadian Council on Animal Care 1989; New York Academy of Sciences 1988; Veterinary Public Health Inspectorate of the Netherlands 1988; Home Office 1990; Wallace *et al.* 1990.) Such ranking schemes may serve as useful initial guidelines for assessing costs, but, since they do not take into account all the features of a given experimental situation, they cannot be used to predict the costs of using particular scientific procedures. Such predictions must take into account the precise details of the proposed experimental technique, the species of animal to be used, as well as its age, sex, and previous experience, what, if any, anaesthetics and analgesics will be used, the quality of the facilities and the skill of those carrying out the procedure, and how the animal will be prepared for and cared for after the procedure. In many cases (particularly where the proposed techniques are novel or where it is difficult to predict the effects of known procedures on the kinds of animal to be used), it will be necessary to monitor especially carefully the first few animals used. The overall costs of the experiment might then be reassessed in the light of these observations.

Applicants for Project Licences under the UK Act, are required to describe in detail the procedures they propose to use, and these descriptions are subjected to close scrutiny by the Home Office Inspectorate. In assessing the overall severity of the procedures account is taken of the likely adverse effects of all procedures (whether regulated or not) applied to each animal or group of animals; the action taken to mitigate these effects; the end-points applying to the procedures and whether these can be refined; the number of animals used; the proportion of animals expected to suffer the most severe adverse effects and the length of time the animals might be exposed to the upper limits of severity (Home Office 1990, paras 4.7 and 4.14, pp. 10 and 11), as well as the quality of the facilities and competence of those carrying out the procedures.

Facilities and personnel

It is, of course, essential for researchers to use appropriate, up to date, high quality and reliable equipment, in hygienic and well-organized conditions, in order to avoid causing animals unnecessary adverse effects. Appropriate training and experience in the proposed techniques should help to reduce the overall costs of the experiment, although there is the problem of gaining manual skills, which, with the exception of training in microsurgery, is not allowed in the UK. If, in any experiment, there is doubt over competence, the advice of more experienced colleagues should be sought (for further discussion see Chapter 9). Sometimes, it may be appropriate for trained animal house staff to carry out scientific procedures on the researchers' behalf, since these staff may be more familiar with the animals (and *vice versa*, the animals may be more familiar with them) and may have more experience in performing particular techniques.

Animal handling

Those handling animals (animal technicians and researchers) should be trained in the methods of handling appropriate for each particular species and should spend time getting to know and gaining the confidence of the individual animals in their care, or which they are using in experiments. This is recognized by the UK Home Office Code of Practice (1989), which states that 'the behaviour of an animal during a procedure depends on the confidence it has in its handler. This confidence is developed through regular human contact and, once established, should be preserved. Where appropriate, time should be set aside for handling and grooming. All staff, both scientific and technical, should be sympathetic, gentle and firm when dealing with the animals' (para. 3.39, pp. 16–17). The overall costs to animals of experiments may be reduced by such sympathetic, skilful handling. In addition, certain kinds of animals (particularly primates and dogs) may be trained to co-operate in some routine procedures, such as injections. This training may also greatly reduce the stress, anxiety and even pain caused to the animals by the procedures (Jaeckel 1989).

Anaesthesia and analgesia

General or local anaesthesia is mandatory for surgical procedures and should be used in all other procedures, except when the giving of an anaesthetic would be more traumatic to the animal than the procedure itself, or when anaesthesia is incompatible with the object of the experiment, as in feeding studies, or in studies of behaviour (Article 8 in the EC Directive). Provision of effective anaesthesia is obviously of great importance in reducing the costs to animals of many scientific procedures. Detailed guidelines are available (Green 1982; Flecknell 1987; Universities Federation for Animal Welfare 1989), but where researchers have any doubts about the best method to be employed, or whether effective anaesthesia has indeed been achieved, they should seek veterinary advice.

Care of the animals before and after procedures is also important in reducing the overall costs to the animals. Acclimatizing the animals to the experimental situation, before carrying out a procedure, can help to reduce the stress caused by the procedure. Where surgery is involved, premedication (tranquilizers, sedatives, antibiotics and/or analgesics) can help to reduce subsequent adverse effects. After-care is equally important. The provision of a warm, quiet environment and, where appropriate, analgesic therapy, can relieve adverse effects and help the animal to recover more quickly from the procedure. Good organization is essential: those using and caring for the animals must be able to recognize adverse effects in the animals and have established, in advance of the experiment, a plan of action, detailing the kinds of adverse effects which might be expected and what is to be done (for example, to give analgesics or humanely kill the animal) should particular effects be observed.

Until recently, the provision of analgesia for laboratory animal species (particularly rodents) had received little attention and there was a lack of information concerning dose rates and efficacy of analgesics (Flecknell 1984). However, there is an increasing awareness of the need to provide effective pain relief, and there is a growing body of knowledge on analgesia in laboratory species (Flecknell 1984, 1987; Universities Federation for Animal Welfare 1989). Researchers have a moral, and, in many countries, a legal, duty, to minimize the adverse effects caused to animals in scientific procedures. This should include administering analgesics where appropriate. In the UK, applicants for project licences are required to give details of the anaesthesia, analgesia or other methods they propose for mitigating or preventing pain and distress. Before granting licences, the Home Office Inspectors check that these proposed methods are appropriate to the animals and procedures involved.

Humane end-points

It may be possible to terminate an experiment before signs of pain and distress are shown by the animals, or before any substantial suffering occurs, without loss to the scientific results. Use of such humane end-points will reduce the overall costs of the experiments. In particular, studies on acceptable dose rates for radiation therapy, toxicity tests and vaccine potency tests have often used death as an end-point (see Silcock 1986). In each case, it should be possible to use a defined clinical effect as an end-point, humanely killing the animals at this stage, rather than waiting until they die, without compromising the scientific usefulness of the study (see further discussion in Chapter 8). Severe endpoints, including death, have also been reported in studies of the pathology and potential therapy of experimentally induced disease. Here again, more humane endpoints may often be possible (see Silcock 1986 for further discussion).

Where animals are to be humanely killed during or after an experiment, the killing should be carried out by a trained, experienced person and the method should be appropriate to the species and weight of the animal concerned. In the UK, 'Schedule 1' of the Animals (Scientific Procedures) Act lists 'standard methods of human killing' for protected animals. Methods other than those listed in Schedule 1 must be licensed under the Act.

Refinement of techniques

So wide is the variety of techniques used on animals in biomedical research, it is impossible to give general guidelines for reducing pain and distress. Particular areas of research involve particular kinds of technique, so that it is possible to develop guidelines for certain research specialisms. Such guidelines have been issued by the International Association for the Study of Pain (Zimmerman 1983—see Chapter 10), the British Psychological Society (1985), the Association for the Study of Animal Behaviour (1986), and the UK Co-ordinating Committee on Cancer Research (1988) amongst others. Amyx (1987) has provided guidance on the control of animal pain and distress in antibody production and infectious disease studies. In the UK, the Home Office has issued guidelines on the conduct of eye irritation/corrosion tests ('Draize' eye tests) and on the use of neuromuscular blocking agents (see Home Office 1988). Guidelines such as these, based on the experience of many researchers, aim to ensure that all researchers design their protocols to follow current 'best practice', so minimizing pain and distress. Where there are no guidelines for a particular technique, researchers may, through experience and advice from colleagues, draw up their own standard protocols, to ensure that all involved in a particular project are aware of, and use, procedures which will cause the least possible animal suffering.

5.2.5 Number of animals used

The greater the number of animals used in an experiment, the greater will be the overall costs, in terms of animal suffering, of that experiment. It is therefore important that, in a given experiment, the minimum number of animals possible is used, consistent with the aims of the experiment. Biological systems are inherently variable and this fact has been used to justify the use of large numbers of animals in order to obtain meaningful results. Proper statistical design and analysis of the results of experiments may make it possible to obtain results of similar accuracy using smaller numbers of animals. Experimenters should decide, in advance of carrying out the work, how statistical analyses will be applied to the results they expect. This prior planning can help to avoid the use of more animals than is statistically necessary. It can also help to ensure that experiments are not invalid (and animal lives wasted) because too few animals are used and the experiment has to be repeated with larger numbers.

5.3 STRATEGIES FOR ASSESSING AND REDUCING COSTS: A SUMMARY

The main object of making the various assessments of pain, distress and anxiety is to arrive at an overall idea of the costs likely to be imposed on the animals used in particular research projects. The overall assessment of cost can then be compared with an assessment of the benefits likely to be derived from the research (see Chapter 3), so that an informed judgement can be made about the ethical acceptability of the work.

In examining the costs which may be imposed on animals in biomedical research, we have identified a number of different features in any given experimental situation, each of which may contribute to the overall costs of that research. These features are drawn together in the Working Party's *Scheme for the assessment of likely cost to animals*, which is found in Chapter 7. The scheme is summarized in Table 5.8. Each of the features should be taken into account whenever an assessment of cost is made.

There are likely to be considerable advantages to being explicit at all stages in arriving at an overall assessment of cost. Because this is a complex area, where both animal lives and professional reputations are at stake, it is unlikely that covert procedures will be tolerated; they certainly will not be trusted. An explicit procedure for assessing cost (which involves working systematically through the various dimensions of suffering likely to arise in a given piece of research and having clear rules for putting judgements about the separate costs together) is likely to help strengthen the dialogue between researchers, those administering the law governing such work, and those who, on ethical grounds, might wish to criticize, or be reassured about, the conduct of research involving animals. At the very least, such a procedure can help all of these people to use a common language when examining

Table 5.8 Summary of the features which should be taken into account in an assessment of the cost to animals of a research project involving animal subjects

1. Quality of facilities and project workers:
   1.1 quality of facilities for housing the animals and for performing the scientific procedures;
   1.2 training, experience and competence of all workers involved in the project.
2. Type of animal used:
   2.1 the animal's capacity for experiencing pain, distress and anxiety;
   2.2 *for wild animals*: the likely adverse effects caused by capture, transport, quarantine (where appropriate), and confinement of the animals in laboratory conditions;
   2.3 *for animals with genetic defects*: the likely adverse effects caused by such defects.
3. Adverse effects caused by housing and husbandry conditions.
4. Adverse effects caused by the scientific procedures, including the proposed end points and method of killing, and taking into account the provision of anaesthesia, if any.
5. Likely severity of adverse effects after the administration of appropriate analgesia.
6. Number of animals involved in the project.

the costs to animals of biomedical research. An explicit procedure, moreover, can help to make such assessments more reliable. In clinical practice, for example, it has proved helpful to make the criteria used in diagnosis highly explicit. When this is done, computers prove to be more reliable than humans as diagnosticians when the outcome can be validated by independent means (Meehl 1954; Dawes *et al.* 1989). This is because humans very often forget to ask crucial questions or forget the weightings previous practice has shown they should use. In this context, our scheme for assessment of cost may serve as a useful reminding list of the features which ought to be taken into account in arriving at judgements about costs.

A further important advantage, is that use of such a scheme can help to pin-point the features of a research proposal, or of research in progress, which might be modified, or 'refined', so that less suffering is caused to the animals involved. Our analysis in section 5.2 suggested a number of strategies which might be used in such refinement, and these are listed in Table 5.9. Such strategies should be considered whenever appropriate, and should be employed wherever possible.

Table 5.9 Strategies for refining research protocols involving animals. These strategies should be considered whenever there is cause for concern in the relevant assessment of cost.

---

1. Facilities and project workers
Provide training for the relevant personnel in the handling and care of the animals, in the proposed scientific procedures, and in the recognition and alleviation of adverse effects in the animals. (Ensure, of course, that the personnel are competent, and that the equipment and facilities are the most appropriate for the experimental purposes).

2. Type of animal
- Use species, strains or individuals with a lower degree of sensitivity and awareness (i.e. a reduced capacity for experiencing adverse effects). If this is not possible, use anaesthesia whenever possible. (Anaesthesia must, of course, be used in all surgical procedures).
- Use purpose-bred animals instead of wild-caught animals. If this is not possible, use species which are easy to obtain, and which do not need to be transported over long distances or subjected to quarantine.

3. Animal housing and husbandry
Develop strategies for improving the care and accommodation of the animals involved. Provide more natural or enriched environments, so that the animals can carry out more of their natural behaviour patterns and fulfil their psychological needs.

4. Scientific procedures
- Find ways of refining the techniques used in the procedures, so that they cause less pain, distress and anxiety to the animals. Examine the literature to find possible refinements, consult experienced colleagues and/or modify the techniques in the light of experience.
- Devise more humane end points for scientific procedures (in particular, use clinical signs rather than death as an end point).

5. Amelioration of adverse effects
Improve pre- and post-operative care, providing analgesia whenever appropriate.

6. Number of animals
Where possible, use fewer animals (a strategy of *reduction*). Take the advice of a statistician to find the minimum group sizes necessary to obtain meaningful results.

REFERENCES

Anderson, G. C. (1990*a*). White House says no. *Nature* **344,** 604.
Anderson, G. C. (1990*b*). Battle lines are drawn. *Nature* **346,** 308.
Anon (1987). Response of the British Home Secretary to the FRAME/CRAE Report on the use of non-human primates as laboratory animals in Great Britain. *Alternatives to Laboratory Animals* **15,** 141–6.

Amyx, H. L. (1987). Control of animal pain and distress in antibody production and infectious disease studies. *J. Amer. Vet. Med. Assoc.* **191,** 1287–9.

Association for the Study of Animal Behaviour (1986). Guidelines for the use of animals in research. *Anim. Behav.* **34,** 315–8.

Barclay, R. J., Herbert, W. J. and Poole, T. B. (1988). Disturbance index method for assessing severity of procedures on rodents. *UFAW animal welfare research report No. 2* Universities Federation for Animal Welfare, Potters Bar.

Barnard, N. (1988). Inherent stress: the tough life in lab routine. *PCRM update March/April 1988* 1–4, Physicians' Committee for Responsible Medicine, Washington DC.

Barnett, S. A. (1963). *A study in behaviour.* Methuen, London.

Beynen, A. C., Baumans, V., Bertens, A. P. M. G., Havenaar, R., Hesp, A. P. M. and van Zutphen L. F. M. (1987). Assessment of discomfort in gall-stone bearing mice: a practical example of problems encountered in an attempt to recognize discomfort in laboratory animals. *Lab. Anim.* **21,** 35–42.

Beynen, A. C., Baumans, V., Bertens, A. P. M. G., Haas, J. W. M., van Hellemond, K. K., van Herck, H. *et al.* (1988). Assessment of discomfort in rats with hepatomegaly. *Lab. Anim.* **22,** 320–5.

Boice, R. (1977). Burrows of wild and albino rats: Effects of domestication, outdoor raising, age, experience and maternal state. *J. Comp. Physiol. Psychol.* **31,** 649–61.

Brain, P. and Brenton, D. (1979). The interpretation of physiological correlates of differential housing in laboratory rats. *Life Sci.* **24,** 99–116.

Brinster, R. L., Chen, H. Y., Trumbauer, M. E., Senear, A. W., Warren, R. and Palmiter, R. D. (1981). Somatic expression of herpes thymidine kinase in mice following injection of a fusion gene into eggs. *Cell* **27,** 223–31.

British Psychological Society, Scientific Affairs Board (1985). Guidelines for the use of animals in research. *Bull. Br. Psych. Soc.* **38,** 289–91.

Broom, D. M. (1988). The scientific assessment of animal welfare. *Appl. Anim. Behav. Sci.* **20,** 5–19.

Bryant, C. E., Rupniak, N. M. J., and Iversen, S. D. (1988). Effects of different environmental enrichment devices on cage stereotypies and autoaggression in captive cynomolgus monkeys. *J. Med. Primatol.* **17,** 257–70.

Bulfield, G. (1990). Genetic manipulation of farm and laboratory animals. In *The biorevolution: cornucopia or Pandora's Box?* (ed. P. Wheale and R. McNally), pp. 18–23. Pluto Press, London.

Canadian Council on Animal Care (1980, 1984). *Guide to the care and use of experimental animals,* 2 Vols. CCAC, Ottawa, Ontario.

Canadian Council on Animal Care (1989). *Ethics of animal experimentation.* CCAC, Ottawa, Ontario.

Chamove, A. S. (1989*a*). Cage design reduces emotionality in mice. *Lab. Anim.* **23,** 215–9.

Chamove, A. S. (1989*b*). Assessing the welfare of captive primates – a critique. In *Laboratory animal welfare research–primates* pp. 39–49. Universities Federation for Animal Welfare, Potters Bar.
Charles River UK Ltd (1990). *VAF/PLUS Price List.* Charles River UK Ltd, Margate, Kent.
Cuthbertson, R. A. and Klintworth, G. K. (1988). Transgenic mice–a goldmine for furthering knowledge in pathobiology. *Lab. Invest.* **58,** 484–502.
Dawes, R. M., Faust, D., and Meehl, P. E. (1989). Clinical versus actuarial judgement. *Science* **243,** 1668–74.
Dawkins, M. S. (1976). Towards an objective method of assessing welfare in domestic fowl. *Appl. Anim. Ethol.* **2,** 245–54.
Dawkins, M. S. (1977). Do hens suffer in battery cages? Environmental preferences and welfare. *Anim. Behav.* **25,** 1034–46.
Dawkins, M. S. (1980). *Animal suffering: the science of animal welfare.* Chapman and Hall, London.
Dawkins, M. S. (1990). From the animal's point of view: motivation, fitness and animal welfare. *Behav. Brain Sci.* **13,** 1–61.
D'Silva, J. (1989). Patenting of animals—a welfare viewpoint. *Paper presented at a conference on 'Patenting Life Forms in Europe',* held at the European Parliament 7–8 February 1989.
Duncan, I. J. H. (1978). The interpretation of preference tests in animal behaviour. *Appl. Anim. Ethol.* **4,** 197–200.
Flecknell, P. A. (1984). The relief of pain in laboratory animals. *Lab. Anim.* **18,** 147–60.
Flecknell, P. A. (1987). *Laboratory Animal Anaesthesia.* Academic Press, London.
Fund for the Replacement of Animals in Medical Experiments and Committee for the Reform of Animal Experimentation (1987). *The use of non-human primates as laboratory animals in Great Britain.* FRAME, Nottingham, and CRAE, Edinburgh.
Gordon, J. W. (1989). State of the art: transgenic animals. *ILAR News* **30,** 8–18.
Green, C. J. (1982). *Animal anaesthesia* (Laboratory Animal Handbook no. 8). Laboratory Animals Limited, London.
Harlan Olac Ltd (1990). *Supporting research worldwide* (brochure). Harlan Olac Ltd, Bicester, Oxon.
Holden, C. (1989). Compromise in sight on animal regulations. *Science* **245,** 124–5.
Home Office (1988). *Report of the Animal Procedures Committee for 1987.* HMSO, London.
Home Office (1989). *Code of practice for the housing and care of animals used in scientific procedures.* HMSO, London.
Home Office (1990). *Guidance on the operation of the Animals (Scientific Procedures) Act, 1986.* HMSO, London.
Jaeckel, J. (1989). The benefits of training rhesus monkeys living under laboratory conditions. In *Laboratory animal welfare research—primates,* pp. 23–5. Universities Federation for Animal Welfare, Potters Bar.

Jaenisch, R. (1988). Transgenic animals. *Science, NY* **240,** 1468–74.
Lawlor, M. (1984). Behavioural approaches to rodent management. In *Standards in laboratory animal management,* pp. 40–9. Universities Federation for Animal Welfare, Potters Bar.
Lawlor, M. (1987). *The effect of caging factors on growth and well-being in laboratory rats: a report to UFAW Council.* Psychology Department, Royal Holloway and Bedford New College, University of London.
McGourty, C. (1989). About-turn on regulations. *Nature* **341,** 6.
Martin, P. and Bateson, P. (1986). *Measuring behaviour: an introductory guide.* Cambridge University Press, Cambridge.
Meehl, P. E. (1954) *Clinical versus statistical prediction: a theoretical analysis and review of the evidence.* University of Minnesota Press, Minneapolis, Minnesota.
Moberg, G. P. (1985). Biological response to stress: Key to the assessment of animal well-being? In *Animal stress* (ed. G. P. Moberg) pp. 27–49. American Physiological Society, Bethedsa, Maryland.
Moberg, G. P. (1987). Problems in defining stress and distress in animals. *J. Amer. Vet. Med. Assoc.* **191,** 1207–11.
Morton, D. B. (1987). Epilogue: Summarization of Colloquium highlights from an international perspective. (Colloquium on recognition and alleviation of animal pain and distress.) *J. Amer. Vet. Med. Assoc.* **191,** 1292–6.
Morton, D. B. and Griffiths, P. H. M. (1985). Guidelines on the recognition of pain, distress and discomfort in experimental animals and an hypothesis for assessment. *Vet. Rec.* **116,** 431–6.
New York Academy of Sciences (1988). *Interdisciplinary principles and guidelines for the use of animals in research, testing and education.* New York Academy of Sciences, New York.
Office of Technology Assessment, US Congress, (1986). *Alternatives to animal use in research, testing and education.* US Government Printing Office, Washington DC.
Oldham, J. (1988). Aspects of farm animal welfare. *Vet. Rec.* **122,** 20.
Orlans, F. B. (1987). Reviews of experimental protocols: classifying animal harm and applying refinements. *Lab. Anim. Sci.* **37,** 50–6.
Palmiter, R. D., Brinster, R. L., Hammer, R. E., Trumbauer, M. E., Rosenfield, M. G., Birnberg, N. C. and Evans, R. M. (1982). Dramatic growth of mice that develop from eggs microinjected with metallothionein-growth hormone fusion genes. *Nature* **300,** 611–15.
Reinhardt, V. (1987). Advantages of housing rhesus monkeys in compatible pairs. *Scientists' Center Newsletter* **9**(3), 3–6.
Rown, A. N. (1984). *Of mice, models, and men.* State University of New York Press, Albany, New York.
Sales, G. D., Wilson, K. J., Spencer, K. E. V. and Milligan, S. R. (1988). Environmental ultrasound in laboratories and animal houses: a possible cause for concern in the welfare and use of laboratory animals. *Lab. Anim.* **22,** 369–75.
Sales, G. D., Evans, J., Milligan, S. R. and Langridge, A. (1989). Effects of environmental ultrasound on behaviour of laboratory rats. In *Labora-*

tory animal welfare research – rodents, pp. 7–16. Universities Federation for Animal Welfare, Potters Bar.

Sanford, J., Ewbank, R., Molony, V., Tavernor, W. D. and Uvarov, O. (1989). *Guidelines for the recognition and assessment of pain in animals.* Universities Federation for Animal Welfare, Potters Bar.

Silcock, S. R. (1986). Refinement of experimental procedures. *Alternatives to Laboratory Animals* **14,** 72–84.

Stewart, T. A., Pattengale, P. K. and Leder, P. (1984). Spontaneous mammary adenocarcinomas in transgenic mice that carry and express MTV/myc fusion genes. *Cell* **38,** 627–37.

UK Co-ordinating Committee on Cancer Research (1988). UKCCCR guidelines for the welfare of animals in experimental neoplasia. *Lab. Anim.* **22,** 195–201.

Universities Federation for Animal Welfare (1989). *Guidelines on the care of laboratory animals and their use for scientific purposes II: pain, analgesia and anaesthesia.* Universities Federation for Animal Welfare, Potters Bar.

Veterinary Public Health Inspectorate in the Netherlands (1988). *Animal experimentation in the Netherlands: statistics 1987.*

Vize, P. D., Michalska, A. E., Ashman, R., Lloyd, B. Stone, B. A., Quinn, P. *et al.* (1988). Introduction or a porcine growth hormone fusion gene into transgenic pigs promotes growth. *J. Cell Sci.* **90,** 295–300.

Wallace, J., Sanford, J., Smith, W., and Spencer, V. (1990). The assessment and control of the severity of scientific procedures on laboratory animals. (Report of the Laboratory Animal Science Association Working Party). *Lab. Anim.* **24,** 97–130.

Wallace, M. E. (1984). The mouse: in residence and in transit. In *Standards in laboratory animal management,* pp. 25–39. Universities Federation for Animal Welfare, Potters Bar.

Watts, S. (1990). Drug industry turns animals into 'bioreactors'. *New Scientist* **126,** (1712), 26.

Weiss, J. and Taylor, G. T. (1984). A new cage type for individually housed laboratory rats. In *Standards in laboratory animal management,* pp. 85–9. Universities Federation for Animal Welfare, Potters Bar.

Wilmut, I., Clark, J., and Simons, P. (1988). A revolution in animal breeding. *New Scientist* **119** (1620), 56–9.

Yamauchi, C., Fujita S., Obara, T. and Ueda, T. (1981). Effects of room temperature on reproduction, body and organ weights, food and water intake, and hematology in rats. *Lab. Anim. Sci.* **31,** 251–8.

Yamauchi, C., Fujita S., Obara, T., and Ueda, T. (1983). Effects of room temperature on reproduction, body and organ weights, food and water intake, and hematology in mice. *Exp. Anim.* **32,** 1–11.

Zimmerman, M. (1983). Ethical guidelines for the investigation of pain in conscious animals (Guest editorial). *Pain* **16,** 109–10.

6

Ethical considerations in the development and use of replacement alternatives to animal experiments

6.1 DEFINITIONS

The term *alternatives* has come to have a special meaning in the context of animal experiments, and the following definition is now widely accepted in Britain, Europe and North America (see Smyth 1978; Balls 1983; Office of Technology Assessment 1986): 'alternatives to animal experiments are procedures which can completely replace the need for animal experiments, reduce the numbers of animals required, or diminish the amount of pain or distress suffered by the animals in meeting the essential needs of man and other animals.' The concept of alternatives thus embraces all of the 'Three Rs' proposed by Russell and Burch (1959), and we can think in terms of *reduction* alternatives, *refinement* alternatives and *replacement* alternatives. Much of the discussion in Chapter 5 was concerned with strategies for diminishing the harm caused to animals in research, by reducing the number of animals and refining the techniques involved in scientific procedures. Now, in this chapter, we consider the third of the Three Rs: replacement alternatives to animal experiments.

First, it is important that we have some definition of what is being (or might be) replaced. Such a definition is provided by Section 1 of the Animals (Scientific Procedures) Act 1986 (and 'animal' will be used according to this definition throughout this chapter).

(1) Subject to the provision of this section, 'a protected animal' for the purposes of the Act means any living vertebrate other than man
(2) Any such vertebrate in its fetal, larval or embryonic form is a protected animal only from the stage of development when –
  (a) in the case of a mammal, bird or reptile, half the gestation or incubation period for the relevant species has elapsed; and
  (b) in any other case it becomes capable of independent feeding.

The 1986 European Community Directive (on the approximation of laws, regulations and administrative provisions of the member states regarding the protection of animals used for experimental and other scientific purposes) does not go as far as this, stating: 'For the purposes of this Directive . . . "animal", unless otherwise qualified, means any other live non-human vertebrate, including any free-living larval and/or reproducing larval forms, but excluding fetal or embryonic forms.'

Thus a replacement alternative could also be defined as one *which does not involve the use of a protected animal*. This is important, since the use of chicken embryos *in ovo* which are more than half-way through the normal incubation period, and *in vitro* cultures of whole rat embryos which are more than half way through the gestation period, although often described as replacement alternatives in other countries, are considered licensed animal procedures in Britain under the terms of the 1986 Act.

By this definition, the range of replacement alternatives includes the following (Balls 1983):

(1) improved storage, exchange and use of information about animal experiments already carried out, so that unnecessary repetition of animal procedures can be avoided;
(2) use of physical and chemical techniques, and predictions based on the physical and chemical properties of molecules;
(3) use of mathematical and computer models, including:
  (a) modelling of quantitative structure–activity relationships (QSAR), that is taking advantage of correlations between molecular structure and biological activity in the prediction of potential desired and undesired effects of chemicals;
  (b) molecular modelling and the use of computer graphics, for example in actively designing drugs and other chemicals for specific purposes;
  (c) modelling of biochemical, physiological, pharmacological, toxicological and behavioural systems and processes;
(4) use of 'lower' organisms not protected by legislation controlling animal experiments, including invertebrates, plants and microorganisms, for example bacteria in genotoxicity tests;
(5) use of the early developmental stages of vertebrates before half-way through gestation (mammals), incubation (birds and reptiles) or the stage when independent feeding occurs (amphibians and fish), for example early chicken embryos in reproductive toxicity tests;
(6) use of *in vitro* methods, including sub-cellular fractions; short-term maintenance of tissue slices, cell suspensions and perfused organs; and tissue culture proper (cell and organ culture), including human tissue culture;

(7) human studies, including the use of human volunteers, post-marketing surveillance and epidemiology, for example skin patch testing in humans and consumer response as an alternative to animals in testing cosmetic products.

Such alternatives can be divided into two main types, according to whether the replacement of animal procedures is direct or indirect. *Direct* replacement is exemplified by the use of the skin of a human volunteer, or an *in vitro* preparation of guinea-pig skin, instead of the skin of a living guinea-pig in testing the effects of a new chemical: the reaction of the skin is tested in both the animal and the alternative methods. *Indirect* replacement is exemplified by the use of the *Limulus* amoebocyte lysate (LAL) test instead of the rabbit pyrogen test for identifying pyrogenic (fever causing) chemicals (see Chapter 8), or of the Wellcome HPLC method (a physico-chemical method) instead of mice in testing batches of insulin for impurities: in both cases the animal method and the alternative method have a different mechanistic basis.

6.2 THE LEGAL REQUIREMENT TO CONSIDER REPLACEMENT ALTERNATIVES

Under UK law, there is a statutory requirement for researchers to consider the possibility of using replacement alternatives to their proposed animal experiments, and Section 5 of the Animals (Scientific Procedures) Act 1986 states that 'The Secretary of State shall not grant a project licence unless he is satisfied that the applicant has given adequate consideration to the feasibility of achieving the purpose of the experiment by means not involving the use of protected animals.' Applicants for project licences must sign a declaration which states that '... in my opinion, no such alternatives will achieve the objectives of the project'.

Such a requirement, in fact, applies to all researchers in the European Community, since Article 7.2 of Directive 86/609/EC, makes the provision that: 'An experiment shall not be performed if another scientifically satisfactory method of obtaining the result sought, not entailing the use of an animal, is reasonably and practically available.' Article 6 of the 1985 European Convention for the Protection of Vertebrate Animals Used for Experimental and Other Scientific Purposes is almost identical to Article 7.2 of the 1986 EC Directive, so that a similar provision applies to researchers in non-EC countries which have ratified the Convention. These clauses are consistent with the underlying principles of the Act, Directive and Convention, that is that experimental procedures necessarily used should be those which cause the least pain, suffering, distress or lasting harm, while also providing satisfactory results. As Article 2.4 of the 1986 EC Directive states, 'All experiments shall be designed to avoid distress and unnecessary pain and suffering to the experimental animals'.

The requirement to consider replacement alternatives is not restricted to European law (see also Chapter 10): in the USA, for example, 1985 amendments to the Animal Welfare Act require that researchers declare that no suitable alternative exists in experiments which might cause pain to animals. Furthermore, the Canadian Council on Animal Care, which administers a national system of voluntary self-regulation in Canada, states that animals should be used in experiments 'only when a researcher's best efforts to find an alternative have failed'. The 1985 CIOMS International Guiding Principles for Research Involving Animals, which are intended to help countries which do not have legislation controlling the use of laboratory animals to draw up regulations, state that 'methods such as mathematical models, computer simulation and *in vitro* biological systems should be used wherever appropriate'.

These legal, and other, provisions indicate clearly the moral duty for all concerned to recognize the concept of replacement alternatives, to inform themselves about the potential uses for particular replacement alternatives, actively to support the development of such alternatives, and to accept scientifically validated alternatives as replacements for animal procedures.

6.3 ADVANTAGES OF REPLACEMENT ALTERNATIVES

There are two principal advantages in the use of replacement alternative methods: a long-established *scientific* advantage and a relatively recently recognized *humanitarian* advantage. In the first place, non-animal methods are often the methods of choice, simply because they are, scientifically, the best methods available for tackling a particular problem or question (see Zbinden 1990). Scientific practice requires the use of the most suitable and most straight-forward means of working on any problem and this has been the overwhelming reason for the trend towards the use of *in vitro* methods in biochemistry, pharmacology, etc. This trend is not new: tissue culture, for example, which is now a basic tool in fundamental and applied research, has been in use for nearly a hundred years. Indeed, much biomedical research is now carried out at the molecular, sub-cellular, cellular and tissue levels.

Furthermore, it is often possible to use human cells, tissues or organs, or human volunteers, so obviating the need for animal–human extrapolation.

In addition to these scientific advantages, the use of replacement alternatives can reduce laboratory animal suffering. This second (humanitarian) motive for the development of replacement alternatives only began to gather momentum in the 1970s, but, as we have seen, it is already recognized in law. In practice, the humanitarian and scientific advantages in the use of replacement alternatives run hand in hand. Some organizations dedicated to the development of alternatives, notably the UK Fund for the Replacements of Animals in Medical Experiments (FRAME), welcome new developments for scientific as well as animal welfare reasons, just as many researchers welcome new developments for animal welfare as well as scientific reasons.

In addition to these principal advantages, non-animal methods may have further benefits. First, such methods can prove more economical than animal methods. Animal experiments tend to be expensive in time, as well as in terms of financial and human resources, and non-animal replacements can sometimes provide savings in both. For example, bacterial tests for mutagenicity provide relatively quick and inexpensive methods for identifying genotoxicity, which can be used in screening certain types of chemicals for carcinogenicity. It should be pointed out, however, that replacement alternatives are sometimes more expensive than the equivalent animal experiments. This, at present, may be the case where *in vitro* production of small amounts of monoclonal antibody is concerned. Except where production is on a very large scale, the use of animals in raising monoclonal antibodies is considerably less expensive than the use of *in vitro* technology.

Second, where laboratory animals are taken from dwindling wild populations, the use of replacement alternatives can help in the conservation of the wild species; and, third, the development and use of replacement alternatives may have political advantages, since animal experiments are unpopular and many members of the public respond warmly to companies, or other establishments, active in developing alternatives or supporting their development.

6.4 ETHICAL CONSIDERATIONS IN THE USE OF REPLACEMENT ALTERNATIVES

In spite of all their advantages, replacement alternatives cannot be said to offer total escape from ethical considerations and problems. As with most other situations, such alternatives have their pros and cons, and the following ethical considerations arise in their use.

6.4.1 General ethical considerations

In general terms, it is essential that putative replacement alternative methods are properly validated and evaluated. *Validation* is an examination of the scientific quality of the proposed replacement method, aimed at assessing its reproducibility and relevance, and whether, scientifically, it is as good as (or better than) the animal procedure it is intended to replace (see Balls and Clothier 1989; Balls *et al.* 1990*a*). Such an assessment involves asking whether or not the method is able to achieve its stated scientific aims, whether it is reliable, and whether reproducible in different laboratories. *Evaluation* of the validated replacement alternative method involves examining wider questions, such as its applicability to real problems, the feasibility of implementing it (including cost and personnel considerations) and its value when compared with other potential alternatives (see Balls *et al.* 1987). Just as it would be unethical to continue to carry out an animal procedure when a suitable replacement alternative exists, so it would be unethical to press for the adoption of a replacement method which had not been properly validated and evaluated. Only when a proposed replacement method has been shown to be valid, and has been favourably evaluated, should it be adopted (Balls *et al.* 1990*b*) However, once a replacement alternative reaches this stage of development, researchers have a moral (and legal) duty to use the alternative, rather than the animal procedure, wherever possible.

By the same token, another general ethical consideration relates to the use of animal procedures for which there are, as yet, no suitable replacement alternatives. Where it is proposed to use such procedures in work which has been shown to have a satisfactory justification in terms of its likely benefits (see Chapter 3), there may in fact be a moral requirement that the animal procedures *should* be carried out. For example, if not to test a batch of polio vaccine in animals thereby increases the risk to children to be vaccinated, there is a moral requirement that the testing be done, provided that no replacement alternative is available, that the smallest possible number of animals is used, and that all possible steps are taken to minimize suffering to the animals concerned.

6.4.2 Ethical aspects of specific types of replacement alternative

In addition to these general ethical considerations, the various types of replacement alternative methods (as listed in section 6.1) may raise their own, special, ethical problems.

Improved use of information

Making information available about the results of animal studies already carried out can help to avoid unnecessary repetition of animal procedures. It is, however, understandable that researchers and companies, who will have invested a great deal of time and money in their research, may be unwilling to give their results away to potential competitors. Animal procedures are considered legally justifiable for biomedical reasons and not because of commercial considerations (such as the protection of market share), but, nevertheless, the interests of individuals or companies doing innovative research must be protected. Here, it might be possible to reach a compromise, whereby data obtained from animal procedures would be made available after a certain interval (see also Chapter 8).

Physical and chemical techniques

Such techniques are especially important in the study of biochemistry in general, and in related specialist fields, such as pharmacology and toxicology. In biochemistry, physical and chemical techniques can, for example, be used to study enzyme structures and mechanisms of action. In pharmacology and toxicology, physico-chemical analysis can, for example, be used in predicting both the likely beneficial and harmful biological effects of chemical substances. EYTEX™, a physico-chemical test based on the breakdown of proteins in solution (Gordon and Bergman 1987) is currently undergoing validation as a replacement alternative for eye irritancy testing (*FRAME News* **24**, 5 (1989)). Physical and chemical techniques may also play a part in other kinds of biomedical research. In dental research, for example, a chemical system can be used to study the formation of dental caries (Office of Technology Assessment 1986).

These techniques usually involve no special ethical problems, unless, in the case of a new chemical entity, laboratory workers are put at risk before the safety of the chemical has been adequately assessed. However, this is also a problem in cases where animal tests have not yet been performed, and the usual practice is to take general safety precautions until an assessment can be made.

Mathematical and computer modelling

In general, the modelling of biological processes and of structure-activity relationships, and the use of computer graphics, also involve no special ethical problems. Various methods and models are all in the course of development and validation and are likely to become more and more useful as experience with them increases and their informational basis improves. They are already being used extensively in education (see Chapter 9), in fundamental research, and for predictive assessment in medicine, pharmacology and toxicology. In pharmacology, for example, mathematical and computer models are used in the evaluation of molecules which it is thought might have useful pharmacological properties, and, on this basis, many such molecules are rejected without ever having been synthesized. Sometimes, as in the educational context, the models may be based on data obtained from real animal experiments, so that some initial use of animals is required. Also, in the research context, the use of such models may result in hypotheses which need to be tested in animals, as a basis for improving the overall quality and predictive value of the models and reducing animal use in future. However, in such cases it is likely that the animal experiments would be better designed, and that fewer animals would be required, than if the computer models had not been used. In this context also, such models can be used in exploring the likely results of employing different combinations of variables in experimental design; so helping researchers to avoid using animals in all but the most salient experiments.

Use of 'lower' organisms

'Lower' organisms (invertebrates, plants and microorganisms), which are not protected under laboratory animal legislation, can sometimes be used instead of protected animals in research, testing and education. An exciting recent development in research, for example, is the possibility that transgenic tobacco plants might be used instead of mice, or instead of expensive *in vitro* alternative systems, in the raising of monoclonal antibodies (Hiatt, Cafferkey and Bowdish 1989). In toxicity testing, bacterial cells, for example, can be used to screen chemicals for mutagenicity; a test involving the invertebrate horseshoe crab (*Limulus*) provides a replacement for some use of rabbits in pyrogenicity testing (see Chapter 8); and the leech brain can be used in testing certain chemicals for neurotoxicity (indeed, a UK company, Neuropharm Ltd, is offering a neurotoxicity screening service based on such a test – see *Alternatives to Laboratory Animals* **17**, 3–4 (1989)). Furthermore, in education, invertebrates may be used instead of vertebrates in the study of animal behaviour and physiology: the 'water flea' (*Daphnia* species) can, for example, be used in studying the influence of temperature on heart-rate, and woodlice can be used in simple studies of animal behaviour.

The use of 'lower' organisms, where possible, rather than vertebrate species, may offer distinct 'economic' advantages. Such organisms are often relatively inexpensive to supply and maintain, and may have a short generation time, producing relatively large numbers of offspring. This latter characteristic means that large numbers of experimental subjects can be obtained quickly, and experiments concerning, say, population genetics, or aspects of development (which would take months or even years using vertebrate species), can be carried out in a matter of weeks or days (or even hours if microorganisms are used). The main practical disadvantage is that there is more difficulty in extrapolating to humans or other vertebrates from data obtained in phylogenetically distant invertebrates, than there is from vertebrate species (see Office of Technology Assessment 1986, for further discussion of these advantages and disadvantages).

The use of lower organisms is not devoid of ethical considerations. In particular, the question of whether invertebrates can suffer (in the vertebrate sense) is, as we have seen in Chapter 4, an open one. As was noted, some invertebrates (particularly the cephalopods) have well developed nervous systems and behavioural characteristics which, in this context, suggest that they should, at least, be afforded the 'benefit of the doubt'. If certain invertebrates can indeed suffer when used instead of vertebrates in biomedical research, they cannot be said to be replacement alternatives in anything other than the current legal sense. The UK Animals (Scientific Procedures) Act includes provision for the Secretary of State to 'extend the definition of protected animal so as to include invertebrates of any description', but, at present, non-vertebrates are only protected by the personal standards (and respect for living material) of those who use them or otherwise come into contact with them. It is worth pointing out that these standards tend to be much higher in scientific establishments than in, say, the farming, fishing or catering industries.

Even when bacteria are used, it may not be possible entirely to avoid causing harm to animals. Tests in which bacterial cells are employed to assess the mutagenic potential of chemicals require the addition of a metabolizing system, in the form of an enzyme containing, cell free fraction obtained from the livers of rats. Before they are killed and their livers removed, the rats must be injected with a chemical (for example, phenobarbitone), to promote the activity of the metabolizing enzymes. The injection of the chemical is a regulated procedure in EC and UK law.

Use of early developmental stages

Early developmental stages, such as the chick embryo or frog tadpole, may also be considered to be replacement alternatives for animals. In toxicity testing, for example, the chorioallantoic membrane of intact chick embryos (before half-way through incubation) is a candidate replacement alternative for the detection of skin or eye irritancy. Early developmental stages, such as rat embryos or fetuses, can be used to screen chemicals for teratogenic potential. As we have seen, provided they are not more than half-way through incubation or gestation, nor capable of independent feeding, these subjects are not protected under the UK Act.

However, Section 1(2) of the 1986 Act, which defines the stages of development at which vertebrates become 'protected animals', seems rather an unsatisfactory compromise and itself raises ethical questions. First, it is not easy to see why a rat fetus should deserve protection from day 10 of gestation and not at day 9 or earlier. Second, mammals differ greatly in the stage of development they have reached half-way through the gestation period—at this time a fetal calf, for example, is relatively much more developed than a fetal rat. The 'half-way' cut-off point seems rather arbitrary. In fact, there is the possibility of changing this cut-off, since the 1986 Act gives the Secretary of State the power to alter the stage of development at which a vertebrate becomes a 'protected animal'. If there has to be a cut-off at all, the selection of a stage more likely to be relevant to suffering through the detection of pain (for example, the stage at which fetal movement is first detectable) would seem a better basis for defining a protected animal.

Furthermore, larval stages and embryos do not appear from nowhere, so their use at stages before they themselves are protected raises ethical questions about the capture, breeding, supply and maintenance of their parents. In the case of mammals, mothers have to be killed in order to obtain their fetuses.

Use of in vitro *methods*

As was noted earlier, *in vitro* methods are now widely used in research, and also in toxicity testing (see Chapter 8). Such methods can involve the use of cultures of unprotected organisms or parts of them, parts of protected animals, or parts of human beings.

Permanent cell cultures are useful for many purposes and, once established, do not require samples of cells or tissues to be obtained from animals or humans. There are, however, some major difficulties to be overcome in such cultures, including finding ways of ensuring the maintenance of tissue-type-specific properties of cells. This is one of the reasons why *in vitro* studies usually require that animals are killed, or that human donors are found, in order to provide fresh tissues. However, there are prospects that it will soon be possible to 'immortalize' many differentiated cell types, so that their special properties and functions are maintained indefinitely.

The use of cultures of unprotected organisms or parts of them raises the same ethical questions as were discussed in relation to 'lower' organisms and early developmental stages. For example, the culture of limb bud and mid-brain cells from rat embryos, although useful as an alternative procedure for screening chemicals for teratogenic activity, is not without ethical concern, since it requires that the pregnant female is killed.

The use of parts of protected animals can reduce animal suffering in research, but will also usually require the death of the animal concerned or, at the very least, surgical removal of, say, skin samples. The killing of animals need not involve suffering, but does involve the shortening of a life and the animal concerned will have had to be bred and raised in captivity, or in some cases captured in the wild, then maintained under laboratory conditions. Tissues from non-human primates, for example, would be ideal for many types of *in vitro* research and testing, but, apart from the fact that the animals would be killed rather than suffering as a result of being used as living subjects in experimental procedures, all the other problems relating to the use of non-human primates would still apply (see Chapter 5). Furthermore, experimental protocols sometimes require that animals are treated in certain ways before they are killed and their cells, tissues or organs are taken for *in vitro* work: the use of cell cultures in cytotoxicity tests, for example, can require the addition of a metabolizing system, the preparation of which (as was noted earlier for bacterial genotoxicity tests) requires rats to be pretreated with chemicals, to induce or increase enzyme activity.

Human tissue culture, the most ideal replacement alternative for many purposes, raises a whole set of problems of its own. Certain tissues (for example, skin, hair and blood) can be taken from healthy volunteers. Others (for example, fragments of liver, muscle and kidney) can be taken as normal tissue biopsies or with diseased tissues during operations. However, significant amounts of important tissues (for example, liver) can only be obtained from deceased humans, such as organ donors. In all three cases, there are ethical questions related to obtaining the consent of the patient or volunteer or the permission of relatives if the donor is deceased. The risk of transmitting infectious diseases, such as AIDS or hepatitis, is a serious problem. For this reason, the use of some human organs, such as the placenta in training in microsurgery, is discouraged in some centres. In the case of organ donors, there is also the ethical problem of deciding precisely when the donor has died, and the practical problems of making sure that surgeons are available shortly after the death, to obtain the organ, of maintaining the organ in a suitable condition and getting it to the researchers whilst it is still useful for their purposes. Cells and tissues obtained from human fetuses may also be very useful in research, but this leads to further ethical problems, not least those resulting from moral questions concerning therapeutic and non-therapeutic abortion. Nevertheless, in spite of all these difficulties, it could also be argued that there is a moral obligation to use human tissue, properly and safely obtained, whenever possible.

Human studies

Much biomedical research involves human patients. These patients, however, are not in general considered to be replacing animals in research: it is more usual to regard animals as replacements for humans in cases where clinical research would be unethical or impractical. Nevertheless, three types of human study are often suggested as potential replacement alternatives for some uses of animals. These are epidemiological studies, post-marketing surveillance and the use of healthy human volunteers. Epidemiological studies may be useful in identifying factors causing or contributing to the prevalence and intensity of human disease, and post-marketing surveillance can be used to monitor the effects of chemicals (pharmaceuticals and other products) after they have been marketed. Such studies can involve questions of confidentiality in some cases (for example, when a particular disease or post-marketing surveillance of new pharmaceuticals is under investigation), but not in others (for example, when experience with many consumer products is under consideration). Healthy human volunteers may take part in the testing of pharmaceuticals and some other products (such as cosmetics), in certain physiological studies, and in testing diagnostic procedures. For example, cosmetic formulations with known ingredients increasingly are being tested directly in human volunteers, without any preceding animal tests.

On the whole, human volunteer studies are fraught with ethical problems (Robertson 1986), including the question of how volunteers should be recruited and selected, whether they should be offered financial inducement, what level of risk of harm to the subject might be acceptable in such work, and how best to ensure that the subject's consent is based on a full understanding of the risks involved (that is, that the consent is valid). In fact, research involving healthy human volunteers stands in contrast to work involving animal subjects, since, in the UK, there is no special legislation relating to the use of healthy human subjects in research; and, furthermore, although principles for the ethical conduct of such work were established with the Nuremberg Code of 1947, and the Declaration of Helsinki (adopted in 1964), specific guidelines have only recently been drawn up (Royal College of Physicians, 1986*a* and 1986*b*, second edition 1990; see also Chapter 10).

6.5 STRATEGIC USE OF REPLACEMENT ALTERNATIVES

Generalizations are usually dangerous, but it might be true to say that, on the one hand, many anti-vivisectionists tend to overstate the current availability of replacement alternatives, while scientists tend to overemphasize their limitations.

Fundamental research on animal subjects involves many different approaches and techniques, so that the possibility of replacing animal procedures with non-animal procedures has to be considered case by case. In toxicity testing, on the other hand, the purpose, procedures and performers of the work are all readily identifiable, easily understandable and relatively small in number, so that the task of developing replacement alternatives may be more manageable. At present, non-animal procedures are used mainly as screens, or adjuncts, in predominantly animal-based testing programmes, leading to better design of any subsequent animal tests and to reductions in the overall numbers of animals used and in any suffering caused to them. However, research on alternatives has already led to genuine replacements for some tests, and such replacement should be regarded as the ultimate goal (Balls 1989; see also Chapter 8). Indeed, acceptance of scientifically validated replacement alternative methods, by regulatory authorities and others, is required by national and international legislation, such as the UK Animals (Scientific Procedures) Act, and the 1986 EC Directive. In the third area of animal use, education and training, there is already the possibility for genuine replacement of many animal procedures (see Chapter 9).

The concept of replacement alternatives is relatively new, and research aimed principally at their development is still in its infancy. At this stage, it is quite impossible to forecast the extent to which it will be possible to replace animal procedures without slowing down the rate of progress in biomedical research. However, it can be said that, whatever certain animal welfarists might want, it would be impossible to go from the current position to total replacement without many intermediate steps. The development of replacement alternatives must be seen as an evolutionary process—it will take time and some promising beginnings will lead to later disappointments. In this development, not only will the various replacement alternatives be interdependent (indeed, the exploitation of their different strengths must be the way forward), but, whilst replacement alternative technology is evolving, there will continue to be, as there already is in many fields, an interdependence of *in vivo* and non-animal methods.

Although it is impossible to predict the likely future extent of replacement of animal procedures, there is also the question of *belief* in the possibility of replacement. Some believe that animal research will always be necessary and that the potential contribution of the alternatives will be limited. Others, however, believe that total replacement will one day be possible, since, they argue, what are currently seen as the limitations of alternatives will be overcome. For the latter group, the integration of alternatives into an overall scheme which includes animal experiments is not enough: the goal is total replacement. Those in the former, more sceptical group, see particular difficulties in developing replacements which can mimic the integrated physiology and behaviour of an intact animal, and which, like whole organisms, can 'reveal the unexpected' (that is, something which could not be predicted by considering cells, organs or systems in relative isolation). However, those in the latter group would argue that, as in other aspects of life, it is only by believing in the possibility of achieving things which, today, might seem impossible, that such difficulties can ever be overcome. Nevertheless, in spite of this contrast in long-term vision and expectation, it is important to recognize that the differences in belief between these two groups need not preclude a shared commitment to recognize and develop replacement alternatives, to use them, and to promote their use wherever possible.

REFERENCES

Balls, M. (1983). Alternatives to animal experimentation. *Alternatives to Laboratory Animals* **11**, 56–62.

Balls, M. (1989). Editorial: Will non-animal tests become genuine replacement alternatives or remain as mere pre-screens? *Alternatives to Laboratory Animals* **16**, 315–16.

Balls, M. and Clothier, R. (1989). Validation of alternative toxicity test systems: lessons learned and to be learned. *Mol. Toxicol.* **1**, 547–59.

Balls, M. Riddell, R. J., Horner, S. A. and Clothier, R. H. (1987). The FRAME approach to the development, validation and evaluation of *in vitro* alternative methods. In *Alternative methods in toxicology* (ed. A. M. Goldberg), Vol. 5, pp. 45–58. Mary Ann Liebert Inc., New York.

Balls, M., Blaauboer, B., Brusick, D., Frazier, J., Lamb, D., Pemberton, M. *et al.* (1990*a*). Report and recommendations of the CAAT/ERGATT workshop on the validation of toxicity test procedures. *Alternatives to Laboratory Animals* **18**, 313–37.

Balls, M., Botham, P., Cordier, A., Fumero, S., Kayser, D., Koeter, H. *et al.* (1990*b*). Report and recommendations of an international workshop on promotion of the regulatory acceptance of validated non-animal toxicity test procedures. *Alternatives to Laboratory Animals* **18**, 339–44.

Gordon, V. C. and Bergman, H. C. (1987). EYTEX™, an *in vitro* method for evaluation of ocular irritancy. In *In vitro toxicology: approaches to validation* (ed. A. M. Goldberg), pp. 293–302. Mary Ann Liebert Inc., New York.

Hiatt, C., Cafferkey, R. and Bowdish, K. (1989). Production of antibodies in transgenic plants. *Nature* **342**, 76–8.

Robertson, A. (1986). What price the human alternative? *Alternatives to Laboratory Animals* **14**, 93–5.

Royal College of Physicians (1986*a*). Research on healthy volunteers. *J. Roy. Coll. Physicians Lond.* **20**, 243–57.

Royal College of Physicians (1986*b*, second edition 1990). *Guidelines on the practice of ethics committees in medical research involving human subjects.* The Royal College of Physicians of London.

Russell, W. M. S. and Burch, R. L. (1959). *The principles of humane experimental technique.* Methuen, London.

Office of Technology Assessment, US Congress (1986). *Alternatives to animal use in research, testing and education.* US Government Printing Office, Washington DC.

Smyth, D. H. (1978). *Alternatives to animal experiments.* Scolar Press, London.

Zbinden, G. (1990). Alternatives to animal experimentation: developing *in vitro* methods and changing legislation. *Trends Pharmacol. Sci.* **11**, 104–7.

The journal *Alternatives to Laboratory Animals* covers all aspects of the development, validation, introduction, and use of alternatives to laboratory animals in biomedical research and toxicity testing. It is published by the Fund for the Replacement of Animals in Medical Experiments (FRAME), Eastgate House, 34, Stoney Street, Nottingham, NG1 1NB, England. FRAME also publishes a quarterly newsletter, *FRAME News*.

7

The assessment and 'weighing' of costs and benefits

INTRODUCTION

So far in our work, we have been concerned to find ways of evaluating the factors and interests which, we believe, should be taken into account when deciding whether particular kinds of biomedical research using animals ought or ought not to proceed. Thus, our discussions have led us to draw up assessment schemes, to help in judging the harm (or cost) likely to be imposed on the animal subjects used biomedical research and in judging the benefits likely to accrue from such research. The detailed assessment schemes are presented in pp. 141–6 of this chapter. The principal features of the schemes, and notes on their potential uses, were given in Chapters 3 and 5. Here, we illustrate how, in practice, the schemes might be applied.

In trying to use these schemes, we are immediately confronted with a difficulty. How should judgements about each of the relatively separate dimensions listed in the schemes be *put together* in order to arrive at overall assessments of cost and benefit? In the scheme for assessing cost, for example, how should the relatively separate features of a given project (the scientific procedure, methods of housing and handling, etc.), which may give rise to several relatively separate dimensions of suffering (pain, distress, anxiety, etc.) be put together in order to arrive at an overall assessment of cost?

In the case of the costs scheme, consideration of this question has lead us to a simple, but essential, rule: 'Higher costs must have greater weighting.' Or, putting it another way, a judgement of low cost in one feature of the experimental situation must not be considered to cancel out a higher cost in another feature. As anybody who has been involved in assessments of human beings in examinations or short listing for a job will know, if an equal weighting procedure is used, a person who scores moderately well on each of, say, three dimensions, can rank higher than somebody else who is outstanding on one of the dimensions and mediocre on the other two. The result can be intuitive nonsense. The same applies in the assessment of cost. If equal weightings were to be given to the low, medium and high judgements in our scheme, a project involving an excruciatingly painful procedure preceded by husbandry conditions involving very little stress or anxiety might be assessed as having a lower overall cost than a project involving a set of conditions that were only moderately painful, moderately stressful and moderately anxiety provoking. The result, again, would not make sense and would be dangerous for those animals.

We therefore recommend that, for a given feature of an experimental situation, if experience on any dimension (pain, suffering or anxiety, etc) is likely to be high, the cost to animals of that feature (husbandry, experimental procedure, handling, etc) should be deemed to be high. Furthermore, whenever a particular feature of the research is judged likely to have a high cost in terms of animal suffering, the overall cost of that research should be deemed to be high. Whilst assessments on different features might be considered to be cumulative, so that, if the costs of many features are assessed as moderate, the overall cost might be considered to be high, a low cost in one feature must never be considered to cancel out a higher cost in another.

Similarly, in the assessment of benefit, the rating of each feature in the assessment scheme, on a scale of low–medium–high, is not intended as a quantitative exercise. The various value judgements do not form part of an equation which can be used to calculate the overall likely benefit. Rather, a judgement of low potential benefit (section 1 of the scheme), and/or a judgement of low quality in any part of the experimental approach (section 2 of the scheme) should give cause for concern about the likely benefits of the project. A rule analogous to that described for the assessment of cost should apply: a 'high' judgement in one feature of the benefits scheme does not somehow cancel out a 'low' judgement in another feature.

Such judgements, about the likely costs and benefits of research involving animal subjects, can form the basis for a decision about the ethical acceptability of the work. If we are to draw conclusions about what ought, or ought not, to be done in particular circumstances, we must, however, find a way of putting together the separate, overall, judgements about cost and benefit, so as to reach a verdict about which is the stronger in a given case. We need to decide, for each particular research project, whether or not the harm likely to be caused to animals can be justified in terms of the outcome of the research. To do this, we need, somehow, to 'weigh' the costs and benefits, one against the other. It has already been noted that such a weighing is, in fact, one of the most crucial requirements of the UK Animals (Scientific Procedures) Act 1986; Section 5(5) of which states that: 'in determining whether and on what terms to grant a project licence the Secretary of State shall weigh the likely adverse effects on the animals concerned against the benefit likely to accrue as a result of the programme to be specified in the licence'.

It might be argued that weighing as suggested here is not possible, since there are no units of human (or animal) benefit and of cost to animals which could make these commensurable. Certainly, if weighing is thought of in terms of a mathematical calculus, this is correct. In everyday life however, personal, professional and political judgements on moral issues normally require the weighing of factors and considerations which cannot be quantified with mathematical precision. A judge, for example, weighing a plea for mitigation of sentence in the 'scales of justice' carries out a procedure of this kind. The fact that morality is not an exact science was recognized many centuries ago, by Aristotle, who observed that in ethics 'we must be satisfied with a rough outline of the truth' and look 'for only so much precision as nature permits' (1955 translation).

Thus, although we can agree on guidelines of the sort illustrated by the 'decision model' in Figure 7.1, we cannot provide universally-applicable rules for weighing costs and benefits. This is because the judgements involved are moral, or value, judgements, which by their nature are *contestable*. If, for instance, we wish to apply the guideline, that 'research involving more than mild animal suffering ought not to proceed unless there is the likelihood that

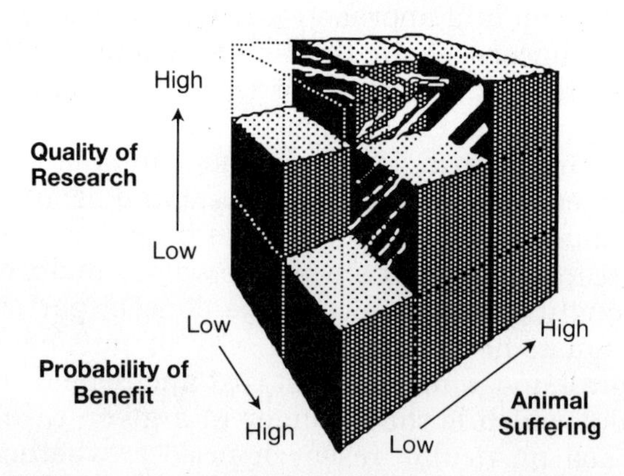

Fig. 7.1 'Decision model', for deciding whether or not a research project should be carried out on animals (see Bateson 1986). 'When a research proposal falls into the opaque part of the "cube", the experimental work should not be done.' This model includes three 'decision parameters': the likely level of animal suffering; the probability of generating biomedically important results, and the quality of the research.

significant benefit will result', then (because there are no agreed quantifiable scales for the axes in the decision model) we must make value judgements about what is to qualify as 'significant' benefit and what is to be considered 'mild' animal suffering.

These value judgements, however, need not be subjective or arbitrary. The possibilities here are not simply objective judgements (of a mathematical calculus kind) and subjective judgements, but also those which can be described as 'inter-subjective'. The latter are morally persuasive because they reflect consensus not on the judgement *per se* but on the procedures which have been used to arrive at that judgement. Confidence in the soundness of such judgements therefore depends largely upon the approach of those who make them: upon whether they have taken into account all the known morally relevant factors, and whether they have shown themselves responsive to all the relevant moral interests.

A SCHEME FOR THE ASSESSMENT OF POTENTIAL AND LIKELY BENEFIT IN RESEARCH INVOLVING ANIMAL SUBJECTS

This assessment scheme is divided into three parts:

Part 1 is an evaluation of the project's *potential* benefits (that is, an assessment of the value of its hoped-for outcome);

Part 2 is an assessment of the quality, validity and necessity of the methods it is proposed to use in achieving the potential benefits; and

Part 3 is an *overall* assessment of the project, to judge the *likelihood* that the potential benefits will be realized, given the proposed scientific approach.

1. Assessment of the *potential* benefits of the project

	N/A*	Low	Medium	High
In terms of the project's				
1.1 Social value Potential contribution towards the improvement of human and/or animal health and welfare (including safeguarding or improving the general environment)	☐			
1.2 Scientific value Potential contribution to the total fund of scientific knowledge	☐			

1. Assessment of the *potential* benefits of the project *(contd.)*

		N/A*	Low	Medium	High
1.3	**Economic value** Potential contribution to employment and profitability in industry and/or conservation of national resources	☐			
1.4	**Educational value** Contribution towards education and training	☐			
1.5	**Other value** Specify	☐			
1.6	**Originality** Necessity in relation to other work in progress				
1.7	**Timeliness** Necessity in relation to the range of problems/questions which might be addressed				
1.8	**Pervasiveness** Links to and implications for other areas of research				
1.9	**Applicability** Potential to lead to further benefits				

***N/A = Not applicable.** Features 1.1–1.5 may not all be relevant in a given project.

2. Assessment of the proposed approach

		Low	Medium	High
2.1	**Assessment of the scientific merit of the project's general approach**			
2.1.1	Relevance of the approach to the stated potential benefits			
2.1.2	Quality of the working hypothesis			
2.1.3	Quality of the experimental design—including statistical aspects			
2.1.4	Appreciation of the relevant background literature and other relevant research in progress			
2.2	**Assessment of the necessity and validity of the scientific procedures to be used in the project**			
2.2.1	Applicability of the scientific procedures to the proposed approach			

2. Assessment of the proposed approach ***(contd.)***

		Low	Medium	High
2.2.2	Necessity to use animals in the procedures			
	(i) at all (in relation to replacement alternatives)			
	(ii) of the species proposed			
2.2.3	Necessity of the proposed procedures in relation to			
	(i) their severity			
	(ii) the number of animals required			
2.2.4	Maximization of information to be obtained from each animal (consistent with welfare constraints)			
2.3	**Assessment of the quality of the project workers and facilities**			
2.3.1	Project workers			
	(i) training			
	(ii) experience, and/or availability of experienced colleagues, to supervise and advise			
	(iii) competence			
2.3.2	Availability of necessary equipment and facilities			
2.3.3	Adequacy of funding			

3. Overall assessment of the project: assessment of *likely* benefits

		Low	Medium	High
3.1	Overall assessment of the potential benefits of the project			
3.2	Likelihood of realization of the potential benefit from the proposed project work			
	(i) at all			
	(ii) within the time available			
3.3	Necessity to use the proposed approach in achieving the likely benefit (can the use of animals be reduced, refined and/or replaced?)			

A SCHEME FOR THE ASSESSMENT OF LIKELY COST TO ANIMALS IN RESEARCH INVOLVING ANIMAL SUBJECTS

This assessment scheme is divided into two parts:

Part A is an examination of the quality of the facilities and personnel involved in caring for and carrying out the proposed procedures on animals;

Part B is an examination of the likely costs of the specific husbandry and scientific procedures to be used on the chosen species of animal.

This second part of the assessment should be set, realistically, in the context of the results of the assessment in Part A. [For example, a particular procedure performed on a particular animal may impose greater costs on that animal if it is done by someone who is poorly trained, than by someone who is skilled and well-practised in that procedure.] The overall costs to animals can then be assessed, in the light of Parts A and B, and in relation to the total number of animals which it is proposed to use.

A. Quality of facilities and project workers

		Low	Medium	High
1.	**Facilities**			
1.1	Quality of facilities for housing animals (in terms of facilities for feeding and watering animals, cleanliness, provision of bedding, cage sizes, environmental enrichment, etc.)	|	|	|
1.2	Quality of facilities for performing scientific procedures (in terms of appropriateness, quality and reliability of equipment, standard of operating theatres, ability to record data, etc.)	|	|	|
2.	**Workers caring for, handling and carrying out procedures on animals**			
2.1	Assessment of workers caring for and handling animals, in relation to their training, experience and competence	|	|	|
2.2	Assessment of workers carrying out procedures, in relation to their training, experience, and competence, and/or availability of more experienced colleagues to advise and supervise	|	|	|

A. Quality of facilities and project workers *(contd.)*

	Low Medium High
2.3 Assessment of all project workers, in terms of their ability to recognise adverse effects (such as pain, distress and anxiety) in the animals and to take appropriate action when and if animals are found to be suffering such adverse effects	|—————|—————|

B. Severity of effects of husbandry and procedures on animals (set in the context of the assessment in part A).

Given that more than one animal will be involved in the project, consider the likely effects on all of the animals to be used, but consider most those animals that will be affected the most (that is, base this assessment on the worst cases).

	Low Medium High
3. Type of animal used	
3.1 Severity in terms of the animal's capacity for experiencing adverse effects such as pain, distress and anxiety and for 'appreciating' what is being done to it during an experiment	|—————|—————|
3.2 If the animal is taken from the wild	
3.2.1 likely severity of adverse effects caused by capture, transport and, where appropriate, quarantine	|—————|—————|
3.2.2 extent of threat to the wild population caused by removal of animals	|—————|—————|
3.2.3 likely severity in terms of novelty of laboratory environment for the animal (taking into account a period of acclimatization)	|—————|—————|
3.3 If the animal has a genetic defect (naturally-occurring or induced), likely severity of adverse effects caused by this defect	|—————|—————|
4. Husbandry and housing conditions (Refer back especially to 1.1 and 2.1)	
4.1 Likely severity of adverse effects (including stress/distress, anxiety, and other effects such as boredom) caused by housing and husbandry conditions. This assessment should take into account the length of time the animals will be maintained in these conditions	|—————|—————|

B. Severity of effects etc. *(contd.)*

	Low	Medium	High
5. Scientific procedures (Refer back especially to 1.2, 2.2 and 2.3)			
5.1 Likely severity of adverse effects (including pain, stress/distress, anxiety and other adverse effects) caused by scientific procedures, including the proposed end points and method of killing. This assessment should take into account the proposed duration of the procedures and the provision, if any, of anaesthesia			
6. Provision for amelioration of adverse effects on animals			
6.1 Severity of adverse effects on animals after providing appropriate analgesia			

How many animals will be used in the project?

Overall assessment of the costs likely to be imposed on animals used in the proposed project

Low Medium High

The Working Party's assessment schemes represent an attempt to set out in detail these morally relevant factors and interests. However, as we have argued, there can be no universally applicable rules for weighing up the various factors, so that we must attempt to develop weighing procedures case by case. We therefore present a number of 'case studies', in order to show how the principles set out in our assessment schemes might be applied. There are five cases covering a range of different types of research, a sixth example relating to the use of animals in education, and a seventh to toxicity testing. The sixth case is a practical exercise drawn from a school textbook. The seventh (toxicity testing) case is a hypothetical, but representative, example. The others are real cases, drawn from published papers. We are aware that this use of published material has its disadvantages and limitations. The alternative, however, is to write fictional cases, which are extremely difficult to make realistic.

The cases are presented without reference to the relevant published material. The chosen examples remain anonymous, because they have been put forward to illustrate the application of our assessment schemes, *not* to criticize or commend the work of particular researchers.

Most of the published work was carried out in the UK, before the Animals (Scientific Procedures) Act came into force. We are aware that, in some of the cases, the decision to licence the work may have been different had the work been assessed under the terms of the 1986 Act, rather than the terms of the 1876 Cruelty to Animals Act, which was in force at the time. Again, however, it must be emphasized that the analysis of the case material is intended to have educational value; to show how our assessment schemes might be used, not just in the UK, but in the broader international context. The cases are *not* intended as a 'test' of UK, or any other, law.

In commenting on the cases, we appreciate that there may well be factors influencing the ethics of a study, which are known to the researchers, but not mentioned in the publication. For this reason, we have been careful to present as 'fact', only those points which can be discerned from the publications. We have raised questions about morally relevant features which are not addressed in the published material, and have tried to show how the answers to these questions would influence the overall judgement. Nevertheless, there may still be factors and interests which have been overlooked.

The importance of these studies, we believe, lies not in the judgements themselves, but in the procedures used to arrive at these judgements. Where judgements are given, these reflect the consensus in the Working Party. In some cases, the commentaries report differences of opinion between members, about the value of the research and about how much suffering might be involved. These differences are not surprising, since we are dealing with ethical judgements, which by their nature are difficult and multifaceted. We hope, however, that by making our procedures explicit, we will provide common ground for furthering the debate about what might, and might not, be judged acceptable practice.

Rather than working through the assessment schemes section by section, the factors and interests which have been judged most relevant in each case have been examined in detail. The relevant points are referred to using the numbers given in the schemes, prefixed by 'C' to refer to the scheme for assessing costs, and 'B' for the scheme for assessing benefits.

CASE STUDY 1. AUTOTRANSPLANTATION OF ISOLATED ISLETS OF LANGERHANS IN THE CYNOMOLGUS MONKEY

1. Summary of the project

The transplantation of islets of Langerhans (insulin-secreting tissue found in the pancreas) has great potential as a treatment for human diabetes. Total removal of the pancreas (total pancreatectomy) is a reliable method of producing experimental diabetes in animals. In this project, cynomolgus monkeys were totally pancreatectomized and the islets of Langerhans isolated from the animals' pancreases. These islets of Langerhans were transplanted back into the same monkeys, so as to test whether sufficient islets could be isolated from a single pancreas to be effective in reversing the experimental diabetes.

2. Potential benefits of the project

The development of a technique for separation of islets of Langerhans from the pancreas of rodents and the subsequent demonstration of reversal of experimental diabetes in rodents has shown the potential of islet transplantation as a treatment for human diabetes. However, until recently, it was not possible to translate these rodent experiments into clinical practice, mainly because of the difficulty of isolating islets from the more fibrous human pancreas. This problem has now been overcome, and islets retrieved from cadaver human pancreases survive and secrete insulin after transplantation into immunologically unresponsive rodents. These results, however, did not indicate whether enough viable islet tissue could be isolated to allow reversal of clinical diabetes. The present project was designed to answer this question. If it could be shown that it was possible to isolate sufficient tissue, it was likely that the therapy would soon after be tried for the first time in human patients. The technique was therefore very close to its clinical application. Non-human primates were used in the study since preliminary work had shown that this was the only group of animals whose pancreas was sufficiently similar in structure to the human pancreas.

3. Personnel, facilities, and funding for the project

The work was carried out by surgeons in a university hospital. The surgical facilities are believed to have been excellent. The work was funded by grants from a government research council and from two diabetes research charities.

4. Procedures employed

Experimental design

Three groups of cynomolgus monkeys were used. Group 1 animals were totally pancreatectomized and islets were autotransplanted into each animal's spleen. The condition of the animals was monitored for several weeks, then their spleens were removed and their condition monitored again. Group 2 animals were totally pancreatectomized and islets were autotransplanted into each animal's liver; their condition was monitored for up to nine months, Group 3 (control) animals were subjected to total pancreatectomy without islet transplantation.

Animals used

The cynomolgus monkeys were imported into the UK. They were purchased from a registered supplier following quarantine isolation. Twelve monkeys were used altogether: four in group 1, five in group 2 and three in group 3.

Techniques used

Total pancreatectomy was performed under full general anaesthesia. Autotransplantation was carried out by injection of isolated islets into either the spleen or the liver. Control animals received no islet transplantation. The animals' abdomens were then closed using the necessary sutures, and they were allowed to recover. At regular intervals after surgery (every two to three days for the first six weeks, and thereafter every one or two weeks), surviving animals were sedated and weighed and blood samples were taken for estimation of plasma glucose and insulin.

Animals which developed diabetes due to total pancreatectomy were expected to have very low levels of plasma insulin and high levels of plasma glucose (the latter being called 'hyperglycaemia'). Diabetic animals would have produced large quantities of urine ('polyuria') and lost weight rapidly. They were expected to become comatose within seven days. However, in order to reduce animal suffering, the researchers intended to humanely kill such animals when they first showed signs of lapsing into a coma (that is, when they were 'precomatose'), rather than waiting until they became fully comatose. When the experimental diabetes was successfully reversed by islet transplantation, no such adverse effects were expected. The animals were expected to have normal serum insulin levels, and to show no hyperglycaemia, polyuria or weight loss.

Animals given islet transplantation into the spleen after total pancreatectomy (group 1 animals) were monitored for five to six weeks. Animals which showed no hyperglycaemia, polyuria or weight loss, then had their spleens removed under general anaesthesia. Before this operation was performed, a glucose tolerance test was carried out, and blood samples for glucose and insulin estimation were taken from the aorta and splenic vein. After the operation, the isolated spleens were studied histologically, and the serum glucose of the animals was measured. Animals given islet transplantation into the liver (group 2 animals) were monitored for up to nine months. Animals which became diabetic were killed after laparotomy under general anaesthesia, and samples of hepatic and portal vein blood were collected for estimation of plasma glucose and insulin.

Commentary on Case Study 1

This research appears to have had great potential benefit in the treatment of human diabetes. Current therapy for diabetes (the administration of insulin) does not deal with the underlying cause of the disease (pancreatic deficiency), and fails to prevent long-term complications, such as loss of eye-sight and arterial disease. Successful transplantation of insulin-secreting tissue obtained from human cadaver pancreases would treat the cause of diabetes, in effect 'curing' patients of the disease and so preventing such complications. The work, therefore, had a high 'social value' (B1.1) in its potential contribution towards improving human health. It was original, timely and applicable (B1.6, 1.7, and 1.9): if it had been shown that sufficient viable tissue might be obtained, the transplant therapy could have been tried for the first time in human patients. The proposed scientific approach was well-suited to answering the question addressed (B2.1), the surgical facilities are believed to have been excellent and the project workers were very experienced in this field, having carried out much of the related background work leading up to the present proposal (B2.3). It was therefore highly likely that the proposed work would realize the stated potential benefits (B3.2).

However, as has been noted in previous chapters in this report, the fact that the work was likely to have a direct and beneficial human application was not, in itself, sufficient justification for the proposed approach. Such a justification should also take into account the likely harm caused to animals in achieving the worthwhile end and, in particular, show that it was not possible to achieve the same end by means which cause less animal harm (B2.2).

The project involved the use of non-human primates in procedures which were expected to produce quite severe adverse effects. The surgery involved total pancreatectomy, followed by injection of islets into the spleen (group 1) or liver (group 2) and, later, splenectomy (group 1). The surgery itself was likely to cause pain and discomfort. Furthermore, in the control group (group 3), in cases where transplantation of islets was not effective, and after splenectomy in group 1, the animals would have developed diabetes, which, if allowed to progress, would have resulted in the development of a series of adverse effects, culminating in coma after about seven days. Cynomolgus monkeys have a relatively well-developed capacity for 'experiencing' the pain and distress likely to be caused by the experiments (C3.1), so that, in the control group especially, the proposed procedures would have been quite severe (C5.1).

The monkeys were imported into the UK and, although there was no threat to the wild population (C3.2.2), an assessment of the overall 'cost' to animals were taken from the wild), transport and quarantine of the animals (C3.2.1). These imported animals were also likely to be stressed by the novelty of the laboratory environment and this stress may have enhanced the severity of the adverse effects caused by the experimental procedures (C3.2).

In this case, it seems that the research was likely to impose quite high costs in terms of animal suffering, in order to achieve a very worthwhile human benefit. The end, it can be agreed, was worth achieving, but could the cost have been reduced? (B2.2).

There would appear to have been no possibility of avoiding the use of animals altogether by employing *in vitro* alternatives (B2.2.2 i). The object was to study the reversal of an experimental disease, which could be shown only by using living animals. Nevertheless, it should be asked whether it was necessary to use this particular species, the cynomolgus monkey, in the work: might a species having less complex psychological abilities and needs have been equally suitable? (B2.2.2 ii). It would seem that, in this project, there was a need to use a species which was anatomically and physiologically as close to man as possible, so that the research could progress as quickly as possible from animal subjects to human patients. A human research ethics committee would be more likely to approve a trial of the transplantation therapy in human patients if the therapy had been shown to be effective in non-human primates, rather than, say, in rats, since in the latter case there would be less confidence in extrapolating to the human situation. In fact, as the researchers stated, the pancreas of the cynomolgus monkey, although smaller, is almost identical in structure to the human pancreas.

Ideally, the animals used would have been purpose-bred in a laboratory, so that they were well adapted to the laboratory environment, including contact with humans. Most of the larger non-human primates available from research suppliers are, however, wild-caught, or have been bred for only a few generations in captive colonies in their country of origin. It may therefore have been impossible for the researchers to use laboratory-bred animals. If this was indeed the case, the extra costs of using wild, as opposed to purpose-bred, animals, could have been kept to a minimum by giving the animals time to adapt to the laboratory environment and to being handled. Furthermore, cynomolgus monkeys, like other non-human primates (whether purpose-bred or wild-caught) have complex social and environmental needs and should be housed in a manner appropriate to these needs. The published paper does not mention any acclimatization, or give details of the accommodation provided for the animals. Acclimatization and the provision of accommodation which meets the monkeys' requirements for space, for social interaction and a varied environment are important steps in minimizing the costs to the animals used in the research (C1.1; C4.1).

There might also have been possibilities for refining the scientific procedures used in the work. The surgical procedures were likely to cause pain and distress, yet the published paper does not mention the use of post-operative analgesia. Good post-operative care, including the use of analgesics where appropriate, is essential in refining the procedures used, since it can reduce the pain and distress caused to the animals and facilitate their recovery from the surgery (C6.1).

Another concern relates to the use of a control group (group 3): was it necessary to use three control animals? Since it is known that removal of the pancreas causes diabetes, from which the animals will die, it might have been possible to avoid the use of the group 3 animals. The cynomolgus monkey's pancreas is quite compact, so total pancreatectomy would have been a relatively straightforward operation, especially for these experienced surgeons. In fact, the animals in group 1 could have served as controls both for themselves and for group 2 animals. The group 1 animals had their pancreases removed, islets were injected into their spleens and, later, their spleens were removed. Once their spleens had been removed, these animals were in the same condition as group 3 animals: they would have been expected to die from diabetes, caused by removal of the pancreas. Thus, after their spleens had been removed, the animals in group 1 could have provided a suitable control for the effectiveness of the pancreatectomy.

A question also arises concerning the 'end point' of the experiments. In the control and other groups, it is important that the end point really was refined. In their published paper, the researchers stated their intention to kill the animals when they were precomatose, rather than waiting until they lapsed into a coma. However, they also recorded that group 3 animals 'all became comatose from hyperglycaemia, polyuria and rapid weight loss within seven days'. This might simply be poor reporting. Nevertheless, the discrepancy leaves open the possibility that the cost imposed on the animals was higher than that necessary to achieve the desired end. There is a need for some reassurance that a precoma end point was indeed used, and that this end point was set as early as possible after the onset of diabetes.

This appears to have been a very worthwhile piece of research, offering the possibility of a direct and highly beneficial application: the relief of human diabetes. The researchers were very experienced and the scientific quality of the work was high. The use of non-human primates appears well-justified. For all of these reasons it can be argued that the work should have proceeded. However, a detailed examination of this high quality work, with reference to the Working Party's assessment schemes, has led to the identification of various possibilities for reducing the costs (in terms of animal suffering) imposed by the research. Opportunities for refinement might have included:

(1) acclimatizing the animals to the laboratory environment and providing them with accommodation which met their behavioural and psychological needs;
(2) providing post-operative analgesia;
(3) refining the end points of the experiments; and
(4) reducing the number of control animals used.

Points (1) and (2) are not mentioned in the published paper, and the paper contains a discrepancy concerning point (3). Therefore, in judging the acceptability of the work, we would wish to be reassured that all three of these refinements were indeed employed; and we would ask, furthermore, whether it might have been possible to have reduced the number of control animals, or to have avoided their use altogether. These refinements, where possible, would have been important in ensuring that the minimum amount of animal suffering was caused in achieving this worthwhile end.

Results of Case Study 1

The results showed that in the cynomolgus monkey sufficient islets could be extracted from a single donor to reverse experimental diabetes. This provided encouragement for the future of pancreatic islet transplantation as a method of treating diabetes.

CASE STUDY 2. EVALUATION OF A NEW THERAPY FOR ACUTE PANCREATITIS

1. Summary of the project

This project was an evaluation of the efficacy of intravenous infusion of a whole, fresh, frozen plasma as a therapy for human pancreatitis. The novel therapy was tested using rats which had been operated so as to induce acute pancreatitis.

2. Potential benefits of the project

The work was aimed at developing an effective therapy for human pancreatitis, which, in the UK, kills around 10 per cent of those affected. The disease takes the form of an acute or chronic inflammation of the pancreas (a gland which secretes digestive enzymes) in which the pancreatic tissue starts to digest itself with its own enzymes. Patients currently are given analgesics and fluid as therapy. Pancreatic inhibitor drugs are also available, but neither drug nor fluid therapy has been effective in reducing the 10 per cent mortality. However, a recent uncontrolled clinical study, involving 239 patients, found that intravenous infusion of fresh frozen plasma (FFP) was effective in reducing mortality from acute pancreatitis during the first five days of the illness (the mortality was reported to be only 3.7 per cent). A possible explanation for this therapeutic effect was that certain enzyme inhibitors, which occur naturally in FFP, helped to eliminate the harmful pancreatic enzymes from the patient's bloodstream. Other substances in the FFP, such as clotting factors, fibronectin, and C-reactive protein, might also have been of benefit in the therapy of acute pancreatitis. The present study was designed to seek more evidence for such a beneficial effect, using an animal model of the disease, before embarking on a controlled clinical trial of FFP therapy.

3. Personnel, facilities and funding for the project

The principal researcher was an experienced surgeon. The surgical facilities are believed to have been excellent. The study was funded by a research grant provided by a Regional Health Authority.

4. Procedures employed

Experimental design

Experiment 1. Three groups of rats were operated on so as to induce acute pancreatitis. After the operation, all of the animals received continuous intravenous infusions containing dextrose and saline, together with an analgesic. Each group of animals also received a fourth substance:

Group 1 received a 'crystalloid' control (sodium chloride);

Group 2 received a 'colloid' control (a commercially-obtained plasma replacement substance called Haemaccel™). The colloid control was similar to FFP, but devoid of enzyme inhibitors and other potentially useful substances found in FFP; and

Group 3 received rat FFP.

Experiment 2. Involved three groups of rats treated in the same manner, except that

Group 1 received the 'colloid' control;
Group 2 received rat FFP in the same volume as that received by group 3, above; and
Group 3 received rat FFP at a lower volume, with the remaining volume being made up with Haemaccel™.

The experiments ran for 72 hours during which time the animals were monitored carefully. The time of any deaths was noted. Any animals surviving after 72 hours were killed, and all of the animals were subjected to *post mortem* examination.

Animals used

Each of the six groups consisted of 30 inbred male rats. In addition, 350 inbred male rats were killed to obtain the FFP needed in the experiments.

Techniques used

Obtaining FFP. Rat FFP was obtained by exsanguination (total bleeding), under general anaesthesia, of 350 male rats from the same inbred colony used for the acute pancreatitis model. The rats died without recovering from the anaesthetic.

Inducing acute pancreatitis. Pancreatitis was induced in each experimental animal by injecting enterokinase and sodium taurocholate into the pancreas, along the pancreatic ducts. The enterokinase and sodium taurocholate damaged the pancreatic tissue in a manner similar to that known to occur in human acute pancreatitis. The operation involved anaesthetizing the animal, making a midline incision and exposing the duodenum, temporarily ligating the bile duct near the liver (*ligature A*, see Fig. 7.2), and then cannulating (that is, inserting a narrow tube into)

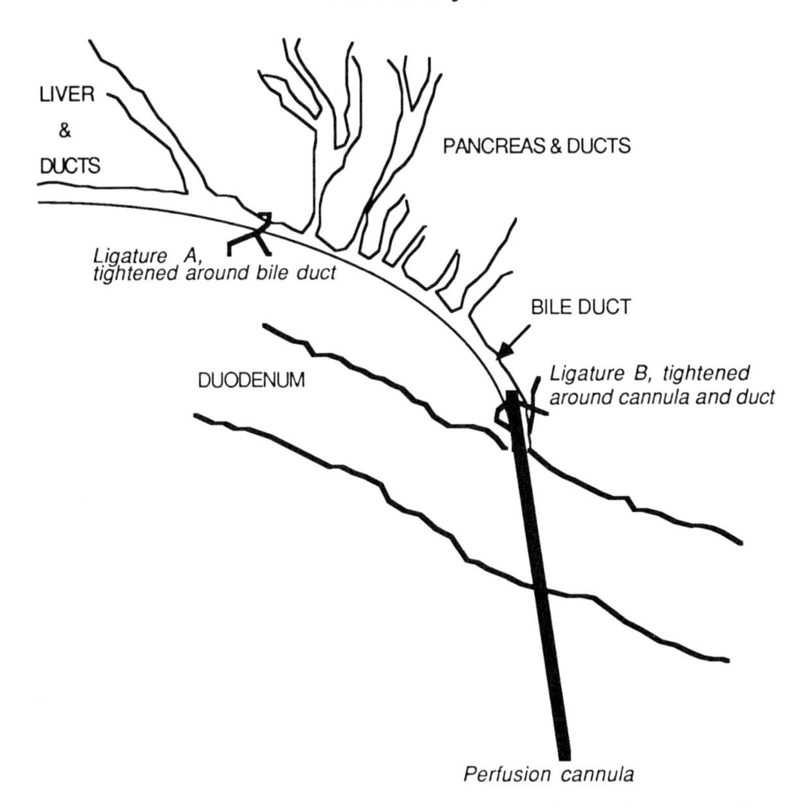

Fig. 7.2 Case Study 2: Diagram to illustrate the induction of experimental pancreatitis.

the duodenal end of the bile duct in order to pass the enterokinase back up the pancreatic ducts, which join the bile duct between the ligature and the duodenum (see Fig. 7.2). A second ligature (*ligature B*) was pulled tight around the cannula and bile duct and the enterokinase and sodium taurocholate injected (via the cannula) into the pancreas. The cannula and ligatures were then removed, and the animal's abdomen closed up with sutures.

Controlled trials of FFP. An infusion line (a narrow tube) was then placed in the jugular vein and connected to a pump by means of a harness, tether, and swivel. This allowed the animals to move about, but inevitably there was some restriction of movement. The therapeutic or control agents, together with dextrose, saline, and an analgesic (buprenorphine), were given to the animals via the infusion line. Mortality in the control groups was high. Between 50 and 70 per cent of these animals died during the 72 hours, with most deaths occurring within 48 hours. The experiments were run for three days because this was found to be the critical period: animals surviving the first three days of the experimental disease were likely to survive.

Commentary on Case Study 2

This project had as its goal a specific human benefit: the relief of acute pancreatitis, a disease which kills about one in ten of those affected. The researchers sought to evaluate the efficacy of a novel therapy, the administration of fresh frozen plasma (FFP), which an uncontrolled clinical trial had suggested was effective in reducing mortality from the disease.

The development of an effective therapy for pancreatitis would certainly be very worthwhile in terms of the relief of human suffering (B1.1). However, examination of the project in the light of the Working Party's assessment schemes reveals some doubt about the benefits likely to be gained by this approach (B2.1). In particular, it can be asked whether new understanding was to be gained from this research, involving animal subjects, which was not already known from the results of the uncontrolled clinical trial (B1.6; B2.1.4). One possible advantage was the validation of this animal model for acute pancreatitis, so as to enable further, detailed, studies of the mode of action of FFP. The primary purpose, however, was to test the efficacy of FFP. Obviously, the experiments on the rats involved the use of controls, enabling comparison between the effects of administration of FFP, containing substances (such as enzyme inhibitors) which the researchers believed might be of specific benefit in the relief of acute pancreatitis, and administration of two control substances (a plasma substitute devoid of these substances, or saline). Nevertheless, offset against the advantages of the controlled nature of the trial, was the difficulty of extrapolating from the animals to man in the clinical situation. It seems reasonable to ask, therefore, whether the investigators could have moved straight into a controlled clinical trial of FFP, without first having carried out a controlled trial using animal subjects (B2.2.2 i). All three substances (fresh frozen plasma, Haemaccel™ and saline) are widely used in clinical practice, so there would seem to have been no ethical objection to their use in a clinical trial. Indeed, at the time of the rat study, routine treatment of acute pancreatitis may have involved administration of saline, or plasma substitutes such as Haemaccel™. These facts, taken together with the results of the uncontrolled clinical trial, which suggested that FFP was efficacious in the treatment of acute pancreatitis, might have provided sufficient evidence to enable a human ethics committee to approve a controlled clinical trial. Indeed, it might be argued that the time taken to carry out the animal experiments could have delayed human patients from benefiting from the work.

It is, however, possible that there might have been some difficulty in setting up such a controlled clinical trial, which would enable reliable conclusions to be drawn concerning the efficacy of the novel therapy. In particular, there might have been difficulty in 'standardizing' the acute pancreatitis condition in a clinical trial, so as to enable meaningful comparison between the treatment groups, since each clinical case would be likely to be different in detail. A controlled clinical trial would have required large numbers of patients. Such difficulties, however, are encountered in most clinical trials. Indeed, at the time of publication of the results of the studies on rats, a controlled clinical trial of FFP was already underway.

There are also concerns about the severity of the procedures used. Death of the animals was used as an end point in the experiments, and a high mortality was observed (50 to 70 per cent of the rats in the crystalloid and colloid control groups, and 27 to 42 per cent of the rats given FFP, died within 72 hours). An analgesic was administered throughout each experiment, representing an important refinement. However, even with this analgesia, the work was likely to have caused the rats considerable pain and distress, since experience with human patients shows that it is very difficult to relieve the pain of pancreatitis (C5.1; C6.1).

There appear to have been two main opportunities for refining the procedures. First, killing the animals when they showed signs of pain and distress which could not be relieved, rather than allowing the experiments to progress in order to find the precise time of death (B2.2.3 i); and, second, avoiding using groups 1 and 2 in experiment 2, since these animals received treatments identical to those given to groups 2 and 3 in experiment 1. That is, it might have been possible to run the study as one experiment, involving one crystalloid and one colloid control group, and two treatment groups, one receiving FFP at the volume given to group 3 in experiment 1, and the other at lower volume. Thus, 60 fewer animals would have been used in the study (B2.2.3 ii). However, even with these modifications, the model would have remained severe and the cost of this work can be said to have been high. Furthermore, in addition to its cost in terms of animal suffering, the study involved the painless killing of 350 rats, to obtain sufficient FFP for the experiments.

In summary, there is doubt about the need for these particular experiments, especially given the high costs of the procedures involved. In considering the ethical justification for this work, we would wish to be reassured that animal studies were indeed required before a human ethics committee could authorize a controlled clinical of the FFP therapy. Given that the substances which would be administered in such a trial were all in use in clinical practice, that an uncontrolled trial had already suggested that FFP might reduce mortality, and that there is, anyway, difficulty in extrapolating from rats to humans, it would seem reasonable to ask whether an ethics committee might have authorized a controlled clinical trial without having the results of these animal studies. Even if this was not the case, and the animal studies *were* required (which is doubtful), there would seem to have been opportunities for reducing the costs of the experiments, by avoiding death as an end point and reducing the numbers of animals involved.

Results of Case Study 2

The experiments showed that intravenous infusion incorporating FFP could improve early survival in a rat model of acute pancreatitis, when compared with crystalloid and colloid controls. The beneficial effect was significant when FFP was given in high volume. Smaller volumes of FFP were also associated with increased survival when compared with colloid controls, but the difference was not significant. The study 'did little to elucidate the mechanism for improved survival in the FFP treated groups'. Taken in conjunction with the results of the previous uncontrolled clinical study, this work provided additional evidence that controlled clinical trials of FFP therapy in acute pancreatitis should be carried out.

CASE STUDY 3. INVESTIGATION OF THE DEVELOPMENT AND NATURE OF IMMUNE SUPPRESSION ACCOMPANYING MALNUTRITION

1. Summary of the project

The cause of the immune suppression which usually accompanies malnutrition in children was investigated using non-human primates. The project involved maintaining animals on a restricted diet until they showed signs of substantial malnutrition. The animals' immune response capabilities were tested in various ways and they were monitored for infectious diseases.

2. Potential benefits of the project

Protein–energy malnutrition in children is usually accompanied by partial suppression of the immune response and severe infection. It is not clear whether it is the infection or the malnutrition, or both, that are responsible for the immune suppression. This study was designed to shed some light on this problem.

3. Personnel, facilities and funding for the project

The project was a collaborative venture, involving members of staff at a university medical school and biochemistry department and at a hospital pathology department. The work is believed to have been supported from departmental funds.

4. Procedures employed

Experimental design

One baboon (the control animal) received a standard primate diet, while two other baboons and a cynomolgus monkey received 20 per cent, 5 per cent and 10 per cent of the control diet respectively, plus vitamins, vegetables and fresh fruit. After four weeks, attempts were made to achieve a steady metabolic state by increasing the amount of food for all the malnourished animals to 35 per cent of the control diet and maintaining it at this level. The nutritional status of all animals was assessed at least twice weekly. Tests for infection and for immune responsiveness were made. After 11 weeks the animals were killed and examined for signs of infection.

Animals used

Three baboons and one cynomolgus monkey, mean age 2½ years, were used.

Techniques used

Blood samples were taken at least twice weekly.

Assessment of nutritional status. The animals were examined daily for any changes in appearance or behaviour. They were weighed twice weekly and the skin-fold thickness of their fore and hind limbs was measured. Levels of serum protein in blood samples were measured.

Tests for infection. Throat and conjunctival swabs were taken at least twice weekly. These swabs, together with samples of blood, were examined by standard microbiological procedures. *Post mortem*, specimens of lungs, liver, spleen, kidneys and lymph nodes were tested in the same way.

Tests for immune responsiveness. A variety of *in vitro* tests were carried out on blood samples (C-reactive protein, IgG and haemoglobin concentrations were measured, and spontaneous rosette formation and PhA stimulation was estimated). The animals were also tested for skin sensitization and rejection of skin grafts.

Commentary on Case Study 3

In this case, there are grave doubts about the purpose and scientific merit of the work, as well as the degree of harm caused to animals in pursuing the aims of the project. At first sight, the work appears to have had both 'social' and 'scientific' value (B1.1 and B1.2), since it sought to explore a question about the nature of immune suppression, which was related to the practical problem of malnutrition in children.

However, examination of the research protocol reveals that the scientific design was very poor (B2.1.3). The study involved only one animal at each dietary level (that is, in each 'treatment'), so that there was little, if any, statistical confidence in the results from the various treatments. In addition, animals of two different species (three baboons and one cynomolgus monkey) and of variable age were used. Both the species and age differences mean that there is little, if any, confidence in conclusions drawn from comparisons between the different treatments. Another problem relates to the animals' exposure to infection. The work sought to establish the time of onset of infection (if any) relative to the time of onset of immune suppression, but the animals' exposure to potential sources of infection was not controlled. It was therefore possible that animals might have failed to become infected during the course of the experiments simply because they were not exposed to infection, or, equally, they might have become infected at different times simply because they were exposed at different times. Here again, there was a major fault in the experimental design. Overall, the project seems to have been of very dubious scientific merit (B2.1) and the likelihood of any significant finding arising from the work was slight. This would have been the case no matter how good the experimental facilities or personnel (B2.3).

In addition, the cost to the animals appears to have been high. The main suffering would have been caused by the dietary restriction. Non-human primates maintained on 1/5 down to as little as 1/20 of the standard diet for four weeks, and thereafter 1/3 of the normal diet for seven weeks (plus vegetables and fruit) were being partially starved and, in the early stages of the experiments at least, were likely to have suffered great distress. The published account of the experiments suggests that this was so, since the animals are reported to have become 'more aggressive' by the end of the first week on the restricted diets, before becoming 'withdrawn and lethargic' by the fourth week.

Monitoring of the animals for infection and immune responsiveness might also have caused adverse effects, but these were likely to have been much less severe than the effects of the dietary restriction. Skin transplants and sensitization tests would have caused discomfort or pain. Blood sampling, weighing and the taking of swabs should have caused minimal stress if carried out skilfully and with the animals' cooperation. However, it is likely that the animals would have had to be restrained for this routine sampling during the first three weeks of the experiment, owing to their increased aggression, and so the sampling may have caused particular stress at this time.

Added to these major doubts about the quality of experimental design and cost to the animals, are doubts about both the originality and timeliness of the study (B1.6 and 1.7). The literature on stress induced immune suppression appears largely to have been ignored (B2.1.4). Furthermore, it seems reasonable to ask whether this same question could have been addressed using epidemiological techniques in studies of malnourished children themselves. Studies involving children would, of course, have raised their own ethical questions, but would have avoided the difficulty of extrapolating the results obtained from laboratory animals, to the practical, human, situation. If such studies could be shown to have been ethically acceptable and scientifically valid, this use of animals might have been unjustified even if the experimental design had been improved (B2.2.2).

Overall, this case is one in which it was unlikely that significant findings would result (because the experimental design was very poor), and in which significant suffering was inflicted on non-human primates. For these reasons, it can be agreed that the work should not have proceeded.

Results of Case Study 3

The results suggested, but did not prove, that severe immune suppression in protein-energy malnutrition may occur long before the onset, and in the absence of, any infection.

CASE STUDY 4. INVESTIGATION OF THE ROUTE OF WATER UPTAKE IN A CEPHALOPOD

1. Summary of the project

The project investigated an aspect of the physiology of *Octopus*: the route via which water lost in excretion is replaced.

2. Potential benefits of the project

The results of the project added to the existing body of knowledge about osmoregulation in cephalopods.
It was known that:

(i) in *Octopus dofleini* the mean volume of fluid lost to the exterior via the kidneys is about 2.6ml/kg/h. No figures were available for the much smaller *Octopus vulgaris*, the subject of this experiment, but, by comparison with *Octopus dofleini*, it was expected that a 0.5 kg animal would lose around 14 per cent of its body weight in water over 24 hours. This water lost via the kidneys must be 'made good' by the animal;
(ii) part of the *Octopus* gut, the digestive gland appendages (see Fig. 7.3) are formed largely from a particular type of cell which has characteristics strongly indicative of bulk fluid transport;
(iii) seawater is taken into the *Octopus* gut at both its anterior and posterior ends: octopuses swallow massive quantities of seawater during and after feeding and can also engulf water and pass it upstream via the rectum.

The project, therefore, was designed to establish the role of the *Octopus* digestive gland appendages in the uptake of seawater. Since the structure of the digestive gland appendages is similar in other cephalopods, it was thought possible to generalize the results to other members of this Class of animals.

3. Personnel and facilities for the project

The work was carried out by university researchers, based at a marine laboratory.

4. Procedures employed

Experimental design

The experiments involved ligation and catheterization of various parts of the animals' digestive system, and monitoring of any loss in weight (indicating the amount of water lost) over a period of 48 hours. The animals were then killed. A variety of physiological measurements was also made, both during the experiments and *post mortem*.

Species and number of animals used
The experiments were conducted using *Octopus vulgaris* collected by trawling or scuba diving. The octopuses were maintained in separate tanks in a marine aquarium and were carefully monitored to check that they were healthy before the experiments were started. Forty-six animals were used.

Techniques used
The basic experiments are listed in Table 7.1, and the sites of ligation and cannulation are shown in Fig. 7.3. All animals were starved for four days prior to surgery. They were anaesthetized using 2.5 per cent ethanol. Mortality following surgery was uncommon, but a 10 per cent loss of animals occurred, due to failure to recover from anaesthetic, or as a result of blood loss. If the peristaltic action of the digestive gland ducts was indeed responsible for water uptake, it was expected that water uptake would be blocked in groups B and C. Where fluid uptake was successfully blocked, a rapid weight loss occurred (around 7–17 per cent of body weight was lost in the first 24 hours after ligation, rising to 12–26 per cent after 48 hours.) Most animals were dead or dying 48 hours after water uptake was blocked. However, experience showed that, until just before their deaths, these animals were indistinguishable in appearance and behaviour from animals in which water uptake had not been blocked. Two to four hours before death, a characteristic set of symptoms emerged,

Table 7.1 Case Study 4: Basic experiments

Group	Number of animals	Procedure
Investigation of the effects of removing the possibility of seawater entering the gut		
A	6	Ligation of oesophagus (blocking the gut anteriorly)
B	7	Ligation of oesophagus and rectum (blocking the gut anteriorly and posteriorly)
Investigation of the effects of removing the possibility of seawater entering the digestive gland appendages		
C	11	Ligation of both digestive ducts
D	3	Control operations (digestive ducts exposed and replaced)
E	5	One digestive duct ligated, other intact

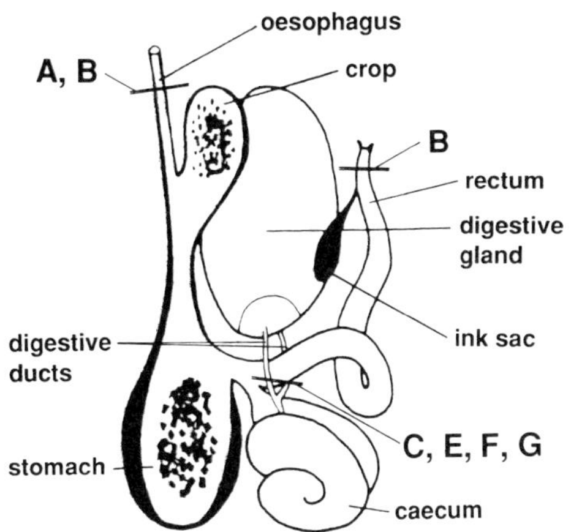

Fig. 7.3 Case Study 4: Diagram to show the sites of ligation used in the experiments. Letters refer to experimental groups shown in Tables 7.1 and 7.2. (Diagram adapted from Laverack and Dando 1979.)

and the researchers killed the animals as soon as these external symptoms became obvious.

The following physiological measurements were also made: osmolality of blood and urine samples, conductivity of these body fluids and copper content of the blood (as a measure of haemocyanin content); muscle tissue water content, using samples of muscle taken from the mantle and second right arm; osmolality of rectal fluid samples; weight of gut and kidney sac contents.

After observations that water was indeed lost in groups B and C, three follow-up experiments were carried out, according to the scheme shown in Table 7.2.

Commentary on Case Study 4

This work was aimed at improving understanding of the physiology of *Octopus*: surgical procedures were carried out on octopuses in order to advance knowledge about octopuses (B1.2). Such work, involving an invertebrate species, would not have been covered by laboratory animal protection law, since most, if not all, such laws—including the UK Animals (Scientific Procedures) Act—refer exclusively to vertebrate species.

The potential benefits of the work appear to have been seen in terms of a gain in scientific knowledge (B1.2), and the project was not expected to have any particular 'social' or 'economic' value (B1.1 and 1.3). The research was directed towards answering a specific question, about the route by which water lost in excretion is 'made good'. The hypothesis that the digestive gland appendages are involved, was based on evidence from microscopical studies and on observations of intact octopuses (B2.1.2;

Table 7.2 Case Study 4: Follow-up experiments

Group	Number of animals	Procedure
Investigation of the possibility that increased urine clearance, brought about by the presence of wastes in the digestive gland, rather than lack of water uptake, was responsible for weight loss observed in experiment C		
F	5	Cannulation of one or both digestive ducts close to the digestive gland, and ligation of the ducts on the caecum side of the cannulae. Cannulae were led into the mantle cavity, so that wastes could be discharged, but water uptake by the digestive appendages was blocked.
Investigation of the role of peristalsis of the digestive ducts in pumping water into the mid-gut gland		
G	7	Cannulation of one or both digestive ducts, close to the caecum, with ligation of both ducts on the caecum side of the cannulae. Cannulae were led to the exterior via the back of the animal's abdomen, so that seawater could be taken in via the cannulae.
Examination of the role of central nervous control in water uptake		
H	2	Removal of gastric ganglion

B2.1.4), and the general approach was relevant in testing this hypothesis (B2.1.1).

It is possible, however, that the effects of the specific procedures may have been severe. Where ligation effectively blocked water uptake, the animals were unable to make good the water lost in excretion. They therefore lost weight rapidly and, in most cases, this led to death within two days. At least 18 animals (groups B and C) were affected in this way. These animals remained apparently normal until two to four hours before their deaths, when a characteristic set of symptoms emerged, and they were killed. There is, however, uncertainty in judging what, if any, suffering was caused to the animals up until the time they were killed. Furthermore, blood loss or failure to recover from the anaesthetic caused mortality in some of the experimental groups, and, again, there is uncertainty about whether the surgery caused the animals any post-operative pain or distress.

Although understanding of what octopuses might experience in the way of pain and distress is limited, the evidence presented in Chapter 4 suggests that we should take seriously the possibility that the animals have some such experiences, and give these animals the benefit of the doubt. On this basis, it can be assumed that the animals suffered at least some adverse effects from the procedures (C5.1). It is reasonable to ask, therefore, whether the approach could have been refined, so as to reduce the suffering presumed to have been caused to the animals used in the experiments (B2.2).

Since the experimental animals themselves were the objects of the study (the octopuses were not being used as 'models' for other animals or humans), it was obviously necessary to use this particular species (B2.2.2 ii). Furthermore, since an integrated physiological process was being studied, the work required the use of whole, living, animals (B2.2.2 i). It might, however, have been possible to refine the scientific procedures. Group A (ligation of the oesophagus) and group B (ligation of oesophagus and rectum) may not have been essential in the experimental design (B2.1.3). Previous work, and direct observation, had shown that water was taken into the gut both anteriorly and posteriorly, suggesting that water uptake occurred via the gut. This, taken together with the observation that the digestive gland appendages have a structure strongly indicative of bulk fluid transport, might have made it possible to move straight into the experiments which investigated the role of the digestive glands. This modification would have reduced both the severity of the procedures (B2.2.3 i) and the number of animals involved in the experiments (B2.2.3 ii). It might also have been possible to use fewer animals in the remaining experiments, particularly in group C.

Nevertheless, even if these refinements were possible, this project involved causing harm to octopuses, in order to improve understanding of cephalopod physiology. The work was not intended to find out something of particular human benefit, nor to benefit the animals under study. The possibility that practical benefits might have resulted was not precluded (it is not impossible, for example, that the octopus could turn out to be a model for mammalian renal physiology), but the researchers were moved to carry out the work even in the absence of any expectation of such benefits. This case, therefore, illustrates the difficulty of deciding where to draw the line in terms of the amount of harm which might be caused to animals in pursuit of such understanding. Some would argue that *no* harm should be caused for such an end. The difficulty, however, lies with those who accept that *some* harm might be justified: the question is, 'How much'?

In this context, it is impossible to provide general rules, and we must judge the question case by case. In the case presented here, there is doubt about whether the understanding gained from the experiments was sufficient to justify the harm caused to the octopuses. Modification of the procedures along the lines suggested would have made the work easier to accept, since this would have avoided the severe effects observed in group B, and would have reduced the number of animals suffering adverse effects in group C. In a general sense, some members of the Working Party consider that this form of curiosity might be a justification for some uses of animals in research, but would wish to be reassured that only the minimum possible amount of harm was caused to the animals involved. As in other biomedical research, there is a price (in terms of animal suffering) which it would be too high to pay for such a cause. In this particular case, from the evidence in the published paper, and given the uncertainty in deciding how much the octopuses might have suffered, it can be argued that if modification had proved impossible, the work should not have proceeded; and furthermore, some members of the Working Party believe that it should not have proceeded at all.

Results of the study

The experiments showed that octopuses lose weight if water is prevented from entering the midgut gland. Seawater normally enters the gut by the mouth and/or the rectum, and the results of this study indicated that the water is passed up into the midgut gland by peristalsis of the digestive ducts. This last process, at least, proceeded in the absence of central nervous control. The experimental observations, taken together with the existing anatomical evidence, suggested that the digestive gland appendages are the principal site of fluid uptake in *Octopus* and other cephalopods.

CASE STUDY 5. LOCALIZATION OF PLANT CHEMICALS

1. Summary

Three studies are described. Each involved the use of antibodies, raised in animals, to localize or isolate specific chemical components of plants. The three studies illustrate the range of animal procedures involved in such work.

2. Potential benefits

In each study, the relevant published paper suggests that the potential benefits were seen in terms of an increase in botanical knowledge.

Study 1. Localization and molecular analysis of actinidin, the cysteine proteinase enzyme of Actinidia chinensis *(the Chinese Gooseberry, or Kiwi, plant)*

The enzyme actinidin belongs to a family of similar enzymes, which includes papain from pawpaw, bromelain from pineapple, ficin from fig and aleurain from barley, as well as enzymes from widely different organisms including mammals and slime mould. The structures of the plant enzymes are known, but, with the exception of aleurain, which has a role in the hydrolysis of storage proteins during seed germination, no clear function has been identified for them.

Actinidin accounts for 50 per cent of the soluble kiwi fruit protein and is a highly active proteinase, so it is clear that the plant must be protected against its activity. It is known that other plant proteinases are synthesized in the form of inactive precursors, or zymogens, which are located inside storage vesicles, within the plant cells. This study was carried out to see if this is the case for actinidin, and to examine further actinidin gene expression in the developing Kiwi fruit. At the time, nothing was known about the physiological function of actinidin in Kiwi fruit.

Study 2. Investigation of the spatial organization of surface antigens of the sperm of Fucus serratus *(a brown seaweed)*

Sexual reproduction in marine brown seaweeds involves the liberation of gametes (large non-motile eggs and small motile sperm) into the sea. There are several species of brown seaweed in UK waters, so there must be some means by which the species-specificity of fertilization is ensured. The eggs secrete a pheromone to attract the sperm, but this is not the basis of the species-specificity of sperm–egg recognition. Several experiments have suggested that the sperm–egg recognition is mediated by saccharides on the surface of the sperm; so the distribution of specific surface antigens is likely to be important in the recognition process.

It is known that specific antigens are localized in particular regions (for example, head, tail) of the sperm cells of many animals and that the molecules in different parts of the sperm often have specific functions in fertilization. The aim of this study was to determine whether surface antigens occurred in discrete regions, or over the entire surface, of *Fucus* sperm; and whether certain antigens were specific to *Fucus serratus*.

Study 3. Localization of prolamine and globulin storage proteins within the developing starchy endosperm of Avena sativa *(oat)*

Developing seeds of angiosperm plants store nitrogen in the form of storage proteins, which accumulate in the seed embryo, endosperm, or both. In many dicotyledonous plants, such as legumes, the principal storage proteins are globulins, which are saline-soluble and synthesized in the cotyledons of the seed. In contrast, most cereal plants have alcohol-soluble storage proteins, called prolamines, synthesized in the endosperm.

Globulins are manufactured in the Rough Endoplasmic Reticulum (RER) of the cotyledon cells and transported (via the Golgi system) to vacuoles in the cell, where they form insoluble aggregates, called 'protein bodies'. Prolamines are also synthesized in the RER, but vary, in that some are transported and some are not, some form protein bodies and some do not.

Oat (along with rice) seeds are unusual amongst cereals in that their endosperms contain relatively small amounts of prolamine, and the predominant storage proteins are like the globulins of dicotyledons, but much less soluble in saline. Electron microscope studies on oat endosperm have found protein bodies located within vacuoles, these bodies having distinct light- and dark-staining regions. Biochemical techniques have shown that the protein bodies contain both prolamine and globulin-like proteins, and it has been suggested that the light-staining regions might be the prolamine component. The present study was carried out to test this hypothesis, to find out whether the prolamine did indeed occur in the light-staining regions, and globulin in the dark-staining regions.

3. Personnel, facilities, and funding

All three studies were carried out by scientists working in university departments of botany or biology. Studies 1 and 2 were supported by grants from UK Research Councils, and study 3 by a grant from the US Department of Agriculture.

4. Procedures employed

Experimental design

Study 1 (Kiwi fruit). Animals were used as part of the initial stage of the project, in the localization of actinidin in the plant. Antibodies to actinidin were raised in rabbits, and used in an *in vitro* test, to detect actinidin in various parts of the plant. The results showed that actinidin was synthesized specifically in the fruit of the plant. Actinidin gene expression in the Kiwi fruit was then investigated using biochemical techniques.

Study 2 (seaweed). Monoclonal antibodies were raised against *Fucus serratus* sperm, using mice. The antibodies were used to localize sperm surface antigens. The monoclonal antibodies were also tested for cross reactivity with *F. serratus* eggs and with sperm from two other species of brown seaweed (*F. vesiculosus* and *Ascophyllum nodosum*).

Study 3 (oat). Anti-globulin antibodies were prepared using rabbits, and anti-prolamine antibodies using mice. The antibodies were used in biochemical studies, to localize the proteins within the oat endosperm.

Species and number of animals used

Study 1 (kiwi fruit). The published paper states that 'antibodies were raised in rabbits', but does not state the number or strain of animals used.

Study 2 (seaweed). The paper states that 'six- to eight-week old Balb-c mice' were immunized in the production of monoclonal antibodies, but the number of animals used is not stated.

Study 3 (oat). The paper states that anti-globulin antibodies were raised according to a method described in a previously published paper. This method used New Zealand White rabbits, but numbers were not stated. Anti-prolamine antibodies were raised using two 30-day old Swiss mice.

Techniques used

Study 1 (kiwi fruit). The paper states only that 'antibodies were raised in rabbits by immunization followed by two booster injections over six weeks with a total of 2 mg actinidin'.

Study 2 (seaweed). Monoclonal antibodies were produced by immunizing the mice, intraperitoneally, with 10^8 live sperm cells of *F. serratus* in 200 μl of seawater, giving a similar booster injection three weeks later, and a final intravenous injection of 10^8 cells in 200 μl seawater one to six months later. The animals were killed four days after the final injection and cells harvested from their spleens. These cells were hybridized with tumour cells and the resulting hybridomas cultured *in vitro*. The culture supernatants were tested using *in vitro* immunological techniques, so as to select cell lines secreting monoclonal antibodies to *Fucus* sperm.

Study 3 (oat). **1. Rabbit anti-globulin**. The previously published paper states that 'antibody synthesis was induced by injection of total oat globulin protein or purified large subunits into New Zealand white rabbits. The initial injection was 1 mg protein in Freund's complete adjuvant followed by weekly booster injections of 1 mg protein in Freund's incomplete adjuvant for three weeks. Beginning two weeks after the last injection, the rabbits were bled from an ear vein.' Anti-globulin antiserum was prepared from the rabbit blood, using *in vitro* techniques. **2. Mouse anti-prolamine**. The paper states that two 30-day old Swiss mice were injected intraperitoneally with 100 μg of the prolamine (avenin) emulsified in Freund's complete adjuvant. The mice received a second injection of 125 μg of avenin emulsified in Ribi adjuvant at 66 days after the initial injection, followed by a final injection of 330 μg of avenin emulsified in Freund's complete adjuvant at 135 days after the initial injection. Sarcoma (tumour) cells were injected intraperitoneally at 155 days after the initial injection (5×10^6 cells per mouse). The resulting hybridoma grew in the peritoneal cavity of each mouse, secreting a fluid rich in monoclonal antibodies to avenin (ascites fluid). A total of 85 ml of ascites fluid was collected over a period of several weeks.

Commentary on Case Study 5

This case differs from those considered previously in an important respect. Here, animals were not used in an experimental manner, but as 'tools' in the production of the antibodies required in the experiments. In the UK, such work has only recently been included under the terms of legislation relating to laboratory animals, with the change from control of the use of animals in 'experiments', under the 1876 Cruelty to Animals Act, to the control of 'scientific procedures', under the 1986 Animals (Scientific Procedures) Act.

This use of animals simply to produce the antibodies, rather than as an integral part of the experiments, may be reflected in the reporting of the studies. In all three published papers the methods used to raise the antibodies were reported in outline only, with the omission of details, such as the number of animals used, and in Study 1, without description of the immunization schedule or mention of the strain of rabbit used. Nevertheless, the basic methods employed in the antibody production were well established (although the details of the protocol involving mice in Study 3 appeared unusual*) and had been applied successfully in similar botanical work (B2.1).

The use of antibodies in botanical research is relatively new, and the techniques are being employed in an increasing number of laboratories (see Wang 1986). Antibodies have proved to be powerful tools for detecting specific plant chemicals, and have considerable advantages over biochemical techniques. In particular, they can be used to probe chemicals whose structure is not known, and can detect very small amounts of specified compounds (see Wang 1986, for further discussion of the use of antibodies in plant science).

However, the procedures used in these studies were not without some cost to the animals involved (C1.5). The raising of polyclonal antibodies in rabbits (Studies 1 and 3) required repeated subcutaneous, intradermal or intramuscular injections (the published papers do not state which), as well as test bleeds, which may have produced local pain as well as stress. Likewise, the intraperitoneal injections used in the production of monoclonal antibodies in mice, may have caused pain and distress. The injection of Freund's complete adjuvant to increase the animals' immune response (Study 3, and possibly Study 1), would have caused local inflammation which may have resulted in painful lesions (see Amyx 1987), and, in Study 3, the accumulation of fluid in the peritoneal cavity (ascites) would have caused discomfort. The report of Study 3 does not mention whether the mice were anaesthetized during removal of ascitic fluid. If they were, the removal of the fluid itself may still have caused physiological shock, and if they were not, the sampling (with a fairly large gauge needle) is likely to have caused pain and distress (see McGuill and Rowan 1989).

There may, however, have been several ways of modifying the procedures so as to have caused less harm to the animals used. The use of animals could not have been avoided altogether, since there are no established methods for raising antibodies entirely *in vitro* (B2.2.2 i). However, in Study 3, it may have been possible to avoid the production of ascitic mice, by growing the hybridomas *in vitro*, as was done in Study 2. (See Home Office 1990, for discussion of the pros and cons of *in vivo* and *in vitro* methods of monoclonal antibody production). The use of an *in vitro*

* Study 3 appears to have involved immunizing and growing hybridomas in the same animals. Whether this was indeed a novel method, or poor reporting, is unclear. However, the usual method would have involved killing the immunized mice after a month or so, harvesting cells from their spleens, hybridizing the spleen cells with myeloma cells *in vitro* and injecting the resulting hybridoma cells into new mice.

culture method would have considerably reduced the costs to the animals involved (B2.2.3 i). If this was not possible, it would have been important to refine the procedures, by removing ascitic fluid under anaesthesia, and by reducing the number of taps, as well as the volume fluid removed in each tap, to the minimum possible, whilst at the same time making sure that the ascitic fluid did not build up beyond a certain maximum volume. If 85 ml of ascitic fluid was indeed obtained from two mice, this would have represented an unacceptably large number of taps and/or an unacceptably large volume of ascites per animal.

In Study 3, and possibly in Study 1, the use of non-inflammatory adjuvant in place of Freund's should have been considered. If Freund's proved to be the only suitable adjuvant for this work, great care should have been taken to minimize its adverse effects. For example, the amount of adjuvant at each site should have been kept to a minimum, so as to reduce the likelihood that lesions developed; the number of injection sites should have been restricted to a maximum of three or four; and the sites should have been widely spaced, so that if lesions developed they could not coalesce (see Amyx 1987). All of these refinements would have reduced the costs to the animals of the procedures (B2.2.3 i). Furthermore, in Studies 1 and 3, it might have been possible to raise sufficient polyclonal antibodies in only one rabbit, rather than the 'rabbits' (of unspecified number) which were used (B2.2.3 ii). We would wish to be reassured that, where possible, all of these modifications were implemented, so as to keep the costs to the animals to a minimum. However, even with such modifications, the techniques would still have caused some adverse effects in the animals involved.

As in the previous case (involving *Octopus*), the fact that the potential benefits of these three studies appear to have been seen solely in terms of an increase in a particular area of botanical knowledge, does not preclude the possibility that the results of the work may eventually have led to some more practical benefit, or may have led to more profound or unifying understanding. The results of Studies 1 and 3, for example, may have been useful in work leading to an improvement in methods or production of these crops, and Study 2 may have had implications in cell development. The researchers, however, appear to have been moved to carry out their work without the expectation of any particular practical benefit or unifying understanding. This, of course, is the nature of much biomedical research: 'breakthroughs' generally come only after much painstaking and detailed work, the usefulness of which may be recognized only with hindsight. Nevertheless, in the absence of any identified practical benefit, we should consider whether the predicted gains in knowledge were, in themselves, sufficient justification for this use of animals.

In this context, the fact that in each case the research had been awarded a grant by a prestigious funding body may not, in itself, have been sufficient justification, since funding bodies often do not take into account the animal interest (leaving this judgement to others).

Most members of the Working Party would be prepared to accept such a use of animals, provided that the procedures, where possible, were indeed refined in the manner described above. Without these refinements, the work would have been unacceptable. There was also some concern amongst the Working Party that the use of animals simply in the production of antibodies, may have led to a lack of detailed consideration by the researchers of the costs in terms of animal suffering of this use, evidenced by the poor reporting of the animal procedures involved.

Results of the studies

Study 1 (Kiwi fruit). The localization study, using antibodies, showed that actinidin was found in the Kiwi fruit, with very low levels detected in other parts of the plant. Further, entirely non-animal, work showed that the difference in the levels of actinidin between fruit and young leaf tissue was due to different amounts of actinidin-encoding mRNA. This mRNA accumulated to very high levels during fruit development, and therefore did not belong to the class of 'ripening-specific' mRNAs described for the tomato fruit. It was suggested that actinidin was encoded by a multigene family of up to ten members. As expected, actinidin was shown to be synthesized in the form of a larger precursor. Analysis of the structure of this precursor enabled the researchers to speculate on the mechanism of inactivation of the proteinase, by analogy with other proteolytic zymogens.

Study 2 (seaweed). Twelve types of monoclonal antibody were raised in mice against surface antigens of *Fucus serratus* sperm. Studies using these antibodies showed that the antigens bound by the antibodies were distributed non-randomly over the cell surface. Seven of the antibodies bound antigens located primarily on the cell body of the sperm, while the remainder bound antigens located primarily on the anterior flagellum of the sperm (*Fucus* sperm are biflagellate). Eight of the antibodies bound to sperm antigens which were not found on *Fucus serratus* eggs, although four antibodies bound to antigens present on both sperm and eggs. Nine antibodies exhibited genus-preferential binding, labelling sperm of *F. serratus* and *F. vesiculosus* more intensively than that of *Ascophyllum nodosum*. Only one antibody showed species-preferential binding, labelling *F. serratus* more intensively than *F. vesiculosus*.

Study 3 (oat). The study demonstrated that globulin and avenin proteins in developing oat endosperm are located within protein bodies contained in a vacuole. The previously observed dark- and light-staining regions contained the globulins and avenins, respectively. The sites of aggregation of globulin and avenin appeared to be spatially distinct; avenin appeared to aggregate within the RER and then migrate to the vacuole, whereas the main aggregation of globulin appeared to take place within the vacuole.

CASE STUDY 6. A SCHOOL BIOLOGY PRACTICAL EXERCISE

1. Summary of the practical exercise

In this practical exercise, school pupils gently handled new-born mice, and assessed the effects on the later behaviour of the mice. The exercise was part of an advanced level school biology course and was described in the pupils' text book.

2. Potential benefits of the practical exercise

The goals of the exercise were purely educational. The aim was to help the pupils to understand the importance of the environment in development and to teach them an experimental technique for investigating animal behaviour. The work was part of a wide-ranging study of development in both plants and animals.

3. Personnel and facilities

The experiment was described as an exercise to be carried out by school pupils, aged around 17–18 years, under the supervision of their class biology teacher. It is most likely that the whole class would have been involved in only one such experiment. If a class was large, it is possible that the pupils would have worked in groups, each group carrying out a separate experiment. The animals would probably have been bred in the school animal house, and brought into the classroom for the period of the experiment.

4. Procedures employed

Species and number of animals

Each experiment involved two litters of inbred new-born mice, and their mothers.

Procedure

The litters of new-born mice, together with their mothers, were maintained in separate cages. In one litter, once a day from the first day after birth, each young mouse was removed from the cage, passed very gently from hand to hand, and replaced in the nest. The treatment was continued until the animals' eyes opened at about the twelfth day. The other litter acted as the control group and was reared normally, with as little disturbance as possible. After the twelfth day, both litters were reared normally and the animals were weaned between 21 and 25 days.

The behaviour of animals from both litters was then examined. Individual mice, selected in random order, were transferred carefully to an 'open field' arena: a box of floor size approximately 45 × 60 cm, with walls approximately 45 cm high, the floor being marked out in squares to enable recording of the animal's movements. The open field was illuminated as evenly as possible, the rest of the room was darkened and the noise level was kept to a minimum during the test.

The activity of each mouse was recorded for three minutes. The amount of movement between squares was used as a measure of the animal's exploratory behaviour, and the number of faecal pellets deposited as a measure of the animal's 'emotionality', or anxiety. After testing each animal was returned to its home cage, the open field arena was cleaned, and the test repeated with another mouse.

The whole test was repeated several times, at daily intervals, in order to obtain a mean activity score for each litter over a period of time. Pupils applied a statistical test in order to find whether there was any significant difference between the litters.

Commentary on Case Study 6

Unlike the other cases considered here, this was not an original piece of research, but an educational exercise (B1.4). The outcome of the pupils' experiment was known and it was not expected that any practical or theoretical benefit would arise from the work (B1.1–1.3). The procedure brought pupils into direct contact with living animals, so that, in addition to the potential educational benefits described in the outline, it might also have helped the pupils to develop an understanding of the needs of the animals (teaching them how to hold the mice carefully, how to care for the mice, and so on). However, it was also possible that this practical exercise involved causing some harm to the animals used.

Provided the mice were properly cared for, breeding and maintaining them in the school animal house should have caused no adverse effect (C4.1). Passing new-born mice from hand to hand, however, may have caused the animals stress (indeed, the object of the expriment was to assess the later effects of this 'handling stress'), and the mother might have been distressed by the removal of her offspring whilst the handling was carried out (C5.1). Although pupils would have been told to handle the new-born mice very gently, there is some doubt whether this would have been possible because new-born mice are so small. The mice might also have been stressed when placed in the open field arena. In fact, again, the object of the experiment was to assess the degree of such stress, by investigating the animals' willingness to explore the novel environment and trying to assess how anxious, or 'emotional', they were when placed in the open field. (It was expected that the handled animals would have been less stressed by the open field than the animals which had not been handled.) The atmosphere in which the open field test was carried out should have helped to ameliorate any adverse effects caused to the animals. With gentle handling, and with noise and other disturbances kept to a minimum, the stress should have remained minimal. Such precautions were necessary from the scientific as well as the welfare point of view, and the teacher would have been able to ensure that these practical instructions were followed.

Overall, then, the practical exercise, if properly carried out, should have caused only minimal stress to the animals involved. However, if the practical instructions were not adhered to strictly, so that the new-born or older animals were handled roughly, or were subjected to noise or other disturbances in the classroom, the animals may had suffered some distress. The pupils themselves were likely to have had very little experience in handling animals (C2.2) and the harm caused to the animals may have depended on how well the teacher (who might also have been inexperienced in working with mice) controlled the practice exercise. In particular, it might have been difficult to ensure that the new-born mice were handled gently.

It can be asked, therefore, whether the educational benefits of this practical exercise were sufficient to justify carrying it out. Some members of the Working Party argued that it was not, stressing the difficulties involved in controlling the practical exercise to ensure that the animals were handled gently. These members suggested that sufficient knowledge about the effects of early development could have been imparted by alternative means (e.g. from the literature, or using films or videos) rather than using animals (B2.2.2); and that development of respect for animals, as well as skills in assessing animal behaviour, could have been achieved in ways which did not run the risk of causing distress to the animals involved. These objectives might, for example, have been achieved simply by observing adult mice and recording their behaviour, rather than by carrying out experiments which involved handling new-born mice (B2.2.3 i). On the other hand, other members argued that the exercise was justified, believing that there was minimal risk that the animals would have been distressed. Provided that the animals were properly cared for and handled gently, these members had no objection to the practical exercise. In this case, it was argued that the exercise would have helped the pupils to develop scientific skills which were an essential part of their biological training. Such skills might have been especially important for pupils studying at advanced level, who were likely to pursue a biological subject at degree level. The study, if carried out properly, would have given pupils experience in the care needed to handle animals, and may have helped to encourage respect for animals.

In summary, this case produced divided opinion in the Working Party, concerning the value of the practical exercise and the amount of harm it caused to the animals. Some saw great difficulty in controlling the experiment so as to ensure that the animals were not distressed. On this basis, it was argued that the practical exercise should not have been carried out, since it offered no educational advantage over other methods which were likely to be less harmful. Others, however, saw the risk of harm as minimal and therefore argued that the educational benefits were sufficient to justify this use of animals.

The issues raised by the use of animals in education and training are examined in more detail in Chapter 9.

Postscript

A recent UK Department of Education and Science Administrative Memorandum (No. 1/89, 9th February 1989) declares that, as a result of the 1986 Act, such practical exercises are illegal in UK schools and can no longer be carried out. The memorandum in fact refers to a 'an experiment involving the tossing of a young mouse from hand to hand', stating that this was 'never ethically acceptable'. Whether the above exercise comes under this description is open to debate. Nevertheless, it is no longer put forward for use in UK schools.

CASE STUDY 7. DEVELOPMENT OF AN INSECTICIDE (HYPOTHETICAL EXAMPLE)

1. Summary of the project

The project is to carry out toxicity tests, according to international guidelines, on a corn insecticide, code-named '123'.

2. Background and potential benefits of the project

The current market of corn insecticides is £500 million a year worldwide. The market is dominated by an insecticide 'XYZ', which holds 70 per cent of the market share. A UK company wishes to develop 123, and aims to place the insecticide on the market in such a way as to take a part share of the market. XYZ is made by a manufacturer based overseas.

The molecule 123 has been patented and is of a similar chemical structure to XYZ. 123 is substantially cheaper to manufacture than XYZ, but has the same efficiency as an insecticide. It is believed that the development of 123 will benefit three parties:

(i) the manufacturer, his shareholders and employees. The new insecticide presents a considerable growth potential with greater profits for the company and the creation of 50 new jobs;
(ii) the country. It will reduce imports of insecticides and increase exports. Hence it will help to improve the balance of payments;
(iii) the consumers. The introduction of competition into the marketplace will reduce the price of the competitor's product, providing the farmer with improved cost effectiveness.

3. Potential disadvantages

The product is similar to XYZ. It is known to be more persistent in the environment. This may be insecticidally beneficial but less acceptable environmentally. It is impossible, at present, to make a firm judgement as to whether 123 will prove environmentally sufficiently benign to be registered. There remains a significant chance (a 25 per cent probability) that in three and a half years time when the environmental studies are completed, the product will not be registered. In addition, from its chemical structure, it is highly likely that 123 will be insecticidal by the same mechanisms as XYZ. Thus, if XYZ induces resistance in insects, it is likely that 123 will also do so.

4. Personnel, facilities, and funding for the project

The company is of international standing. The facilities are excellent, and all personnel involved in the project have been trained to a high standard.

5. Procedures to be employed

Toxicity studies and environmental evaluations will be carried out in parallel, in a carefully planned order. This is necessary to protect workers involved in the environmental studies, to avoid delays in marketing and to enhance the product's patent life.

Experimental design

Thirty-seven different toxicity studies will be carried out, as laid out in the international protocols according to the most recent OECD guidelines. These include

(i) *acute toxicity tests*: eye and skin irritation, oral and dermal acute toxicity, LD50 estimation in rats and mice, and skin sensitization tests;
(ii) *sub-acute toxicity tests*: 28 and 90 day repeat dose studies in rats and mice; and chronic studies (one year studies in dogs and lifetime studies in rats and mice);
(iii) *reproductive toxicity tests*: teratogenicity tests in rats and rabbits and multigeneration studies in rats;
(iv) *mutagenicity tests*: bone marrow micronucleus tests in mice and cytogenetic studies in rats.

Metabolism studies will also be required. Additionally, acute toxicity studies are required for each formulation. Some toxicity studies of similar design will be required for the principal metabolites found in plants.

The design of the studies is laid down in the regulatory guidelines, which, it is argued, must be followed. The guidelines include the need for LD50 estimations, the administration of doses close to the LD50 for the mutagenicity assays, and the top dose in the lifetime studies to be the Maximum Tolerated Dose (MTD). The MTD is selected to produce slight signs of toxicity but no mortality in the lifetime studies.

All experiments will be conducted according to the guidelines for Good Laboratory Practice (GLP), which require that all data are verified by a Study Director and audited by a Quality Assurance Unit. The laboratory is subject to regular government inspections to ensure compliance with GLP.

Animals

Approximately 1400 rodents, 20 dogs, 48 rabbits, and 20 guinea-pigs are to be used.

Techniques

The insecticide will be administered by gavage in the acute oral studies and by addition to the diet in the sub-acute, chronic, and reproductive studies. There will also be studies requiring dermal application and application to the eye.

Commentary on Case Study 7

This is a hypothetical example. The commentary is presented as a dialogue, in order to illustrate the wide variety of considerations which are important in deciding whether such a project ought or ought not to proceed. These issues are examined in more detail in Chapter 8.

Potential benefits of the product

Commentator A. The potential benefits of the introduction of this new product are seen mainly in economic terms (B1.3). In particular, there is likely to be improved cost effectiveness of insecticide application, due to the introduction of a cheaper product onto the market, which will force the competitor to reduce his price. The company will also receive benefits, with the expansion of its product range and hence its profitability. Fifty new jobs are likely to be created by this project.

Set against these potential benefits, however, are several potential disadvantages, which, if they emerge during the development of 123, could lead to the abandonment of the project. If 123 is judged not to be environmentally acceptable, it will not be registered. Furthermore, if XYZ induces resistance in insects it is likely that 123 will also do so; hence the agricultural utility, and thus the commercial viability, of the novel insecticide could be limited.

One of the critical issues in the development of a pesticide of this type is that all of the studies must be carried out in a carefully controlled sequence in order to minimize the interval between the granting of a patent on the product and the time at which all the information is available for registration. Even when this is achieved the remaining patent life on such a product at present is only between seven and ten years, and the manufacturer needs to sell his product with sufficient profit to recoup the costs of development, which may total around £10 million. Taking into account the factors which might lead to the studies being terminated, it seems that the company considers that there is sufficient chance of success to justify the investment of considerable resources.

Commentator B. This appears to be a 'me-too' insecticide, since it is similar to the market leader in its insecticidal effects and in its likely toxicity. If insects resistant to XYZ were not resistant to 123, this would be a major advantage, since the two insecticides could then be used strategically. This, however, seems unlikely to be the case and it looks as if the usefulness of 123 would be limited. 123 appears to be less environmentally acceptable than XYZ and may not be marketable after the animal testing has been completed. This further weakens the benefits case.

Whilst having successful companies, creating new jobs and reducing the balance of payments deficit, are undoubtedly important, this is only one part of the case. Difficult though it may be, we need our companies to develop insecticides which will be better than those already on the market; which are safer, and less damaging to the environment, and which, for these reasons, will capture more of the market share. The value of this new product, in terms of its originality (B1.6) and its value in safeguarding the environment (B1.1) is therefore in doubt.

Benefits and costs of testing the product

Commentator A. It is likely that XYZ and ABC will cause toxicity by the same mechanism. However, it is not possible to predict, without further investigations, that unexpected toxic effects will not be discovered. The product must be tested for toxicity, in order to safeguard the health of humans and animals who may come into contact with the insecticide, and to protect the environment from unforeseen damage (B1.1). The testing will be done to satisfy the company's own requirement for a safe product and in order to achieve registration by the regulatory authorities.

There is no doubt that some of the animals (probably less than 1 per cent) will die as a consequence of chemical toxicity and a larger number (perhaps about 5 per cent), will be killed for humane reasons (C5.1). From the company's point of view, the number of animals and the detailed design of the studies to be used are the subject of Government guidelines and may not be altered without prior discussion with the regulatory authorities. As this product is planned to be marketed in 60 countries it will be virtually impossible to negotiate variations in any of the studies. Thus it is necessary to perform tests according to the most recent OECD guidelines, including all of the studies listed in the outline (B2.1.1). Insecticides of this type require different formulations for application to different crops in different climatic areas. The company and the regulatory authorities require testing of the formulations in the acute topical and systemic toxicity studies. Thus, if the product formulations are to be registered, there is a need to carry out these tests (B2.2). The facilities in which the tests are conducted will be of very high quality, carefully inspected and serviced by a fully trained staff (B2.3).

It is likely that some of the countries in which it is planned to market the product will require acute toxicity testing of a type which is not permissible under the UK Animals (Scientific Procedures) Act 1986. It is likely, therefore, that these studies will be carried out in the country which requires them for registration of the product.

Commentator B. If new pesticides are to be produced, then, since they are highly biologically active, and designed to be toxic to *something*, it is clear that potential toxic hazards must be identified, so that risk assessment can be properly carried out. Thus it can be agreed, as commentator A has noted, that if the product is to be marketed, it must be tested, for the protection of workers, the population in general, and the environment (B1.1; B2.1.1).

In the project outline, however, the only justification given for the planned studies is that they are required by regulatory bodies and are to be carried out according to internationally accepted guidelines. This view is understandable from the company's point of view, but it does not necessarily imply that the tests are acceptable ethically. In examining the benefits of the proposed testing approach, it should be asked what tests and how many animals would be needed to test 123 according to the company's *own* toxicological and risk assessment experience, rather than to meet the demands of the regulatory authorities (B2.1 & 2.2). Furthermore, questions should be asked about the morality of having testing done abroad, on behalf of a UK company, which could cause more suffering than is acceptable under the UK Act.

From the project outline, there is no evidence that the company has any commitment to the rationalization of toxicity tests through active support for the Three Rs—Reduction, Refinement, and Replacement of animal use (B2.2.2–4), nor to the promotion of more-scientific approaches to hazard identification (B2.1). The proposed approach has not been shown to be necessary in terms of its scientific aim—which is to identify the potential hazard posed by the chemical.

Conclusions

This hypothetical case study has been carefully selected to illustrate the uncertainty involved in commercial invention and development, and also the ethical problems raised when such development involves the use of experimental animals. It illustrates how difficult it can be to reach consensus about ethical matters.

There are several reasons why the development of a new product—in this case an insecticide—might be terminated. As explained, studies using animals must run in parallel with other studies necessary for the successful development of the new product. Information will be generated on environmental and toxicological effects, the efficacy of the insecticide in the field, and the ability of the company to manufacture the chemical economically. In each of these areas, it may be found that the chemical has unacceptable properties for an insecticide. In that case, it could be said, in retrospect, that the animal studies had been carried out unnecessarily, although new scientific information will have been generated. Even though there are many points at which the development of the insecticide can fail (in common with the development of most new products), it is clear that the company considers that there is a sufficiently large chance of successful development of this insecticide to justify the investment of considerable resources in assessing its safety and usefulness. The opinion that the development of such a product is unjustified is partly based on the view that it is a 'me-too' insecticide similar to others in the field, and may not offer any advantage over these similar products. Proponents of this view would argue that the discomfort, distress and pain caused to some of the animals is therefore too high, and that development of 123 should not be permitted. It should be noted, however, that although 123 is similar in many respects to XYZ, it will not be known whether significant differences will emerge, conferring advantages to 123 over XYZ, until the field trials are completed. This will happen only when the majority of the toxicology is complete or well under way. In other situations 'me-too' products have been found to be sufficiently different when actually used in the field, and they have subsequently superseded the original product.

There are governmental organizations involved in most OECD countries which could prevent the development of such an insecticide even being started. The drive for improvement of most chemical products and the development of new products is to large extent in the hands of the industrial companies. In response to a request from an industrial company, the governmental organizations decide whether such development should take place and, in making their decision, are influenced by the current attitudes in society, not least about the use of animals in the development of pesticides. The overall aim is to maintain the drive to compete by the improvement or development of new products, while at the same time avoiding causing unnecessary suffering to animals. If development of the product is not refused then, from the company's point of view, the protocols for the existing battery of animal tests required by the national and international regulatory bodies must be followed. There are those who believe that the test systems approved for national and international bodies should be modified so as to reduce the number of animals and the suffering they may experience – that is, by the application of the Three Rs. Inevitably, there is a tension between these two views, and, in order to stimulate the acceptance and validation of new protocols, it is essential to have a continuous dialogue between those primarily concerned with animal welfare and those primarily concerned with the production of new chemicals to which humans will be exposed. The development of new scientific information to support the Three Rs and continuous dialogue with the regulatory bodies are the ways in which substantial reduction in animal use can be effected without impairing the validity of decisions about hazard and risk.

Finally, the case study illustrates an important facet of the consequences of stringent national control of animal experiments. While some countries may demand that certain toxicity tests should be carried out, others may prohibit the conduct of those studies. This puts the industrial companies in a difficult practical and ethical position. Again, this illustrates the importance of international dialogue, not only with regard to toxicity testing protocols, but also with regard to the controls which are appropriate to protect experimental animals.

REFERENCES

Amyx, H. L. (1987). Control of animal pain and distress in antibody production and infectious disease studies. *J. Amer. Vet. Med. Assoc.* **191,** 1287–9.

Aristotle (1955 translation). *Nichomachean ethics,* Book 1, Chapters 1–3. Penguin, Harmondsworth.

Bateson, P. (1986). When to experiment on animals. *New Scientist* **109** (1496), 30–2.

Home Office (1990). *Report of the Animal Procedures Committee for 1989* (pp. 25–9, on the production of monoclonal antibodies by *in vivo* and *in vitro* methods). HMSO, London.

Laverack, M. S. and Dando, J. (1979). *Lecture notes on invertebrate zoology* (2nd edn). Blackwell Scientific Publications, Oxford.

McGuill, M. W. and Rowan, A. N. (1989). Refinement of monoclonal antibody production and animal well-being. *ILAR News* **31,** 7–10.

Wang, T. L. (1986) (ed.). *Immunology in plant science.* Society for Experimental Biology Seminar Series, **29**. Cambridge University Press, Cambridge.

8

Ethical considerations in the use of animals in toxicology and toxicity testing

8.1 INTRODUCTORY REMARKS

Toxicology is the study of the toxic effects produced by chemicals. *Research in toxicology* serves the dual but complementary purposes of throwing light on the mechanisms of toxicity and furthering understanding of the biological systems affected. As Bernard wrote in 1875; '... the poison [a chemical] becomes an instrument which dissociates and analyses the most delicate phenomena of living structures and by attending carefully to their mechanisms we can learn much about the physiological processes of life'. The orderly function of complex organisms, such as vertebrates, requires continuous 'communication' between cells and tissues, in order to maintain the functions necessary for life. Much of this control is mediated by chemically-specific messengers (hormones, transmitters), interacting with specific targets on or within cells, or by chemical modification of enzymes (the biological catalysts) at specific places in their structure. From the existence of such specified control mechanisms it follows that there is the potential for chemicals, which interact with these systems, to disrupt normal function. When particular chemicals have been studied in detail, it has been shown that this is the case; toxicity caused by the chemical produces defined types of toxicity (signs and symptoms or histological changes) by interaction with specific cells and/or macromolecules, such as enzymes (targets). Of course this is a generalization, since the mechanisms of many forms of toxicity remain to be elucidated. However, except when substances are so chemically reactive and corrosive that they indiscriminately damage all cells they touch, general toxicity appears to be rare.

Toxicology research, in general, is similar in kind to the other types of biomedical research described in Chapter 2. *Toxicity testing*, on the other hand, raises different ethical questions and so will be the focus of this chapter. Toxicity testing is concerned with identifying the potential of chemicals to cause adverse effects (toxicity). As practised, much toxicity testing *requires* adverse effects to be produced in animals, often leading to considerable pain and distress. This aim leads to a special ethical problem which does not arise in many other kinds of research, in which suffering arises only indirectly, as an unwanted side effect of the experimental procedure used. Toxicity testing is also one of only a very few kinds of work in which there is a legal obligation to use animals. Other examples include the testing of vaccines for potency, the use of sentinel animals to detect disease in animals held in quarantine, and efficacy testing of new pharmaceuticals. The legal obligation to use animals reflects an ethical concern for ensuring human health, the health of other animals and the protection of the environment. A conflict arises because, whilst some laws and regulations demand the use of animals in testing (reflecting public concern about the safety of chemicals), others seek to reduce the use of animals in precisely this sort of activity (reflecting public concern about animal experiments).

8.2 WHAT IS TESTED AND WHY

8.2.1 Exposure to chemicals

We cannot avoid being exposed to chemicals. They are all around us. Everything we touch, eat or breathe is made of chemicals. Around seven million chemical substances have so far been described, representing only the top of a much larger iceberg of chemical possibilities. Most of the chemicals to which man and animals are likely to be exposed are natural chemicals, being found in, or extracted from, plants and animals and the general environment. Others are synthetic (man-made) chemicals. Both synthetic chemicals and naturally-occurring chemicals are used in the manufacture of myriad different products, including medical and veterinary drugs, clothing, buildings, furniture, other household and industrial products, pesticides, foods and cosmetics. Exposure to chemicals may occur during the manufacture, use and dissemination of such products; and a variety of different types of contact with chemicals can be identified (see Table 8.1).

8.2.2 Reasons for testing chemicals

Toxicity testing (by whatever method, whether employing animals or non-animal alternative systems) is aimed at providing information which can be used to attempt to protect society and the natural environment against the harmful effects of chemicals. Testing is

Table 8.1 Types of contact between humans (or animals) and synthetic chemicals

Type of contact	Examples of chemicals
Contact which is therapeutic	Human and veterinary medicines
Contact which is implicit when using a product	Food colours; clothes dyes; cosmetics and toiletries, e.g. toothpaste
Contact which should be avoided, where exposure is accidental or adventitious	Household chemicals, such as bleach, and industrial chemicals; toxins produced by microbial contamination of food
Contact which should be reduced, but some exposure cannot be avoided	Residues of pesticides in food chain; car exhaust fumes
Contact as a by-product of some other primary purpose	Dioxins in nappies and coffee filters; migration of chemicals into food from wrapping materials; chemicals produced by incineration of waste; insufficiently cooked beans
Contact results from abuse of the substance	Drugs; solvents

used to provide information about the potential toxic effects (that is, potential hazards) of particular chemicals. Such data are put together with information about the likely conditions of exposure to the chemical, so that an assessment of the risks posed by the chemical can be made. This risk assessment is used to judge the harmful effects which are likely to be incurred when the chemical is used in particular circumstances. Assuming that toxicity testing does what is demanded of it (that is, provides information which can be used to make reliable assessments of risk), then potentially harmful products can be identified and their use controlled or prevented altogether.

It is important to recognize that the effects that chemicals produce on biological systems are dependent both on the toxicity of the chemical and the amount present in the biological system. Some chemicals may be beneficial at one dose but toxic at another. For example, vitamin A is essential for life, but produces malformations if administered at high doses during early pregnancy. The importance of dose toxicity was recognized by Paracelsus, in 1538, who concluded: 'What is there that is not a poison? All things are poison and nothing is without poison. Solely the dose determines that a thing is not a poison'. There is no such thing as absolute safety of chemicals and the use of any chemical will carry some, albeit in some cases a vanishingly small, risk of harm.

Toxicity testing provides a basis for the quantification of risks from exposure to chemicals, allowing government regulatory authorities and the general public to judge whether these risks are acceptable in the light of the benefits likely to accrue from their use (see section 8.2.3). Testing (by whatever method) is carried out to protect both the workers manufacturing a chemical and also those humans, animals and, perhaps, parts of the general environment which may later be exposed to the product. Information derived from toxicity tests can be used to assess risk in each of the types of contact listed in Table 8.1.

Manufacturers are under a legal obligation to test new chemicals before putting them on the market (see section 8.3.1), and companies are also under increasing pressure to carry out more extensive testing of existing chemicals. People are becoming more aware of safety issues and have a greater expectation of safety in the chemical products they use. Legal requirements for safety are becoming more stringent, and this may force manufacturers to test chemicals and products that have been on the market for considerable periods of time, even though there is no evidence of human hazard.

Manufacturers and suppliers also have an ethical responsibility for assessing the safety of their products: it would be immoral to cause harm to people, animals or the environment by placing unsafe products on the market. This responsibility is supported by product liability legislation, which imposes a strict obligation on producers to ensure that products are safe when they are introduced into the market. In the past, a consumer who had been harmed by a product would have to show that a company had been negligent in not establishing or warning him of the possible harm. However, as a result of the 1985 EC Directive on product liability (85/374/EEC), the UK Consumer Protection Act 1987 makes producers responsible for all damage caused by defective products, subject to a special exemption if a producer can show that he has taken all reasonable care in the design of a new, 'state of the art', product. It is not known whether this change in product liability legislation has lead to increased testing of products before their entry into the market.

8.2.3 Benefits of synthetic chemicals

Chemicals produced by the chemical industry are used in a multitude of ways in modern society. Indeed, it is now almost impossible for most of us to imagine life without the use of at least some synthetic chemicals. Such chemicals have brought benefits to both humans and animals and, in some cases, to the general environment. Use of synthetic chemicals has contributed to improvements in health care, in quality and productivity in agriculture, and in food, housing, communications, household products, clothing and many other aspects of life. The manufacture of chemicals has also brought economic benefits, including employment for many people.

In spite of these overall benefits, however, it remains the case that each person's impression of the benefits of a particular product depends on that person's own circumstances. The benefits (and indeed the risks) of, say, a particular pesticide might be perceived differently by a UK farmer, a farmer in the Third World (where crop diseases and food shortages are a grave problem), a manufacturer of the pesticide, a member of an organic farming organization, and a UK or a Third World consumer of the food product on which the pesticide had been used. For this reason, it is very difficult to evaluate the costs to society of doing without particular kinds of products.

Here, however, we are concerned with the ethics of using animals in the testing of such products. It is therefore important that the benefits of synthetic chemical products, however they may be perceived, are considered in terms that will enable them to be judged against the costs imposed on the animals used in testing them.

8.3 THE USE OF ANIMALS IN TOXICITY TESTING

Widespread, systematic use of toxicity testing to collect information about the adverse effects of chemicals has developed only relatively recently, particularly over the past 30 years or so. This testing has relied almost exclusively on the use of whole animals, generally as models for humans and sometimes as models for other animals.

8.3.1 Regulations

There are a large number of national and international regulations which demand that certain animal toxicology studies are carried out before new chemicals are introduced on to the market. In the UK, for example, these controls include: Medicines Act (1968); Agriculture Act (1970); Poisons Act (1972); Health and Safety at Work Act (1974); Biological Standards Act (1975);

Food and Environment Protection Act (1985); Consumer Protection Act (1987); Environmental Protection Act (1990); Food Safety Act (1990). The information required from the tests and the ways in which the tests are to be carried out are laid down in a variety of guidelines and regulations, such as the *Guidance Notes on Applications for Product Licences* under the Medicines Act (1968), issued by the Department of Health. As a member of the European Community (EC), the UK regulations are also subject to the EC Directives governing the testing of chemical products such as pharmaceuticals, veterinary medicines, cosmetics, industrial chemicals and pesticides. Companies wishing to export their chemical products outside the EC must also meet the demands of the relevant national regulations and international transport regulations.

The pressure for further testing of existing chemical products is being felt in the EC and elsewhere, and regulations are being introduced which will require producers to carry out relevant animal studies (for example, EC Existing Chemicals Regulations and amendments to the EC Cosmetics Directive, which are under discussion; and US Pesticides Regulations, requiring re-registration of old products).

8.3.2 Toxicity test procedures

A wide variety of test procedures is used in toxicity testing. In the development of these procedures the aim has been to provide a comprehensive range of tests, capable of identifying as many toxic effects as possible. This aim has been fulfilled in part, since the present testing system (which has been developed gradually over many years) seems to work in practice. However, it should be noted that very few, if any, tests have received formal validation against relevant end points in man, although the performance of many tests has been evaluated retrospectively.

Table 8.2 lists the main types of toxicity test and indicates the circumstances for which each type of test would be considered appropriate. The following notes give brief details of the *in vivo* methods used in each type of test.

Acute systemic tests

These tests are used to examine the adverse effects caused by a single dose of a test substance, or in some cases by multiple doses given within a period of 24 hours. The substance under test is administered by an appropriate route (for example, orally, by in-

Table 8.2 Main types of toxicity test and their use

Type of test	When appropriate
Acute systemic tests	
Examine the adverse effects occurring within a short time after the administration of a single dose of a substance or multiple doses given within 24 hours	For chemicals which might be ingested or absorbed into the body either directly, or indirectly via the general environment
Eye and skin irritancy tests	
Test for irritant (reversible tissue damage) and/or corrosive (irreversible tissue damage) effects of substances on skin and in eyes	For substances and products directly applied to skin or eyes or to which people may be exposed at the workplace
Sub-acute and chronic tests	
Examine the effects of repeated exposure to small doses of a substance	For substances to which people or animals may be directly or indirectly exposed over a long period (e.g. during long-term medical treatment or industrial exposure)
Carcinogenicity and mutagenicity tests	
Examine whether substances cause cancer (carcinogens) or cause permanent changes in genes which are passed on to descendant cells (mutagens)	As above and in other cases where the chemical structure of a substance suggests that it may be a carcinogen and/or a mutagen
Reproductive toxicity tests	
Examine the effects of chemicals on reproduction at any stage from the production of sperm and eggs (fertility studies) to development of the fetus (teratogenicity studies) and the fully-formed fetus and newborn (pre- and post-natal studies)	For substances to which people or animals may be directly or indirectly exposed over a short or long period
Sensitization tests	
Examine hypersensitivity/allergic reactions to substances, in single or multiple exposures	Any substance, any exposure—an expanding area of testing, as understanding of sensitivity/allergy is growing

Based on information in Commission of the European Communities (1988).

halation, or by injection) to several groups of experimental animals (often of two different species, usually both rodents), a different dose being used for each group. The animals are then observed and the effects of the test substance are recorded.

LD50 tests are used to determine the dose or concentration of test substance (the 'lethal dose') which will kill half of the animals tested.

Classical LD50 tests. The classical LD50 test was introduced by Trevan in 1927, as a means of establishing the potency of medicines, such as insulin and digitalis extracts, which have a very narrow difference between the therapeutic and toxic doses. The test later gained acceptance as a method of assessing the acute toxicity of a wide variety of chemical products, and it came to have special uses in the detection of undue toxicity (due to impurities) in product batches and in the classification and labelling of hazardous substances. Large numbers of animals were used in classical LD50 tests, in an attempt to gain a high degree of precision in the LD50 values obtained. There was no upper limit to the amount of test substance which could be administered to the animals (Table 8.3).

Table 8.3 Comparison of acute systemic toxicity test procedures

Test	Reference	Number of animals per species	Dose limit (mg/kg)	End point
Classical LD50 test	based on Trevan (1927)	60–80	none	death
Modified LD50 tests				
Approximate LD50	OECD (1981)	30	5000	death
Approximate LD50	OECD (1987)	20	2000	death
Step-wise	BGA (1985)	12–18	2000	death
Up-and-down	Bruce (1985)	6–8	2000	death
Non-lethal test				
Fixed dose procedure	BTS (1984, 1987)	10–20	2000	clear toxicity

In the past two decades, such classical tests have been strongly criticized. It has been argued that the tests cause a great deal of animal suffering for dubious toxicological ends, since a high degree of precision in the LD50 value is not necessary in risk assessment and, even if it were, it is not attainable technically (see for example, Home Office 1979; Zbinden and Flurey-Roversi 1981; Zbinden 1986). A variety of modified LD50 tests have now been proposed.

Modified LD50 tests. In such modified tests, information about the mortality caused by a test substance is obtained using far fewer animals and is used to calculate an *approximate* LD50 value. In addition, a limit is set for the maximum dose which can be administered to the animals under test. Such tests can be used to provide information, not only about lethality, but also about the nature and time of onset of any non-lethal toxic effects.

OECD test procedures. Internationally agreed test guidelines are published by the Organization for Economic Co-operation and Development (OECD). In 1981, the OECD adopted a 'modified' LD50 test procedure, which was revised in 1987 in order to reduce further the suffering caused to animals (see Table 8.3).

If it is expected that the highest possible (limit) dose will cause no mortality, then this one dose (2000 mg/kg in 1987 guidelines) can be administered to a single group of animals and the toxic effects observed. This is called a limit test. If the test chemical is expected to produce mortality at doses below the limit dose, or if mortality occurs in the limit test, a full test is carried out. In a full test, each of three groups of animals is given a different dose of the test substance, the three doses being selected to span the range in which the LD50 dose is expected to fall. The top dose is therefore expected to produce a high level of mortality. The animals are observed daily for at least 14 days, and toxic effects, including any deaths, are recorded. In addition to mortality, observations may include changes in the condition of the animal's skin, fur, eyes or mucous membranes, changes in weight, and changes in breathing rate, circulation, co-ordination or behaviour pattern. Particular attention is paid to observation of tremors, convulsions, salivation, diarrhoea, lethargy, unusual sleep patterns and coma.

Other modified tests. A variety of other modified LD50 test procedures have been developed, including the 'step-wise' procedure put forward by the German Ministry of Health (BGA) in 1985 and the 'up-and-down' test described by Bruce (1985). As shown in Table 8.3, these methods require fewer animals than the current OECD guideline, but, like the OECD guideline, unless the substance under test is not toxic at the limit dose, they both require that a proportion of the animals die during the study.

Non-lethal tests. A test scheme put forward by a Working Party from the British Toxicology Society (BTS) in 1984 offers a radically different approach to the assessment of acute toxicity. The test approach was validated in an inter-laboratory trial within the UK (Van den Heuvel, Dayan and Shillaker 1987) and is now known as the BTS Fixed Dose Procedure (FDP).

This test procedure does not set the death of the animals as its objective. Rather than establishing a lethal dose, as would a classical or modified LD50 test, the BTS FDP aims to establish the dose at which clear signs of toxicity are detectable. An initial dose is selected on the basis of the predicted toxicity of the test substance, the aim being to cause toxic signs, but no mortality, in the animals used in the test. This dose is administered to a group of ten animals (five male, five female) and the toxic signs observed for up to 14 days. If the initial dose has been selected correctly and toxic signs but no deaths are observed, no further testing is required. However, if no toxic signs are observed, the test may have to be repeated at a higher dose level using a second group of animals (the limit dose is set at 2000 mg/kg). If the initial dose is found to cause mortality, testing with a lower dose may be required.

An extensive validation study (funded principally by the EC) has judged the BTS FDP acceptable as an alternative to lethal dose procedures. The EC has stated its intention to accept the FDP as a standard procedure for chemicals classification purposes and will call on all OECD nations to accept data generated by the test (*FRAME News* **24**, 7 1989).

Skin and eye irritation tests (Draize tests)

Albino rabbits are usually used in skin and eye irritation tests and, in general, at least three adult animals are used to test each sample. Most test guidelines recommend that an assessment of skin irritancy should precede assessment of eye irritancy, since substances that are corrosive or severe skin irritants can generally be assumed to be severe eye irritants.

In skin testing, the usual method is to apply a small quantity of the test substance to a small area (approximately 6 cm^2) of shaved skin (which sometimes has to be abraided or scarified), holding it in place with a gauze patch. The patch is normally left in place for four hours. Longer exposures are used when human contact with the chemical is likely to be prolonged. The animals are examined for signs of skin irritation, and the response is classified according to the degree of redness and swelling observed. This formal examination takes place 30–60 minutes after the patch is applied, then daily for three days after the patch has been removed. The observation period may be extended for up to 14 days, in order to determine whether or not the effects are reversible and hence to decide whether the substance has ‘irritant’ (reversible) or ‘corrosive’ (irreversible) effects.

In eye testing, the normal procedure is to instil 0.1 ml of the substance into the conjunctival sac of the eye of a rabbit. Observation of the other eye of the animal serves as a control. The reactions of the eye to the test substance are observed at one, 24, 48 and 72 hours. The effects on the cornea, iris and conjunctivae are noted and graded. If there is no evidence of irritation after three days, the study may be ended. If irritation is found, the study may be extended for up to 21 days, in order to determine whether or not the changes are reversible.

Sub-acute and chronic toxicity tests

These tests are used to examine the effects of long term/repeated exposure to chemicals. A test substance is normally administered to the animals daily, seven days a week, in order to expose them to the chemical and its metabolites (substances produced as a result of the body's normal modification of the test chemical). Several different dose levels are required, with the highest dose being sufficient to produce some toxic effects in the animals. The duration of the tests is normally correlated with the likely duration of exposure to the chemical in man. Animals may be exposed to the test chemical for, say, seven, 28 or 90 day periods or, where rats and mice are used, for a 'life time' of up to two years. The effects of the substance on the animals' behaviour and blood and tissue biochemistry are investigated during the study, and at the end of the study (or at suitable intervals during the study) animals are killed and the organ systems (for example, the nervous system, cardiovascular system, digestive system, etc.). are examined for gross and microscopic effects.

Carcinogenicity and mutagenicity tests

Many carcinogens cause mutations (alterations in the chemical structure of genetic material). Such chemicals may be detected using *in vitro* tests (such as the Ames test), involving bacteria or cells grown in culture. Other chemicals, however, may cause cancer by different mechanisms, which are not shown up in such *in vitro* tests. Life time studies in whole animals are used to detect such carcinogens. Large numbers of animals are used in these studies, in order to show up the small incidence of carcinogenic effects of the chemical against the natural background of cancer in the species (see Commission of the European Communities 1988).

Reproductive toxicity tests

Examination of the effects of chemicals on reproductive processes involves dosing animals with test substances before mating, and during and after pregnancy. The tests are prolonged, continuing over several (usually two or three) generations, and generally taking around a year to complete. A large number of animals is used. In general, three different dose levels are used, and are selected so that the top dose causes some maternal toxicity, such as a 10 per cent reduction in body weight gain. Three kinds of study may be carried out:

(a) fertility studies, that is, studies of the effects of chemicals on the production and function of eggs and sperm;
(b) teratogenicity studies, that is, studies of the effects on the early development of embryos and fetuses;
(c) pre- and post-natal studies, that is, studies of the effects on later development of fetuses and on the new-born, from birth and then through the period of lactation to weaning.

Sensitization tests

These tests are used to examine the sensitizing (that is, allergenic) potential of substances. The major difficulty with such testing arises because, in humans and animals alike, sensitization is likely to occur in only a small proportion of subjects. In testing, therefore, various experimental conditions are used to try to improve the chances of identifying potential sensitizing substances. The guinea-pig is usually the chosen species since these animals are particularly susceptible to allergens.

The tests are conducted in two parts.

(a) Induction phase. This first part involves attempting to induce allergy in the animals, by repeatedly applying the test substance either onto the surface of the skin (epidermally) or by injection into the skin (intradermally). The substance is applied at the maximum concentration which does not cause any skin reaction. In some tests, the animals are also given one or more injections of an adjuvant (a substance designed to increase the immune response), so as to 'maximize' any immunological reaction occurring in the second part of the test.

(b) Challenge phase. After an interval of 10 to 14 days, the animals are challenged with a single re-exposure to the test substance. The animals are then monitored for signs of any sensitizing (allergic) skin reaction to the substance. A numerical scale is used to score the intensity of any observed reaction.

8.3.3 Harm caused to animals in toxicity testing

As mentioned at the beginning of the chapter, toxicity testing poses a special ethical problem, since the aim is to establish the presence (or absence), nature and extent of any toxic effects of the chemicals under test. Techniques such as feeding, injecting, oral dosing or inhalation are used to administer chemical substances to the animals. Provided that they are properly carried out, these techniques usually cause little distress (although repeated injections may cause considerable local pain). It is, however, the *effects* of the chemicals administered which are the major determinants of the costs imposed on animals used in testing.

The adverse effects of chemicals on animals may range in severity from a small weight loss or a small change in a physiological variable (such as the amount of a particular hormone circulating in the blood), through to loss of function in major organs of the body and, in some cases, the death of the animal concerned. The most minor effects are unlikely to cause pain or distress to the animals, and so, in these terms, will impose relatively low costs. Other effects, particularly those that result in lethality or severe tissue damage, may impose high costs in terms of animal pain and distress. As Zbinden (1988) has written,

> considerable suffering must be assumed in animals bearing large tumors or afflicted with organ damage, for example, perforated gastrointestinal ulcers, myocardial infarctions, liver necrosis and muscle wasting. Functional disturbances, such as paralysis, excessive central nervous stimulation, diarrhoea, polyuria, hypotension, and sensory organ dysfunction, cause stress and anxiety.

Although the potential costs of all toxicity tests are high, the actual costs imposed on the animals will depend on the purpose of a study and on the particular type of chemical or formulation tested. For example, a test which is designed to identify the cause of a particular type of toxicity, requiring the toxic effect to be reproduced, may cause more animal suffering than a test which is designed to confirm the absence of toxic effects.

The overall cost in terms of animal suffering imposed by toxicity testing will also depend on the number of animals used in the tests and on the proportion of these animals which suffer toxic effects. Table 8.4 shows the approximate numbers of animals required in each of the basic tests in an application for a licence to market a typical pharmaceutical product. These figures can be taken to represent the minimum requirements for such a product licence application.

Information on the scale of animal use in toxicity testing in the UK is recorded in the Home Office *Statistics of scientific procedures on living animals*. The *Statistics* record that 591 483 scientific procedures were carried out on animals for the purposes of toxicity testing (including screening for carcinogenicity) during 1989. This represents 18 per cent of the total number of scientific procedures

Table 8.4 Animals used in tests required in a product-licence application for a pharmaceutical product (1982 figures)

Toxicity test	Number of animals	Species
Acute toxicity	40	all rodents
Chronic toxicity	290	240 rodents; 50 non-rodents
Carcinogenicity	500	all rodents
Teratology	188	48 rabbits; 140 rats
Fertility	180–400	usually all rodents = initial number + offspring for 2 or 3 generations

From a report by the Economic Development Committee (1987).

carried out on animals in the UK during 1989 (Home Office 1990). Table 8.5 illustrates the scale of animal use and lists the main species of animal used in each of the different kinds of toxicity test.

The *Statistics* also give information on the reasons for carrying out the toxicity tests (although carcinogenicity screening tests are not included in this analysis). Figures 8.1 and 8.2 summarize this information. Figure 8.1 shows that over three-quarters (76 per cent) of toxicity test procedures were carried out for legislative reasons: to fulfil the regulatory requirements for product licence applications. The remaining procedures were performed for non-

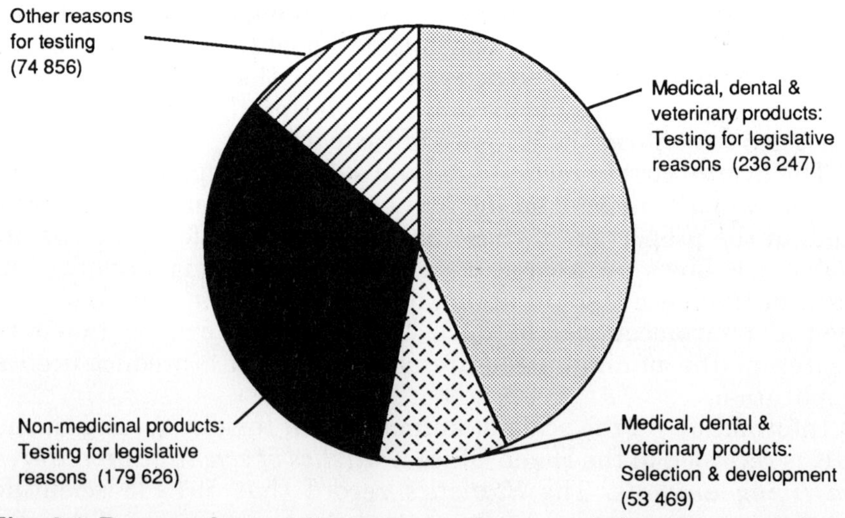

Fig. 8.1 Reasons for toxicity test procedures carried out on animals in the UK during 1989 (number of procedures in brackets; data from Home Office 1990).

Table 8.5 Animal use in toxicity tests in the UK in 1989 (data from Home Office 1990)

Type of test	Total scientific procedures	Scientific procedures by species of animal: Mice and rats	Other rodents[c]	Rabbits	Carnivores[b]	Ungulates	Primates	Birds	Reptiles/ Amphibians	Fish
ACUTE AND SUBACUTE TESTS[a]										
1 **Systemic tests**										
Formal LD50 tests	73 130	41 017	953	90	—	—	—	2706	—	28 364
Ranging or limit setting lethal or clinical sign tests	102 879	84 788	2844	570	1420	52	79	5056	—	8070
Non-lethal clinical sign tests	143 428	74 031	7087	55 770	2001	874	776	855	—	2034
2 **Skin irritancy and sensitization tests**	41 599	4743	29 888	6879	13	76	—	—	—	—
3 **Eye irritancy tests**	4921	431	107	4349	1	33	—	—	—	—
CHRONIC TESTS[a]										
1 **Systemic tests** (other than reproductive toxicity, mutagenicity or carcinogenicity tests)	132 765	124 766	1061	164	2514	—	1320	1172	—	1768
2 **Teratogen & mutagen tests**	45 476	38 507	92	6737	1	121	—	—	18	—
3 **Carcinogenicity screening**	47 285	*No species data recorded*								

The Statistics record the number of scientific procedures carried out, not the number of animals used (see Chapter 2).

[a] The Home Office Statistics divide tests into two main groups: 'acute & subacute' and 'chronic' tests. In the first category, 'acute' refers to single-dose tests and 'sub-acute' to multiple-dose tests of less than one month's duration. 'Chronic' tests are multiple-dose tests of more than one month's duration.

[b] Mainly dogs.

[c] Mainly guinea-pigs (which are used especially in sensitivity testing).

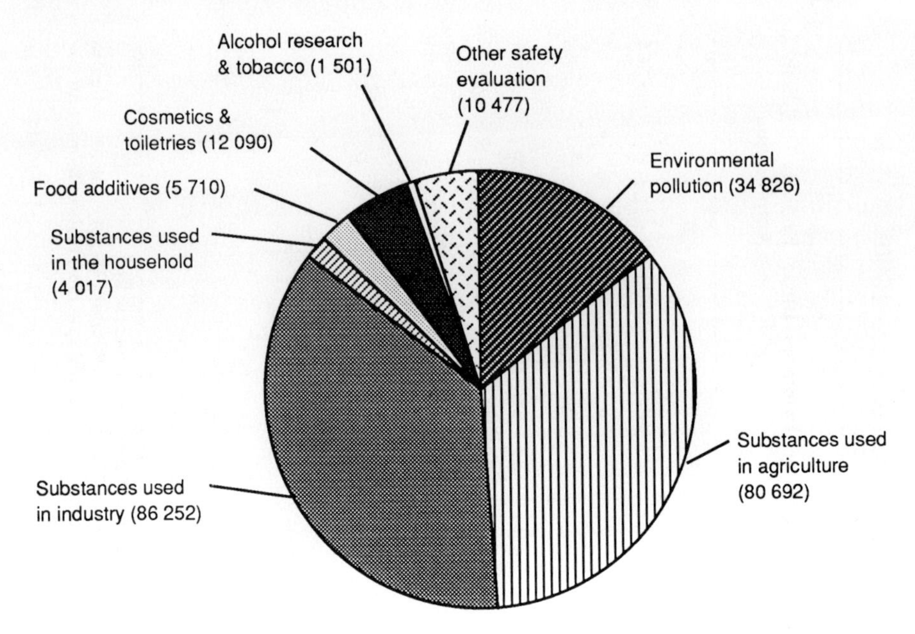

Fig. 8.2 Non-medicinal products: Safety evaluation procedures* involving animals, performed in the UK during 1989.

legislative reasons, including toxicological research and the selection and development of products. The figure also shows that over one half (53 per cent) of all toxicity test procedures were for medical, dental or veterinary products. The majority of the remaining toxicological procedures involved non-medicinal products, and Fig. 8.2 shows the variety of non-medicinal products for which safety evaluation procedures were recorded.

It is impossible to establish from any of the published statistics the proportion of animals which suffered pain or distress in the tests. It should be noted, however, that although large numbers of animals are used in a given test, or in testing a given type of product, it is unlikely that all of these animals will have suffered from toxic effects of the test substance. In most tests a range of dose levels is used, so that significant toxic effects may often only be produced at the higher dose levels.

8.4 ETHICAL ANALYSIS

8.4.1 Introduction

We have seen that in toxicity testing there is a conflict between the

*Procedures for the 'protection of man, animal or the environment by toxicology or other safety evaluation' (Home Office 1990).

moral claims of, on the one hand, the interests of human and animal health and environmental safety, which the testing is intended to protect, and, on the other hand, the well-being of the laboratory animals used, which the testing harms, the ethical justification for toxicity testing practices depends on some form of moral weighing of these two claims, one against the other. As with other uses of laboratory animals, such weighing examines whether the beneficial outcome of the tests, in terms of predicting the risks involved in using chemical products, is sufficient to justify the harm caused to the animals used in the tests. It is generally accepted that some form of safety evaluation is necessary if such products are to be introduced into, or kept on, the market. The reasons were outlined in section 8.2.2. The question, therefore, relates to whether *present* testing practices, which may harm animals, are justified.

If, at the outset, we were to accept a popular criticism of present practice, which says that many products themselves are unnecessary, so that by implication the use of animals in testing them is unnecessary and therefore unjustified, we might be able to avoid the moral conflict in at least some instances. This possibility is examined in section 8.4.2. If, however, we accept that at least some chemical products *are* necessary and must be tested, we should consider whether, and if so how, conditions for the ethical acceptability of testing practices can be improved. In this context, the resolution or diminution of the moral conflict outlined above will depend on:

(a) reducing the harm caused to animals in testing; and
(b) improving the validity, relevance and necessity of the tests in the assessment of risk.

The following account examines these strategies in more detail, focusing on criticisms of present testing practices. Section 8.5 then examines the possibility of eliminating animal suffering altogether, by replacing the animals used in the tests with non-animal alternatives.

8.4.2 Necessity of chemical products

As noted above, it is often argued that some animal toxicity tests are unnecessary, and hence unethical, because the products being tested are themselves unnecessary. The problem with this line of reasoning, however, lies in deciding what is meant by 'unnecessary'. There is a tendency for such arguments to be based on the tacit assumption that 'What I want is necessary, but what other people want, and I don't want, is unnecessary'. As discussed in section 8.2.3, an individual's perception of the benefits of a particular chemical will depend on that person's own particular circumstances. There is therefore a great deal of uncertainty in attempting to judge which products are to be considered necessary or essential and which are to be considered unnecessary or trivial.

An example frequently debated in the media, and elsewhere, is that of beautifying cosmetics, which are often described as 'trivial' products.* It is argued by some that it is unethical to use animals to test new ingredients or formulations of beautifying cosmetics, 'Because the use of these products is not essential', (that is, is trivial) and there are already enough such products on the market. On the basis of public demand for the products and size of the market, however, cosmetics would appear to be no more trivial than many other chemicals. Furthermore, in terms of the use to which the chemicals are put, many other products, such as clothing dyes or food colourings, would seem to be no less trivial than beautifying cosmetic products.

If there is legitimate demand for a particular product, as a result of which it is manufactured and used, adequate safety assessments, which may involve the use of animals, must be carried out in order to protect both the manufacturing workers and the consumers of such products. 'Triviality' issues are too uncertain and too complex to permit a generalized judgement concerning the acceptability of animal use in testing products to be made: this judgement has to be made case by case. In practice, the judgement is best made by the consumers of such products, who, given *accurate* information about the use of animals in testing, can exert their considerable buying power to influence the size of markets.

* This example may frequently be misrepresented. The following points should be borne in mind when considering the ethics of using animals in testing beautifying cosmetics. **1.** The term cosmetic may be applied to a wide variety of products: EC Directive 76/768 defines *cosmetic* as 'any substance or preparation intended to be applied to any part of the external surface of the human body (that is to say, the epidermis, hair system, nails, lips and external genital organs) or to the teeth, or buccal mucosa [mucous membranes of the mouth] with a view to cleaning, perfuming or protecting them or keeping them in good condition or changing their appearance or combating body odour or perspiration except when such cleaning, perfuming, protecting, keeping, changing or combating is wholly for the purpose of treating or preventing disease'. **2.** Only a small proportion of the animals used in testing cosmetics and toiletries are used to test *beautifying* cosmetics (that is, cosmetics designed to 'enhance' appearance, rather than to clean or protect the body). **3.** Most new ingredients for cosmetic products have multiple uses, so that they would be tested anyway, even if they were not used for cosmetics; furthermore, testing of such ingredients is carried out to protect the workers manufacturing the chemical as well as the consumer of the final cosmetic product. **4.** The adverse effects caused by such testing are usually very mild: by far the majority of tests are for skin sensitization and are not Draize eye or skin irritancy tests.

8.4.3 Reduction and refinement of practice

As was outlined above, two considerations are particularly important in the design of toxicity tests involving animals. First, the end points of the tests may involve unavoidable animal suffering: hence tests should be designed so as to minimize the numbers of animals used and to minimize the suffering caused to those animals. Second, the information from the tests in practice is usually applied to humans: this imposes demands over and above those of good scientific design, since it must be possible to extrapolate from the animal to the human with as much confidence as possible. Balancing these design criteria is often difficult. Nevertheless, there is inter-laboratory variation in how toxicity tests using animals are carried out, both within the UK and across the world. There would seem to be room for refinement of much of this testing, so as to cause less suffering to the animals used, whilst at the same time maintaining or improving the scientific validity of the tests. Possible reduction and refinement steps include:

Use of the least number of animals consistent with the scientific objectives

Good experimental design of toxicity tests, as of other biomedical experiments using animals, is essential (see also Chapter 5). Marsh (1985), for example, has described a statistical test (the randomization test) which may be used to decide on the smallest group size which will give statistically significant results in a given experiment. Healey (1983) has described a variety of design techniques which can help to avoid the need to repeat tests and can ensure that the maximum amount of useful information (consistent with welfare) is obtained from the least number of animals. A case in point is the reduction of the number of animals used in acute toxicity testing. It is now recognized that it is neither possible nor necessary to find 'precise' LD50 values. Internationally accepted testing procedures have been refined so that small numbers of animals are used in tests which generate data which can be analysed statistically, in order to provide approximate LD50 values (OECD 1981, 1987). However, such tests still require that at least some of the animals die, and here there is an immediate opportunity for further refinement.

Refinement of end points used in testing

It should be possible to refine the end points of toxicity tests without compromising the validity and usefulness of the information obtained from the tests. This is particularly so in acute toxicity testing, where, as has been described, lethality is often used as an end point. Humanely killing an animal when evident signs of toxicity are shown, rather than continuing the test until the animal dies, will reduce animal suffering and perhaps also improve the quality of the information obtained *post mortem*. There will, for example, be no autolysis of tissues. With the validation in the EC of the BTS Fixed Dose Procedure, which does not require the death of animals as its end point, there should now be little, or no, need for acute systemic toxicity tests to be designed so as to produce lethality (but see also section 8.4.5.).

Refinement of measurements made in testing

There are often opportunities for designing tests which measure graded physiological variables (such as levels of tissue-specific serum enzymes, blood glucose or composition of urine), rather than all-or-none end points such as lethality. Not only will such measurements be more humane (as described above), but they may also be quantitatively more useful, since they can indicate a graded response to the test chemical.

Refinement of the techniques used in testing

Although, as was noted in section 8.3.3, most of the effects of the techniques used in toxicity testing are relatively mild, there are ways in which techniques can be refined so as to reduce the adverse effects caused to the animals: for example, the use of indwelling cannulae for body fluid collection, and the avoidance, where possible, of repeated injections or gavaging (oral dosing by passing a tube down the oesophagus), which may cause pain and distress.

Limiting the maximum dose administered in testing

A limit should be set for the maximum amount of a substance (per kilogram body weight) which can be administered to the animals under test. In the past, the amount of test substance administered to animals was sometimes taken to extreme levels: limiting the dose can both reduce animal suffering and improve the validity of the test as a model for human exposure to the test substance. For both of these reasons, the recent lowering of the recommended maximum oral dose for acute toxicity testing in rodents from 5000 mg/kg body weight to 2000 mg/kg body weight, represents an important refinement (OECD 1987, *cf.* OECD 1981).

Use of the least sentient species possible

Where several species might equally well be used to achieve the toxicological aims of a particular test, the least sentient species should be used (as in biomedical research generally – see Chapter 5).

Reduction of the duration of tests where scientifically practical

There is variation in the duration of chronic toxicity tests required by the different regulatory authorities, and there has been much debate concerning the necessity of carrying out toxicity tests of longer than six months duration. If the maximum duration of chronic toxicity tests could be reduced to six months without compromising their scientific validity, this would help to improve both the humaneness and cost-effectiveness of the tests. The general consensus now is that tests longer than six months are not necessary (see, for example Bass, Grosdanoff and Lehnert 1985; and Lumley and Walker 1986). This view is supported by a study involving retrospective analysis of the results of 88 long-term studies for which data were available both before and after six months. The results of the study showed that no significant new effects were observed after six months (Lumley and Walker 1986).

Use of a hierarchical approach to testing

This approach involves the use of a series of non-animal screening methods, considered sequentially, to identify toxic chemicals at an early stage in the testing process. If such screens suggest that the chemical is likely to be very toxic, whole animal tests should not be carried out at all, or should be carried out using only a very few animals. Whole animal tests should therefore be used only to confirm the absence of significant toxicity, or to grade low or moderate levels of toxicity.

Hierarchical approaches to testing are especially useful in the assessment of skin and eye irritancy, and Fig. 8.3 shows such an approach to the assessment of skin irritancy (Fielder *et al.* 1987). Evaluation of what is already known about the irritant potential of a chemical (by literature review), examination of the physical and chemical properties of the substance and *in vitro* studies on isolated skin, can all be used to identify severe irritant or corrosive chemicals without the need to use whole animals. If these screening tests suggest that the substance is unlikely to be a severe irritant, a carefully controlled test may be carried out on a single animal. In this case, the substance should be tested first as a dilute solution, the concentration being gradually increased until irritancy is observed, and the skin patches should not be left on for a full four hours, but inspected more frequently, from as early as three minutes after application of the substance. Once irritancy is observed, the experiment can be terminated. Only if this single

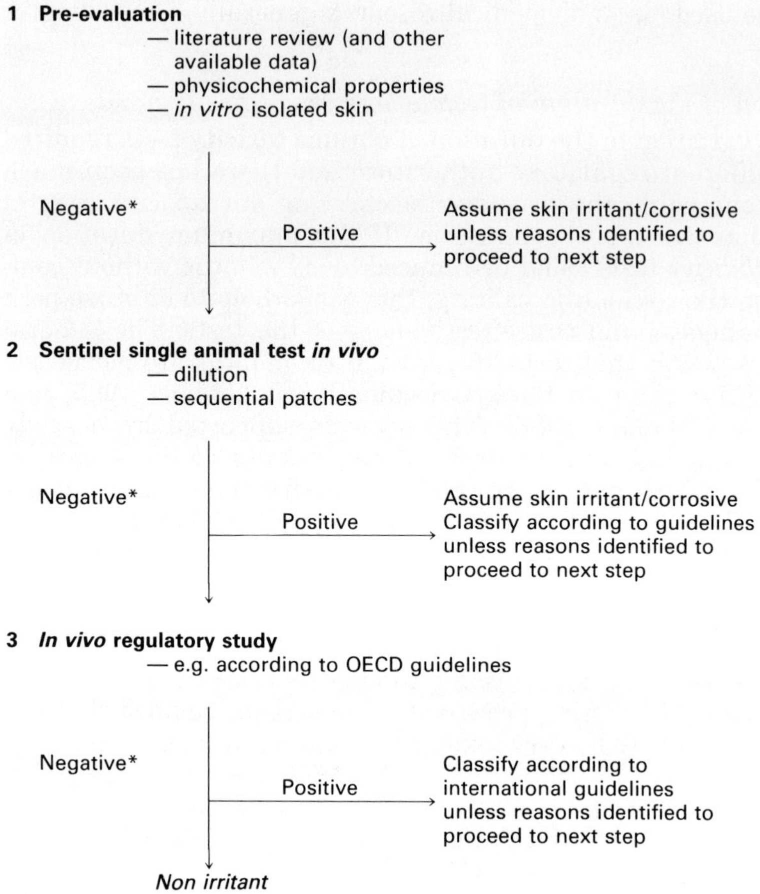

1 Pre-evaluation
— literature review (and other available data)
— physicochemical properties
— *in vitro* isolated skin

Negative* | Positive → Assume skin irritant/corrosive unless reasons identified to proceed to next step

2 Sentinel single animal test *in vivo*
— dilution
— sequential patches

Negative* | Positive → Assume skin irritant/corrosive Classify according to guidelines unless reasons identified to proceed to next step

3 *In vivo* regulatory study
— e.g. according to OECD guidelines

Negative* | Positive → Classify according to international guidelines unless reasons identified to proceed to next step

Non irritant

* Except for those established by international guidelines, the negative and positive criteria are those established by the individual laboratory.

Fig. 8.3 A hierarchical approach to the assessment of skin irritancy (taken from Fielder *et al.* 1987).

animal test indicates that the substance is unlikely to be a severe irritant should a full, regulatory, study be carried out.

A similar approach may be used in the evaluation of eye irritancy. In this case, the 'pre-evaluation stage' (as shown in Fig. 8.3) includes assessment of the results of skin irritancy studies. Unless there is a clear reason to suspect otherwise, substances which are skin irritants can also be assumed to be eye irritants, and there should be no need to perform eye tests. UK Home Office Guidelines, issued in 1987, aim to ensure that such a hierarchical approach is adopted in the assessment of eye irritancy.

8.4.4 Use of the information generated by the tests

As we have seen, toxicity tests generate information which is used to assess the risk posed by the use of chemical products. The data provided by the tests therefore have a direct practical application. Several aspects of the practical use of such data may affect both the validity and necessity of testing practices.

Extrapolation to human hazard assessment

When animals are used to test substances in order to provide a basis for the assessment of risk to humans, the animals are used as 'models'. The rationale is that it is probable that many toxic effects of substances under test will be similar in the chosen laboratory mammal and in man, because much basic physiology is similar among all mammalian species. It should be remembered, however, that biological models (whether they be, say, rainbow trout used as models for all the different kinds of fish in environmental toxicity testing, or rats used as models for man in safety testing of drugs) are, by their very nature, imperfect. It is therefore essential for toxicologists to have a clear understanding of the physiology of the species of animal used in testing, paying special attention to any physiological differences between the test species and the species which the tests are intended to protect. Research into the physiology and mechanisms of toxicity in the various species must take place, since it plays a vital role in improving the validity of extrapolation from the model to the real subject.

Designing tests to suit their purpose

Whilst toxicity tests must aim to achieve scientific validity, the design of the tests should also take into account the purpose for which the results of the tests will be used. Sometimes, testing procedures seem to be over-concerned with numerical precision, so that more animals are used, and more suffering and distress is caused, than is necessary for the purposes of the test. A particular criticism is that animal toxicity tests may use sophisticated procedures in order to obtain detailed quantitative information which is then used only in low-grade classification systems. For example, detailed results from LD50 tests are used to classify substances into only four broad categories: 'no effect', 'harmful', 'toxic', or 'very toxic'.

Avoiding unnecessary duplication of tests

If information about the toxicity of chemicals already tested is not made available to other workers, this can lead to the unnecessary duplication of animal tests. This may be a problem in some areas of testing, and in these areas efforts should be made to avoid duplication. Computerized databanks are now being used to increase the accessibility of information about the toxic effects of chemicals, and about 20 such databanks are already in existence. A general problem, however, is that companies usually wish to keep their toxicity reports, on which product safety assessments are based, confidential. Such testing information might take a manufacturer many years to obtain, at considerable cost, so it is understandable that the manufacturer might wish to protect this information, rather than give it away to a third party without financial recompense. If possible, a balance should be struck between protecting the commercial interests of the companies involved, by keeping the testing information secret, and avoiding duplication of animal tests, by making data on toxic effects more widely available. Such a balance might be achieved, in the case of generic (or me-too) drugs for example, by:

(a) setting a limit for the duration of confidentiality of reports of toxicity tests on the original product, after which the information would be made available in support of generic product licence applications; and,
(b) in the meantime, encouraging originating companies to give other companies access to their toxicity information, in return for appropriate financial recompense.

A further strategy would be to encourage companies with a common interest in a particular chemical to collaborate in testing the chemical (so-called consortium testing).

A precedent has been set in respect of (a). A legal case, involving the company SmithKline and French, ruled that the UK product licensing authority could use safety data for SmithKline and French's drug Tagamet (cimetidine), which was out of patent, when processing generic cimetidine applications (Reg. v. Licensing Authority ex parte SmithKline and French Laboratories Ltd. (1990) 1 A.C. 64). The judges considered that if the licensing authority could not have recourse to originator's data in assessing second applications this would entail extra and unnecessary tests on both animal and human subjects (*Scrip*, July 6, 1988, No. 1323, p. 1).

8.4.5 Effects of regulations on testing practices

Test guidelines

A great deal of effort may be put into developing and validating tests which cause less animal suffering, but if the improved tests are not incorporated into the relevant regulatory authority guidelines they cannot be used routinely to fulfil the regulatory and legal requirements for product licence applications.

It is, of course, important that new or changed tests, aimed at improving animal welfare, are fully validated in terms of human welfare. Many putative alternative schemes have proved to be poorly validated and stand no chance of incorporation into the regulatory guidelines. Nevertheless, over the years there has been inertia in the regulatory system, leading to long delays between acceptance of an improved test by toxicologists and acceptance of the test by the regulatory authorities. The Ames test, for instance, was developed in 1973, but did not find its way into the regulatory guidelines until 1981–2.

One possible cause of delay in acceptance of modified procedures arises from a lack of standards against which to judge the results of the new tests. If testing has been done in a particular way for a long time and a lot of data generated, a new or modified test will generate a new kind of data which may be difficult to compare with the old data. An example occurs with the labelling of hazardous substances: it is argued that changes in the test protocols will necessitate changes in the interpretation of the labelling. Another example concerns the use of a reduced volume of test substance in eye irritation testing. Although studies have shown that reducing the volume from 0.1 to 0.01 ml increases the predictive value of the tests for man, this change has not been incorporated into the regulatory guidelines. It is argued that new data, generated by the changed tests, would not be comparable with data on existing chemicals.

However, even when improved tests are accepted into regulatory guidelines, they may not be accepted by the regulatory authorities of all countries. Thus, there may be considerable variation in the regulations themselves or in the interpretation of regulations by regulatory authorities. This can lead to companies having to repeat tests for different authorities, or having to carry out extra tests, which are not a part of their home regulations, in order to be able to export their products.

Once an improved test procedure has been shown to be valid scientifically, it would seem unethical to continue to carry out testing which involves more suffering than can be justified scientifically. In fact, it might not only be unethical, but also unlawful. As Zbinden (1988) has pointed out, the animal protection laws of many countries now require researchers to demonstrate the necessity of the procedures in which they propose to use animals. The 1986 EC Directive, for example, requires that 'no pain, suffering, distress or lasting harm is inflicted *unnecessarily*' (our italics). Zbinden has expressed the opinion that such animal protection laws may require 'more justification for an animal experiment than the simple statement that a proposed test is necessary because it is required by a regulatory authority guideline'. It will be particularly difficult to fulfil this requirement 'if the country in which the study will be conducted does not require the proposed test, or is satisfied with an experiment involving fewer animals or a shorter duration of treatment'.

Difficult moral and legal dilemmas are posed by the conflicting demands of the animal protection legislation and some regulatory authority guidelines. A particular example occurs in acute toxicity testing. In the 1980s, such testing was largely carried out according to the 1981 OECD Guideline (see section 3.4.1). However, a new OECD Guideline, involving the use of fewer animals and reducing the limit dose, was agreed in 1986 and published in 1987. Nevertheless, the regulators of some OECD members (notably Japan) still insist on the submission of data according to the 1981 Guideline, so companies selling internationally are forced to test to the most demanding guideline. Now, as Zbinden has put it, the requirement to demonstrate the necessity of particular animal tests may mean that 'the easy way out . . . i.e. to conduct toxicity studies always according to the most demanding national guideline, will, in the future, often not be possible'.

It is often suggested that, in order to avoid such problems, regulations should be harmonized to a common standard across the world. A particular kind of harmonization, which has been successful, is in the generalized conduct of toxicity tests. In 1975, OECD published guidelines for the conduct of tests to conform to Good Laboratory Practice (GLP): that is, in such a way that retrospective auditing allows confirmation that test reports reflect test conduct accurately. With this regulation in place, and with mutually acceptable national GLP inspection procedures agreed, reports are acceptable to all OECD registration authorities and cannot be rejected on the basis of quality of conduct.

Harmonization of test guidelines, however, could have both advantages and disadvantages. On the one hand, harmonization could result in less repetition of tests, so that fewer animals would be used. This would both reduce animal suffering and also reduce the costs incurred in breeding or buying in and maintaining the animals. Further economic benefits would also result, since companies would be able to perform a single battery of tests for the world wide market. On the other hand, once harmonized, the tests would be fixed for a particular class of products: hence, even if toxicologists had special concerns they would not be able to bypass the harmonized test. It is also possible that such harmonized regulations could be based on the wrong protocol, so that there could be scientific advantages in variation between regulations.

Whilst harmonization, by and large, should reduce the number of animals used in testing, it will also reduce flexibility in the interpretation of test guidelines. On this basis, we believe that *rationalization* of guidelines is better than harmonization. Rationalization would involve flexibility in interpretation of tests and mutual acceptance, by the various regulatory authorities, of data from approved test procedures. Such a strategy would help to reduce repetitive testing and would also ensure that toxicologists would not always have to employ the most demanding tests, in terms of animal use. It would mean that toxicologists would not be bound to use a particular set of tests, but would be able to use the procedures which they consider will best fulfil the objectives of the regulatory requirements whilst causing the least possible animal suffering. A continuing dialogue between regulators and toxicologists, including preregistration discussion (at the test design stage), would play an important part in this process.

It is likely that such rationalization, involving flexibility in interpretation of guidelines and in the precise test protocol used, will benefit not only the animals, but also the science of toxicology. Since the currently prescribed battery of toxicity tests has not been validated scientifically against relevant end points in man (see section 8.3.1) it does not make sense to regard it as being cast in tablets of stone. Instead of being forced into adopting a checklist approach in determining the toxicological characteristics of a test compound, rationalization would give toxicologists more freedom to treat chemicals case by case, using the test procedures which they consider best suited to the characteristics and likely conditions of use of the particular substances under consideration.

This kind of approach has been suggested by the European Commission in the adoption of the now validated BTS Fixed Dose Procedure. At the time of writing, the Commission has proposed that the FDP be accepted as a means of classification for labelling and risk assessment, and is amending the EC test guidelines accordingly. It is also proposing that all members of the OECD who are not EC members be urged to accept FDP data according to the agreed principle of mutual acceptance of data, while, on the same principle, the EC will continue to accept data from non-EC countries submitted according to the 1987 OECD Guideline. Furthermore, it is to be recommended that all LD50 procedures which make greater demands than the 1987 Guideline should no longer be considered acceptable (*FRAME News* **24**, 7 1989).

Testing existing chemicals

At present, there is much international discussion concerning the potential hazard posed by existing chemical substances, in particular those which were in use before regulations controlling the introduction of new chemicals came into force. In the EC, a preliminary draft proposal for an amendment to Directive 76/768 on cosmetic products is under discussion. The Commission proposes to draw up an inventory of substances used in cosmetic products. Member States would be required to send existing toxicological information on cosmetic ingredients to the Commission, and the Commission would draw up a list of substances for which further toxicity studies would have to be provided. A draft Directive on the testing of existing chemical products generally is also being considered in the EC.

It is reasonable to assume that a considerable amount of the required data would not be available for substances currently in use, and so both of these Directives, if eventually implemented, would require the collection by Member States of toxicity data on existing chemicals. The collection of such data would be likely to involve the use of large numbers of animals, at considerable economic cost. It has been estimated that to provide data on only 3000 of the 6–8000 existing cosmetic ingredients would require the use of around two million animals, at a cost of at least £1.2 billion; and to complete a comprehensive toxicity database on the 12 860 chemicals in commerce listed by the US National Research Council as being produced in quantities greater than 450 tonnes per annum, would require the use of nearly 11 million animals, at a cost of around £5.8 billion (*FRAME News* **25**, 203 1990).

It is not unreasonable for the EC to ask for such inventories: the fact that substances have been in use for a long time does not necessarily mean that they are 'safe'. However, apart from the problems of economic cost and the increased use of animals, it is likely that the majority of the data obtained would be negative. Since most of the chemicals would be expected to be non-toxic, testing all of them in a full battery of standardized tests would provide only reassurance, and would not greatly improve human safety in the use of chemicals (*FRAME News* **25**, 203 1990). If any such Directives were to be adopted there would be a need for prioritization of chemicals according to their scale of production and the likely hazards posed by their use. A common sense approach would be required in testing, *not* a standardized, checklist testing of all chemicals (see *FRAME News* **25**, 203 (1990) for further discussion of this issue).

8.5 ALTERNATIVE STRATEGIES: REPLACEMENT OF ANIMALS IN TESTS

Chapter 6 has reviewed the range of replacement alternatives available and the ethical questions associated with the development and use of such alternatives to animals. Here, we consider the possibilities for replacing animals in toxicity testing.

Section 8.4 has shown that there are immediate opportunities for the reduction and refinement of animal use in toxicity testing, so as to reduce the harm caused to the animals used and to improve the toxicological validity of the tests. Some companies have been using such reduction and refinement methods routinely, as part of their overall testing programme, and some of these alternative strategies have been, or are now being, introduced into the regulatory authority guidelines. (There is, however, a need for a system which promotes the rapid incorporation of improved, validated tests into the guidelines—see section 8.4.5.)

The third of the 'Three Rs' objectives, the complete *replacement* of animal use in testing (while at the same time maintaining or improving the quality of data for risk assessment), although unlikely to be attained in the foreseeable future, should be regarded as the ultimate goal. This was recently affirmed at the EC Seminar on the LD50 test, see *FRAME News* **24**, 7 (1989). Indeed, regulatory authorities, and others, have both legal and moral obligations to accept scientifically-validated replacement alternative procedures (Balls *et al.* 1990; see also Chapter 6).

At present, non-animal methods (*in vitro* techniques and computer modelling) are used mainly as screens or adjuncts in predominantly animal-based testing programmes. The methods can be used in the initial stages of product development, to identify potentially efficacious new chemicals and so to establish priorities for further efficacy and toxicological evaluation. They can also be used as part of a hierarchical approach in toxicity testing, helping to improve the design of subsequent animal studies, should they be required (Fig. 8.3). *In vitro* methods are also used to establish specific mechanisms of toxicity or toxic effects on particular target organs. The use of non-animal screening methods can lead to a reduction in the number of animals used and a refinement of the tests, so that animal suffering is minimized. However, such methods are not truly replacements since the use of animals, although reduced and refined, is not avoided altogether.

There are a few areas in which complete replacement of particular whole animal tests with *in vitro* alternatives is likely to occur in the foreseeable future. Examples include:

Use of the Limulus *amoebocyte-lysate assay as a replacement for rabbits in pyrogenicity testing*

Some pharmaceuticals have to be tested for the presence of bacterial endotoxins (substances found in the cell walls of bacteria) which, amongst other effects, can cause fever (endotoxins are said to be 'pyrogens'). Traditionally, rabbits are injected with the test substance and then observed for signs of fever. However, the *Limulus* amoebocyte-lysate (LAL) assay now provides a more sensitive, more reliable and less expensive *in vitro* alternative test.

LAL is an extract of the blood cells (amoebocytes) of the horseshoe crab, *Limulus*. In the assay, small amounts of the test substance are incubated with equal volumes of LAL in glass tubes. When bacterial endotoxins are present solid clots form in the tubes, but when there is no toxin present there is no clot formation. The test does not harm the crabs, since up to 30 per cent of the blood of a horseshoe crab can be collected without adverse effect. After blood has been collected, the crabs are returned alive to the sea (McCartney 1986).

The LAL test was accepted in 1977 by the US Food and Drugs Administration (FDA) for testing biological products and medical devices, and rabbit tests are no longer required. However, the test has yet to be accepted by the European Pharmacopoeia. Once this happens, it will automatically be acceptable under the relevant European Community pharmaceuticals legislation.

Use of in vitro *tests as replacements for animals in certain specific neurotoxicity studies*

Certain organophosphorus compounds cause delayed neurotoxic effects, usually manifested as a dying back of the long nerve axons in the arms and legs and in the spinal cord. If the damage is in the spinal cord, the condition is permanent. Humans are known to be a sensitive species, since there have been several large-scale poisoning episodes due to criminal adulteration of food and drink. It is therefore essential to have a routine test which can be used to screen potential organophosphorus pesticides for this undesirable property.

Such testing, until recently, has had to be carried out on living hens. A few hens are dosed repeatedly and the effects noted. However, basic research has now established the biochemical mechanism for the initiation of the delayed neurotoxic effects, making possible the development of *in vitro* alternative tests. If the structure of the active agent is known, it is now possible to predict its likely neurotoxic effects from *in vitro* measurements made using nervous tissue obtained from hens or, if available, human *post mortem* tissue.

Use of in vitro *skin and eye irritation tests*

As was discussed in section 8.4.3, *in vitro* tests are commonly used as screens in hierarchical approaches to the assessment of skin and eye irritancy. Now, especially in eye irritancy testing, *in vitro* and isolated eye tests are 'being developed towards regulatory acceptance' as replacement alternatives (Commission of the European Communities 1988).

Potential alternatives which are currently undergoing development and validation in Europe and North America fall into four categories:

(a) *in vitro* methods which involve killing animals, including the use of isolated, enucleated rabbit eyes, isolated bovine corneas and isolated sections of rabbit intestine;
(b) the use of the extra-embryonic membranes of chick embryos (classed as animal procedures in the UK if the embryos are more than halfway through their incubation period);
(c) physico-chemical tests, for example the EYTEX™ method which is based on the breakdown of proteins in solution; and
(d) cytotoxicity tests, including the assessment of the effects of chemicals on cell viability, cell morphology, cell adherence and/or detachment, cell membrane integrity or cell proliferation (*FRAME News* **24**, 5 1989).

However, in spite of these advances, the question whether alternative methods can ever replace the full battery of animal tests remains controversial. There are some who doubt that such complete replacement will ever be possible, arguing that, whilst non-animal tests are very useful as screens or adjuncts, they can never be used to predict all possible kinds of toxic effect. Such an argument is usually based on the fact that some toxicologically significant effects (such as behavioural or hormonal changes) depend on the integrated response of a multicellular organism and thus can never be shown except by using a whole animal. Others believe that complete replacement will one day be possible, arguing that, with more understanding of mechanisms of toxicity, it should be possible to design alternative testing strategies capable of predicting the full range of toxicologically significant effects. Such alternative strategies, it is argued, would not only employ different kinds of tests, but would also generate different kinds of evidence on which to base assessments of risk.

In spite of these differences in long-term views, all would agree that if we are to work towards the goal of replacing animals in some or all of the current battery of tests, continued and increased funding is needed for research aimed at improving understanding of the molecular and cellular mechanisms of toxicity and at developing, validating and evaluating alternative testing procedures.

8.6 CONCLUSIONS

The use of animals in toxicity studies poses difficult moral questions. In the light of these questions, we have reviewed the purposes, methods and costs to animals of toxicity testing. We have attempted to evaluate the ethical acceptablity of current practice and, on this basis, to identify strategies for improving such practice.

We have been particularly concerned to present accurate information about the ways in which animals are used in testing, so as to provide a common factual foundation, which we hope will help readers to enter into the debate. The conclusions from our deliberations are presented here as answers to several questions.

What is the need fulfilled by toxicity testing and toxicology?

If we are to have the benefits of using new chemical products, some form of safety assessment of these products is required. Toxicity testing provides a basis for such safety assessment, being used to fulfil practical, ethical and legal duties to try to ensure product safety and to improve knowledge of the risks posed by exposure to chemicals. Toxicology research (the study of the mechanisms of action of toxic chemicals) may be linked with the practical aims of safety assessment and the treatment of accidental poisoning. Studies of mechanisms of toxicity can help in the design of more definitive *in vivo* toxicity tests and can lead to the development of *in vitro* methods for the prediction of toxicity. In a more general context, the use of chemicals or toxins for the controlled derangement of biological systems has revealed much of significance in the understanding of the normal physiology of mammals and other species. In this respect, toxicology research has similar aims to biological research in general.

Both toxicology research and toxicity testing involve the use of animals. The latter activity, in particular, relies on the use of animals. We are therefore faced with a conflict between duties: on the one hand, to try to ensure the safety of chemicals for humans, animals and the general environment (for which reason testing is carried out) and, on the other hand, to try to safeguard the welfare of laboratory animals (which the testing, by its nature, harms).

Are some chemical products unnecessary?

In at least some instances, we could avoid this moral conflict if we agreed with the argument which says that particular chemical products are themselves unnecessary and hence that their development, and with it the use of animals in their testing, should cease. However, whilst it is legitimate for individuals in society to decide that they do not wish certain products to be developed and produced if this requires animal testing, we have had great difficulty in developing a logical argument which permits the rational identification of necessity of products. Our conclusion is that, if society wishes to use certain products, there is a moral and legal responsibility on the provider of those products to do all possible to ensure their safety. Whilst safety assessment (in terms of toxicity) relies on animal toxicity tests, it is inevitable that those products will require to be tested in animals.

How good are the animal tests for their designated purpose?

Standard tests used in compliance with international guidelines for the development of new chemical products (and the testing of existing chemicals) have been developed by an iterative process, based on the experience of testing and evaluation internationally. This process aims to achieve international consensus on testing protocols, taking into account scientific objectives and economic, legal and, sometimes, humane constraints. For the majority of tests, it is not practical or ethical to validate them using human data, because of either the paucity of such data or the nature of the toxicity being studied. The success of testing by existing methods relies on many basic similarities in physiology and response amongst mammalian species and the conservative methods used in interpreting the results and assessing human hazard.

Anomalies remain, particularly where there are interspecies differences in toxic response, which pose problems for the prediction of effects in humans. Such anomalies may require further non-standard mechanistic studies for their resolution. Similarly, mechanistic studies, using both *in vivo* and *in vitro* methods, have been shown to be an effective means of developing new knowledge of mammalian physiology, although there are, again, inter-species differences in physiological processes which require careful interpretation of experimental results.

What are the possibilities for reducing and refining the use of animals in practice?

We suggest that, in the immediate future, the most moral way forward is to minimize to an essential core of tests, consistent with the objectives of the testing process. Strategies for achieving this include *refinement* of the testing procedures, *reduction* in the numbers of the animals used in the tests, *rationalization* of the testing guidelines, and *investment* in the development and validation of potential *replacement alternatives*.

Both the scientific validity and humaneness of the tests can be improved by allowing for flexibility in interpretation of the regulations which demand testing. Tests can then be designed individually, tailored to the particular chemical under test, rather than being carried out by rote, simply following the guidelines set out in the regulations. The intelligent design of tests can improve their predictive value for man (and, perhaps, other animals): more appropriate species, routes of administration of the substance, and measurements of the effects of the substance can be used. Use of hierarchical (step-wise) approaches, limit doses and humane end points can all help to reduce the suffering caused to the animals used in toxicity testing, without compromising the utility of the information gained from the tests.

Rationalization of regulatory guidelines will help to ensure that toxicologists need not always apply the most demanding guidelines (in terms of animal use) when testing products intended for the world market, and it will eliminate the need to repeat tests for different regulatory authorities, so reducing the number of animals used in testing. Making information on the toxicology of chemicals readily available can also reduce the need to carry out repeat studies, so saving animal lives. However, this has to be balanced against the need to protect the interests of the companies which carried out the original work.

Increasing attention has been paid to the issue of animal use in toxicity studies, and some changes have already occurred. We have noted, for example, that concern about the conduct of acute toxicity studies has resulted in the acceptance by several regulatory agencies (including the EEC) of revised study protocols which aim to reduce and refine the use of animals in such testing (OECD 1981, *cf.* OECD 1987). Refinement of acute toxicity testing practice now looks set to proceed further, with the acceptance in the EC of the

British Toxicology Society's Fixed Dose Procedure, which does not have lethality as an end point.

All such changes require painstaking work. Careful discussion must take place among interested parties, to ensure commitment and that the changes do not compromise the utility of the studies in human hazard assessment. There is much more to be achieved by, for example, further rationalization of regulatory guidelines and wider acceptance of the improvements in reduction and refinement already available. The moral issues, nevertheless, remain acute, since, set against these concerns for animals, there are increasing concerns for the protection of human health, which have resulted in additional studies (for example, in immunotoxicity and neurotoxicity) on more chemicals (especially existing chemicals), leading to a trend towards increasing animal usage.

What are the possibilities for replacing animals with non-animal alternatives?

Mammalian physiology, particularly in respect of the delicately balanced and complex co-ordination of cellular and organ function, poses formidable problems in the development of 'non-animal' toxicological techniques. The result is that, so far, non-animal methods (such as computer simulation and *in vitro* studies) have found their greatest utility in the study of specific mechanisms or defined toxicological end points. In the latter context, such techniques have been developed for screening purposes, permitting the selection and prioritization of chemicals for further testing or providing information leading to improved design of subsequent *in vivo* tests.

There is no doubt that non-animal techniques will increasingly become part of the armamentarium of the toxicologist. We can look forward to development of an increasng range of non-animal screening methods. However, such techniques still have as their limitation the difficulty of modelling the complex homeostasis of an animal. There is particular difficulty in the use of non-animal techniques in risk assessment, in establishing the target organ, and in the study of some obvious end points such as the behavioural component of neurotoxicity. Furthermore, as has been pointed out, most *in vivo* studies have not been validated against human toxicological experience of equivalent severity, so that the development of non-animal studies may still require the use of animal data for validation.

While some believe that developments in science will allow non-animal techniques to replace whole animal toxicology in the long run, others believe that the complexity of mammalian physiology and the increasingly broad demands of regulatory toxicology will make it impossible to replace animals for all toxicity testing. This difference in belief, however, should not prevent collaboration and co-operation between those who hold these disparate views, or alter the shared commitment to try to implement change. Progress will be made only if both time and money are invested in the search for alternatives. Changes will occur step by step, with each step being carefully evaluated before it is accepted and implemented. Unfortunately, there are no instantaneous solutions.

REFERENCES

Balls, M., Botham, P., Cordier, A., Fumero, S., Kayser, D., Koeter, H. *et al.* (1990*b*). Report and recommendations of an international workshop on promotion of the regulatory acceptance of validated non-animal toxicity test procedures. *Alternatives to Laboratory Animals* **18,** 339–44.

Bass, R., Grosdanoff, P. and Lehnert, T. (1985). Alternatives to chronic toxicity studies. In *Drugs between research and regulations* (ed. S. Steichele, U. Abshagen, and J. Kock-Weser), pp. 21–6. Steinkopff Verlag, Dormstadt.

Bernard, C. (1875). *La science experimentale.* Bailliere, Paris.

Bruce, R. D. (1985). An up-and-down procedure for acute toxicity testing. *Fund. Appl. Toxicol.* **5,** 151–7.

British Toxicology Society Working Party (1984). A new approach to the classification of substances and preparations on the basis of their acute toxicity. *Human Toxicol.* **3,** 85–92.

Commission of the European Communities (1988). *Report on the possibility of modifying tests and guidelines laid down in existing Community legislation in compliance with Article 23 of Council Directive 86/609/EEC on the approximation of laws, regulations and administrative provisions of the Member States regarding the protection of animals used for experimental and other scientific purposes,* COM (88)243 final. Brussels, 29 April 1988.

Economic Development Committee (1987). *Pharmaceuticals: focus on R&D.* National Economic Development Office, Millbank, London.

Fielder, R. J., Gaunt, I. F., Rhodes, C., Sullivan, F. M. and Swanston, D. W. (1987). A hierarchical approach to the assessment of dermal and ocular irritancy: A report by the British Toxicology Society Working Party on irritancy. *Human Toxicol.* **6,** 269–78.

FRAME News, published by Fund for the Replacement of Animals in Medical Experiments, Eastgate House, 34, Stoney Street, Nottingham NG1 1NB.

Healey, G. F. (1983). Statistical contributions to experimental design. In *Animals and alternatives in toxicity testing* (eds. M. Balls, R. J. Riddell and A. N. Worden) pp. 167–78. Academic Press, London.

Home Office (1979). *The LD50 test.* Home Office, London.
Home Office (1987). *Guidelines on eye irritation/corrosion tests ('Draize' eye test).* Home Office, London.
Home Office (1990). *Statistics of scientific procedures on living animals, Great Britain 1989,* Command 1152. HMSO, London.
Lumley, C. E. and Walker, S. R. (1986). The questionable value of long-term animal toxicology studies: a regulatory dilemma. *Arch. Toxicol. Suppl.* **9,** 237–9.
Marsh, N. W. A. (1985). How few animals may we use in an experiment? Logical constraints upon lower limits for group sizes. In *Animal experimentation: improvements and alternatives* (ed. N. Marsh and S. Haywood), pp. 79–86. ATLA Supplement, FRAME, Nottingham.
McCartney, A. C. (1986). The *Limulus* amoebocyte lysate assay for bacterial endotoxins. *Alternatives to Laboratory Animals* **13,** 180–92.
OECD (1981). *Guidelines for the testing of chemicals–guideline 401,* Decision of the Council, C(81)30 Final. OECD, Paris.
OECD (1987). *Guidelines for the testing of chemicals–third addendum, revised guideline 401.* OECD, Paris.
Paracelsus (Theophrastus Bombastus von Hohenheim) gen. Epistola dedicatora St Veit/Karnten, Sieben Defensionen oder Sieben Schutz-, Shirm- und Trutzreden. Dritte Defension. August 24, 1538.
Trevan, J. W. (1927). The error of determination of toxicity. *Proc. Roy. Soc.* **B101,** 483–514.
Van den Heuvel, M. J., Dayan, A. D., and Shillaker, R. O. (1987). Evaluation of the BTS approach to the testing of substances and preparations for their acute toxicity. *Human Toxicol.* **6,** 279–91.
Zbinden, G. (1986). Invited contribution: acute toxicity testing, public responsibility and scientific challenges. *Cell Biol. Toxicol.* **2,** 325–35.
Zbinden, G. (1988). A look behind drug regulatory guidelines. In *National and international drug safety guidelines* (ed. S. Alder and G. Zbinden), pp. 7–19. MTC Verlag Zollikon, Switzerland.
Zbinden, G. and Flurey-Roversi, M. (1981). Significance of the LD50 test for the toxicological evaluation of chemical substances. *Arch. Toxicol.* **47,** 79–99.

9

Ethical considerations in the use of animals in education and training

A separate chapter has been devoted to discussion of the issues raised by the use of animals in education and training, since this use may be considered to be different in kind from the use of animals in biomedical research *per se*. In education and training, animals are used not, as in research, to try to discover, prove or develop *novel* facts, ideas or techniques, but to teach or demonstrate *known* facts, ideas or techniques. The difference means that this use of animals poses a distinct set of morally-relevant questions which are not encountered when animals are used in research or in testing.

In the most general terms, three such morally relevant questions may be identified:

(a) is it acceptable to use animals to teach facts and ideas which are already known?
(b) is it acceptable to use animals in order to gain manual skills which may be used to benefit humans or other animals?; and
(c) how can educational experiences with animals be used to encourage the development of balanced, humane attitudes towards animals? How best can we educate the consciences of present and future biomedical researchers, and of animal keepers and users in general?

We have considered these questions in relation to the use of animals in primary and secondary schools, in tertiary and postgraduate education and in the training of biomedical researchers and technicians.

9.1 USE OF ANIMALS IN PRIMARY AND SECONDARY SCHOOLS

9.1.1 Possible effects of using animals in schools

Table 9.1 sets out in detail some of the perceived educational aims and benefits of keeping animals in schools. In general, it is believed that school experiences with animals can help to develop

Table 9.1 Educational benefits provided by animals in schools

- Opportunities for detailed observation and investigation of animal behaviour, structure and function, growth and life cycles.
- Contributions to the social education of students through observation and discussion of reproduction, social interactions, and death, leading to an appreciation of the material and social needs of animals—including human beings.
- Increased motivation to study living animals and to use and develop skills of literacy and numeracy in describing and recording patterns of animal behaviour or productivity, and planning management programmes.
- Stimulation to do creative work and encouragement of the aesthetic appreciation of animals.
- Identification and investigation of the normal range of environmental factors influencing living animals leading to an appreciation of the importance of protecting the environment.
- Contributions to the personal development of students by shared responsibility for animal welfare, establishment of human–animal bonds and caring attitudes, and introduction to potential out-of-school interests involving animals, such as bird-watching.

From a joint statement by the Association for Science Education, Institute of Biology and the Universities Federation for Animal Welfare (1986).

pupils' understanding of the physiology and behaviour of living organisms and may also play a role in shaping their 'attitudes' towards animals. On this second point, such experiences may determine how children and, later, adults treat animals, as well as how they perceive the issues surrounding man's use of animals.

The manner in which animals are used in schools might therefore have important knock-on effects for biomedical research in general. It might, for instance, influence whether or not pupils choose to pursue degrees and/or careers in the biological sciences; it might help to shape public opinion concerning the use of animals in research; and, since experiences with animals in school form part of the early training of researchers and animal technicians, it might help to determine the attitudes of those engaged in such research towards the animals in their charge.

9.1.2 Permitted uses of animals in schools

In the UK, procedures likely to cause pain, suffering, distress or lasting harm may not be performed on living vertebrate animals in schools. Such procedures are regulated under the Animals (Scientific Procedures) Act of 1986, which states that licences may be issued only for 'education and training *otherwise* than in primary and secondary schools'. This prohibition applies not only to the use of adult vertebrates, but also to the use of vertebrate embryos, including hen's eggs, from halfway through their incubation or gestation period, and to larval forms, such as tadpoles, from the time they become capable of feeding independently.

Vertebrate animals may, however, be used by schools in a number of ways which are *not* likely to cause them any pain, suffering, distress or lasting harm. Such uses may include simply 'keeping' animals, perhaps as pets, in the classroom; carrying out observational, behavioural or breeding experiments (involving procedures which do not cause any adverse effects) and examining and dissecting dead animals or parts of dead animals. Invertebrates, which fall outside the scope of the UK law, may also be used in experiments in schools; although the fact that this use is legal does not necessarily mean that it is also morally acceptable. The Protection of Animals Act, 1911 (1912 in Scotland) remains the most important legislation pertaining to vertebrate animals used in UK schools (O'Donoghue 1988). This Act makes it an offence to cause unnecessary suffering to a domesticated or captive animal, applying to suffering caused both by deliberate cruelty and by neglect (see Cooper 1987, for further details).

Many European countries have laws which limit the use of animals in schools in a similar manner, though such restrictions may not apply throughout Europe. The 1986 EC Directive on the use of animals for scientific purposes (Directive 86/609/EC) does not cover use in education and training. Article 25 of the European Convention is intended to restrict such use of animals to higher education and continuing professional training, although the wording of the Article rather leaves it open to interpretation:

1. Procedures carried out for the purpose of education, training or further training for professions or other occupations, including the care of animals being used or intended for use in such procedures, must be notified to the responsible authority and shall be carried out by or under the supervision of a competent person, who will be responsible for ensuring that procedures concerned comply with national legislation under the terms of this Convention.
2. Procedures within the scope of education, training or further training for purposes other than those referred to in paragraph 1 above, shall not be permitted.
3. Procedures referred to in paragraph 1 of this Article shall be restricted to the minimum measures absolutely necessary for the purpose of the education or training concerned and be permitted only if their objective cannot be achieved by comparably effective audio-visual or any other suitable methods.

In contrast, the situation in the USA is quite different. Although several national organizations and local school systems have issued policy statements which state that experiments which subject vertebrates to pain, distress or discomfort should not be performed in US schools none of these policies can be enforced in law: they are merely guidelines and not rules (US Congress Office of Technology Assessment 1986). Morton (1987) has expressed concern about the type of work that is carried out using animals in schools in the US, particularly in view of the work displayed at Science Fairs (see section 9.1.4).

9.1.3 Animals in primary schools

It seems that in most countries, animals in primary (elementary/junior) schools are usually kept purely as pets, giving pupils the opportunity to learn about the animals' husbandry and care and to make observations of their behaviour. This exercise can result in the pupils developing a sensitive, caring and responsible attitude towards, and relationship with, the animals. Such an attitude will in no small part be determined by the teacher's own attitude towards the animals concerned. The psychologists Zahn-Waxler, Hollenbeck and Radke-Yarrow (1985) have suggested that basic humane attitudes and behaviours are laid down in the first years of life and that the teaching of a sense of responsibility for animals, as well as for other people, should begin as early as possible, preferably from a child's first birthday onwards. It is therefore important that any use of animals in primary schools is carefully thought out, so as to be sure that the experience really will enhance the childrens' knowledge and respect for the animals in their care.

The main cause for concern over such use of animals lies in the adequacy of the husbandry and general care provided. Particular problems are posed by the need to care for the animals during holiday periods and in cases where there is insufficient knowledge (amongst teachers, children and/or parents) about how to handle the animals, what to give in their diets and so on, so that inadvertent mistakes may be made in their husbandry. In the UK, the RSPCA (1986*a*) has produced a comprehensive set of guidelines on the keeping of small mammals in schools, aimed at helping schools to avoid this kind of problem.

9.1.4 Animals in secondary schools

In addition to keeping animals as described for primary schools, UK secondary (high/senior) schools may also use dead animals, or parts of animals, in order to teach biological principles in a more exam-orientated fashion. As discussed in section 9.1.2, it is illegal for 'invasive' procedures to be carried out on living vertebrates in UK schools. Living vertebrates may, however, be used in recognized husbandry procedures so that, for example, breeding experiments may be carried out. Experiments may also be carried out on invertebrates, although as a Nuffield 'A' Level Guide to Projects in Biological Science (Tricker and Dowdeswell 1970) suggests, these should be afforded the same consideration as that provided by legislation for vertebrates.

Four main areas of concern over such use of animals in secondary schools can be identified.

Standards of animal handling, husbandry, breeding and care

The breeding of animals in schools is not subject to legal control and inspection in the same way as the breeding of animals for scientific purposes is controlled, and so standards of husbandry and care in some cases may not be adequate. Again, the RSPCA's Guidelines on the keeping of animals in schools are helpful. When keeping animals for any purpose, schools should be required to register with a local veterinary surgeon, and should be encouraged to seek his or her advice on any matter relating to the welfare of the animals.

Methods of euthanasia of animals used in schools

In secondary schools, teachers or laboratory technicians may kill animals in order to provide material for dissection and/or experimental work, in order to dispose of surplus animals from breeding programmes, and to 'put down' pet animals when these are sick or injured and unlikely to recover. Any method may be used provided there is no cruelty involved (that is, provided the method is acceptable under the Protection of Animals Act 1911). The standard methods of humane killing listed in Schedule 1 of the Animals (Scientific Procedures) Act may all be used in schools, but at least some initial training in these techniques is desirable. The method of cervical dislocation, with or without initial concussion, requires particular skill. Remfry (1987, in a Universities Federation for Animal Welfare booklet) points out that whilst physical methods of euthanasia, such as cervical dislocation, stunning and pithing are 'recognised as the quickest and most humane', 'these methods require skills which are rarely taught'. Remfry lists the methods of euthanasia of vertebrates and invertebrates which are currently used in schools and which 'are believed by most authorities to be humane' (see Table 9.2).

Two main questions arise. First, is it indeed true that the

Table 9.2 Methods of euthanasia for vertebrates and invertebrates used in schools

Vertebrates	
Rodents	CO_2 (may be followed by dislocation of the neck or decapitation).
Rabbits	Take to veterinary surgeon. In emergency, CO_2 or dislocation of neck.
Birds	CO_2.
Fish	*Cold water*: MS-222, CO_2 (followed by dislocation of 'neck'). *Warm water*: Refrigeration, MS-222, CO_2.
Amphibians	Refrigeration (followed by pithing or freezing), CO_2, MS-222.*
Reptiles	Freezing, chloroform or ether.*
Invertebrates	
Crustacea	Freezing, boiling water.
Insects	Chloroform, ether, CO_2, freezing.
Molluscs	MS-222, freezing, boiling water.
Helminths	Chloroform, ether, boiling water.

From Remfry (1987).
* Further information about methods of euthanasia for amphibians and reptiles is included in: Report of Universities Federation for Animal Welfare/World Society for the Protection of Animals Working Party (1989). *Euthanasia of amphibians and reptiles.* Universities Federation for Animal Welfare, Potters Bar.

methods listed in Table 9.2 are the most humane? None of the methods of killing reptiles, for example, is listed as a 'standard method of humane killing' under Schedule 1 of the Animals (Scientific Procedures) Act 1986 and there are no weight limits given for the different species and methods, as are laid out in Schedule 1. Second, are school laboratory technicians adequately trained in the standard methods of humane killing?

If the use of animals in secondary schools is to be morally defensible, it is important that the animals are not subjected to any unnecessary suffering. If school teachers or technicians have doubts about the euthanasia techniques to be used, or about their competence to perform techniques, they should take the animals to a veterinary surgeon or other suitably qualified person. It is good practice to ensure that animals are not killed in front of students and that animals kept as pets are never killed for dissection.

Dissection in schools

This use of animals in schools has received a great deal of attention in recent years. In the USA, especially, attention has been focused by a court case involving a Californian High School student, Jennifer Graham, who claimed moral objections to dissecting a frog. Having had her marks in class docked because of her refusal to dissect, she took her case to court. Several months later, in August 1988, the case was dismissed and Jennifer was offered a compromise: she could be tested on frog anatomy using a frog that had died of natural causes! Although this particular 'solution' was rather impractical, it took into account Jennifer's objection that she did not wish to participate, even indirectly, in the death of an animal. The case resulted in the passing of a new Californian State law, which upholds the 'right' of school pupils under 18 years old to refuse to dissect (see Orlans 1988).

In the UK, an Association for Science Education, Institute of Biology and Universities Federation for Animal Welfare statement on the place of animals in education (1986) suggests three specific educational benefits of dissection:

(a) active involvement in an important method of biological enquiry and discovery;
(b) obtaining first-hand knowledge and understanding of internal structures and their inter-relationships;
(c) gaining direct personal experience of both the fragility and strength of fresh tissues.

It should be noted that these are perceived benefits, and doubt has been expressed that they may indeed be gained from animal dissection in schools. From this viewpoint, it has been argued that there should be no dissection in schools at all.

Some members of the Working Party are agreed, however, that dissection may have some such educational benefits. In considering the ethics of this use of animals, it seems reasonable to begin by asking whether the benefits listed above could be achieved without the use of whole animals, killed especially for dissection. Possible alternatives to whole animals include the use of preserved 'museum mounts' so that students can observe dissected material, the use of animal organs obtained from the abattoir, and the use of slides, films or videos. There is some debate about the value of dissection in schools and about the usefulness of such alternative systems. On a simplistic analysis it would seem that, whilst each of these alternatives might achieve some of the benefits listed above, none can fulfil all of the aims listed. The use of animal organs obtained from the abattoir might achieve all the benefits except promoting an understanding of inter-relationships between internal structures of the body. The other methods can teach the principles of the body's organization, but the pupils will lack the first hand experience of fresh tissues.

If whole animals *are* to be used, questions of how the animals are kept during their lives and then killed for dissection must be addressed. As Cochrane and Dockerty (1984) have pointed out, 'it is somewhat trite to talk of organisms used for dissection as "dead material". The dead material referred to is, in fact, living organisms, bred and killed for dissection.' Any animals which are to be used for dissection should be kept in the best possible conditions and, when the time comes, killed humanely. The number of animals used should be reduced to the minimum needed to achieve the educational objectives of the practical work.

The effects on the pupils of such use of animals must also be considered. It is possible that dissection could be counterproductive in achieving the aims listed above, either because it upsets the pupils and so puts them off biology, or conversely, because it numbs the pupils so as to give them an indifferent attitude towards the use of animals. As a recent RSPCA policy document has noted, dissection can also 'offend the sensibilities or consciences of some pupils. Rightly or wrongly, some see it as "cruel" ... Many pupils find it upsetting or distasteful' (RSPCA 1986*b*). Since there *are* alternatives available, it would seem realistic to make dissection a voluntary activity, allowing pupils to choose to study non-animal alternatives. Before animals are used, class discussion of the reasons for using animals in this way and of the issues surrounding such use, may help to ensure that the value of animal life is not seen as being cheapened, and may also give pupils who object a chance to discuss their feelings.

In practice, however, the decision about whether or not to include dissection in class rests with the teacher of that class. This decision depends on the teacher's perception of the value of alternative systems and whether or not the teacher believes that the likely educational outcome can justify such a use of animals. The teacher also has to take into account the demands of the various syllabuses for public examinations.

In the past, dissection was a compulsory part of many UK Advanced Level practical examinations (taken by pupils around 18 years old). This requirement sometimes led to animals being used simply so that pupils could gain the manual skills necessary to pass the exams. Pupils who refused to dissect lost the marks. Recently, however, the 'A' Level Examining Boards seem to have become more sensitive to the dissection issue. Where there is a practical examination at 'A' level, most Boards now do not include a dissection. Where the teacher is required to assess the students' practical abilities, most Boards suggest dissection as a suitable practical task but give special consideration to students who object. The 'A' Level Boards may also include dissection as part of the teaching syllabus requirements. The London University Schools Examination Board has become the first of the eight 'A' Level Boards to remove animal dissection from its syllabus. Instead, the Board's 1990–91 syllabus requires pupils to study a computer alternative (*Daily Telegraph*, 27 March 1990).

If animals are to be dissected in schools, they should be kept in the best possible conditions and killed humanely. Class discussion of issues surrounding such use should be encouraged and pupils who object should be allowed to study alternative systems. It is important to ensure that dissection is not carried out simply in order to gain the manual skills necessary to pass a practical examination.

Teaching about the use of animals

Teachers' attitudes towards the use of animals in, for example, medical research, are important in shaping the pupils' attitudes. It is therefore important to ensure that a balanced perspective is maintained and that accurate information is passed on to the pupils. It is, however, difficult for teachers to obtain material which presents such a balanced and accurate perspective (stating arguments both for and against the various uses of animals), since little is available. The UK Association for Science Education (1988), in response to the need for discussion material concerning the use of animals in schools, has produced a resource pack for teachers covering the 'issues surrounding the use of animals in science lessons'. This booklet includes an outline of the ethical issues involved in animal use in schools, together with a series of statements written from a variety of standpoints, which may be used to help stimulate discussion by the pupils. The material shows an awareness of the need to discuss such issues in an informed and sensitive manner. Further such discussion material, especially concerning the use of animals in research, would be welcomed.

Use of animals in US schools

In addition to the four areas of concern discussed above, which apply in secondary schools throughout the world, a particular concern arises over the use of animals in schools in the USA. As noted under 9.1.2, North American High School students *may* use living vertebrates in invasive studies, particularly in projects submitted to local, state or national competitions – the student Science Fairs.

Although some Science Fairs allow only observational studies to be carried out on vertebrate animals, others allow a much wider range of procedures. For example, the International Science and Engineering Fair (ISEF), which is entered by students aged between 14 and 18, allows students to perform experiments involving 'anesthetics, drugs, thermal procedures, physical stress, pathogens, ionising radiation, carcinogens or surgical procedures', provided these are carried out 'under the direct supervision of an experienced and qualified scientist or designated adult supervisor'. The rules of the Fair state that 'surgical procedures may not be done at home'! (Office of Technology Assessment 1986). Morton (1987) has expressed concern that such work, carried out in schools, may be done by untrained students, supervised by teachers who are insufficiently knowledgeable to ensure humane treatment of the animals concerned. It is indeed difficult to see how the educational benefits of such experiments performed by school children could possibly justify the harm inflicted upon the animals used. For this reason, such use of animals should be discouraged and, if necessary, prohibited in law. In this context, UK legislation, and the legislation of several other countries, (which prohibits the use in schools of scientific procedures likely to cause pain, distress or lasting harm to vertebrate animals) is to be welcomed.

9.2 USE OF ANIMALS IN TERTIARY EDUCATION

9.2.1 Permitted uses of animals in tertiary education

In the UK, animals may be used in undergraduate courses in Institutes of Higher Education, Polytechnics and Universities, not only in all the ways described for schools but also in ways which are regulated under the Animals (Scientific Procedures) Act 1986 and which therefore require the appropriate Home Office licences. In such higher education and training, licences may be obtained in order to use animals for the 'demonstration of known facts' (a use which previously required a certificate C under the Cruelty to Animals Act, 1876) and for experimental purposes, for example for undergraduate projects. Animals may *not* be used to enable undergraduate students to acquire any form of manual skill. Home Office guidance on the operation of the 1986 Act (1990) states that, in applying for a project licence for education and training, 'applicants must show that they have carefully considered alternatives, such as video material and computer simulations, and that none is suitable'. The procedures used in such projects must normally be carried out under terminal anaesthesia, or (on a scale of mild, to moderate, to substantial) should not cause more than moderate animal suffering (see Home Office 1990, para. 4.40, p. 14).

Other European countries also permit scientific procedures to be carried out on living vertebrate animals in tertiary education. However, the extent and manner of *control* of such use may vary considerably between the different countries, especially since the 1986 EC Directive on the use of animals for scientific purposes does not cover the use of animals in education and training. The UK is the only country in the world specifically to prohibit training in manual skills, other than in microsurgery (see section 9.4).

9.2.2 Scale of animal use in tertiary education

Unfortunately, there is little objective information available concerning the extent and variety of animal use in tertiary education. The Home Office Statistics, however, can be used to paint a very general picture of the scale of animal use in tertiary education in the UK. Figure 9.1 shows the number of experiments carried out between 1978 and 1986 which were licensed by Certificate C under the Cruelty to Animals Act 1876, together with 1987, 1988 and

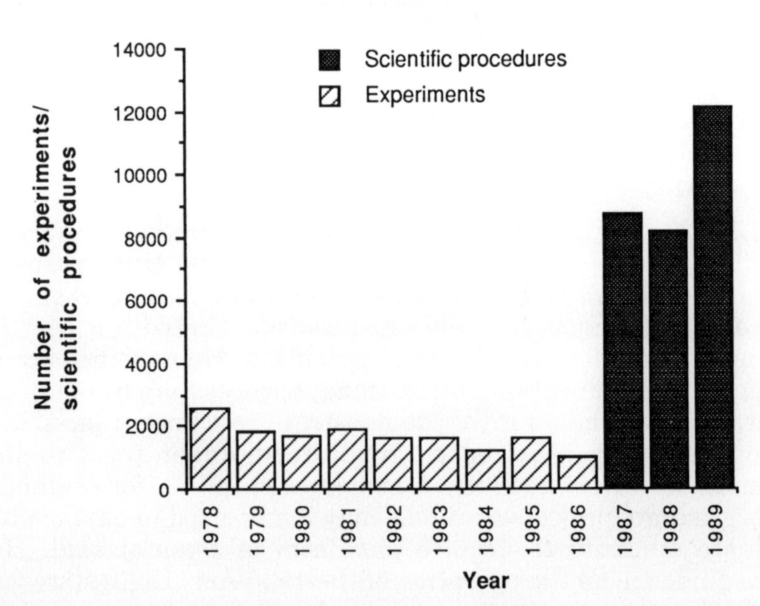

Fig. 9.1 Use of animals to demonstrate known facts (1978 to 1986) and in scientific procedures for the purposes of education and training (1987 to 1989) (Home Office Statistics).

1989 figures for the number of procedures carried out under the Animals (Scientific Procedures) Act for the purpose of education and training. These data include experiments or procedures carried out in undergraduate courses, postgraduate courses and in continuing professional training.

Certificate C, under the Cruelty to Animals Act, licensed work which was to 'demonstrate known facts'. There is a jump between the 1986 and 1987 figures since the statistics collected under the 1986 Act put together *all* scientific procedures carried out for the purposes of education and training: that is, the data include scientific procedures whose outcome was known (requiring Certificate C under the old 1876 Act) and those whose outcome was not known (requiring different certificates under the 1876 Act). The 1986 Act also encompasses procedures which were not covered under the 1876 Act, for example, the use of decerebrated animals and the use of animals in training for microsurgery.

In each year since 1978, the use of animals in education and training has constituted less than 0.4 per cent of the total use of animals under the relevant legislation.

9.2.3 Variety of animal use in tertiary education

The ways in which animals are used and the aims of this use vary between the different types of biomedical degree course. Broadly speaking, such courses can be divided into two groups:

(a) those, such as veterinary, medical and dental courses, which provide a 'professional training', and
(b) those courses in the biological sciences which provide more of a 'research training', for example, degree courses in biology, zoology, biochemistry (including genetics and pharmacology), toxicology, psychology, etc.

The study of physiology could be considered to link these two groups, being a common theme in veterinary, medical and basic biological education.

The following notes give just a general indication of the ways in which animals may be used in tertiary education.

Use of animals in the teaching of physiology

A University Lecturer involved in teaching physiology to students of medicine and biology has informed the Working Party that in both of his courses 75–80 per cent of the practical work is now carried out on human subjects—the students themselves—and furthermore, neither of the courses now uses whole animals at all.

Table 9.3 Changing pattern of animal use in a physiology course for medical students: practicals using animals in 1978 and 1988

1978

1. Student practicals

- Rat pituitary function (used 50–60 rats)*
- Dissection of female rats treated with oestrogen
- Effect of thyroxin on metabolic rate in mice
- Mice vaginal smears
- Frog nerve-muscle preparation
- Frog isolated heart
- Sensory nerve recording from frogs
- Pharmacology of rabbit ileum preparation

2. Demonstrations

- Cardiovascular reflexes in cat
- Observation of posture in a decerebrate pigeon, and in an intact cat and rabbit

1988

Student practicals

- Pharmacology of rabbit ileum preparation (four rabbits/year)
- Effect of calcium ions on isolated frog ventricles (16 frogs/year)

---

* These experiments were rapidly removed from the course—they were expensive and the course teachers and students were unhappy about the taking of so much animal life.

The changing pattern of animal use in the course for medical students is shown in Table 9.3. The whole animal work which was a part of the 1978 course is no longer carried out and the use of parts of animals has decreased dramatically over the last ten years. Although more animals may be used in other universities it seems that in the UK there has been a countrywide decrease in the use of animals in the teaching of physiology, prompted by both economic and ethical concerns.

Use of animals in medical education

In medical studies, animals are used as models for human subjects and courses aim to study human subjects as far as possible. Most anatomy, for example, is learnt by dissection of human corpses. As much physiology as possible is also taught using human subjects (the students themselves), although there are, of course, limits to what can be done. As mentioned earlier, in the UK, animals may not be used for gaining manual skills in clinical subjects. The practice in the USA, however, is different: animals are used in teaching a wide range of surgical, anaesthetic and other techniques. According to the Office of Technology Assessment (1986), around 10 per cent of animals used in surgical procedures in US medical schools are allowed to recover, and a proportion of these animals (the report does not say what proportion) may be subjected to multiple recovery procedures.

Use of animals in veterinary education

In veterinary training, animals are used as models for other animals. Veterinary courses use some animals in the teaching of physiology, but the majority of animals are used in dissection. Students dissect a variety of animals in order to study their anatomy and function and so provide a foundation for gaining clinical competence. In the UK, it is illegal for such students to practise surgical techniques on living animals. These techniques have to be practised on the job, under supervision, in real clinical practice. The rule is again different in the USA, where undergraduate veterinary courses may include practice in survival surgery. A report in the *Veterinary Record* of 23 April 1988 described an example of such survival surgery as follows:

> Two goats are operated on for six days on alternate days by students under strict supervision. The procedures include jugular catheterisation, plaster casing a leg, rumenotomy, eye enucleation, digit amputation abomasopexy, dehorning, teat surgery, urethrotomy and cerebrospinal fluid collection. The goats are only allowed to survive one major surgical procedure such as rumenotomy and are finally euthanased. [Report of a paper given at the Annual Congress of the Association of Veterinary Teachers and Research Workers. Paper by Dr A. D. Weaver of the US University of Missouri College of Veterinary Medicine.]

Such use of animals in veterinary education may not be confined to the USA: a group of Dutch veterinary students has recently drawn attention to the use of survival surgery on a pig in the undergraduate surgery course at the University of Utrecht veterinary school (EURONICHE 1989).

Use of animals in basic biological science courses

In some basic biological science courses, perhaps particularly zoology courses, a range of different animals (vertebrate and invertebrate) may be dissected. Some courses may also involve the use of animals in experimental work, designed to teach the students the principles of experimental design and to illustrate biological concepts. Further generalization is very difficult, however, since the particular use of animals may vary considerably from course to course.

Information about the use of animals in undergraduate teaching is difficult to obtain: universities, perhaps understandably, seem reluctant to provide even general information about the use of animals in their biological science courses. A general enquiry from the Working Party, circulated to all UK universities by the Committee of Vice-Chancellors and Principals, received only one reply.

9.2.4 Alternatives to the use of animals

Again, perhaps the first question to be asked when considering the acceptability of these uses of animals is, 'can the same educational benefits be achieved without using whole animals?'. The use of slides, museum mounts, prosected specimens, films, videos and interactive videos and computer programmes can all help to reduce the numbers of animals used in teaching biomedical science. The latter provide particularly useful models for biological systems. A wide range of sophisticated computer programmes is now available. The 'Mac' family of physiological models (developed at MacMaster University in Canada), for example, may be used in clinical, physiological and pharmacological teaching and research (Dickinson *et al.* 1985). Several programmes are available, each simulating a different physiological system:

MacMan: A model for the complete systemic circulation.
MacPuf: Simulates the lungs and airways, pulmonary circulation and gas exchange.
MacPee: Models the systemic circulation and kidneys (incorporates MacMan).
MacDopex: A model for the absorption, distribution, and excretion of drugs in a multi-compartment system.

In each programme, students can monitor important physiological variables, which are displayed graphically at the computer terminal, and can study the effects on the physiological system of altering one or more factors. Computer programmes may also simulate particular experiments commonly performed in undergraduate practical classes. For instance, Clarke (1987, 1988) has developed programmes to simulate a frog sciatic nerve preparation and the mechanical properties of skeletal muscle. In each case the programme was produced by recording from a real preparation. Students can study the effects of altering the relevant experimental variables and a 'biological' variability is built into the results provided by the computer. Use of such programmes means that the students spend less time trying to make viable experimental preparations, leaving them with more time, and enthusiasm, for mastering the appropriate neurophysiological concepts (Clarke 1987). Some programmes also enable the students to chart their own progress, providing questions to test comprehension before passing on to the next stage of the experiment.

Sharpe (1988) has described the 'biovideograph', which enables students to watch experiments on video and also provides a record of the experimental results on a pen or chart recorder. As Sharpe suggests, the biovideograph may be useful for teaching topics in physiology or pharmacology where similar demonstration experiments are carried out year after year. A further development is 'interactive video', which combines interactive computer programmes with moving or still video pictures (Computers in Teaching Initiative Centre for Medicine 1989).

However, where such non-animal alternatives are used, the students are denied the experience of dissecting the animal and making experimental preparations for themselves. Hence they may not appreciate all of the difficulties involved in carrying out such techniques, nor develop any expertise in the techniques. Nevertheless, as Clarke (1987) has pointed out, many students will enter careers which will not require skills in animal dissection or experiment and so it may be desirable to spend less time developing such skills and more time fulfilling other teaching objectives. If it is thought especially important for students to see a dissection or a particular experimental preparation then it might be argued that one demonstration dissection or preparation would suffice. In veterinary and medical circles, there is some debate about the value of using prosected (predissected) specimens as against the students dissecting for themselves. Prosection is time-saving and, perhaps for this reason alone, is becoming more popular in an ever-increasing curriculum. Prosected specimens also offer a high quality of dissection so that all students can learn from them (not only those who manage to do a good dissection). In addition, such demonstration dissections do not consume animals, offering savings in animal life and in money. On the other hand, it is argued, students learn better when they experience dissecting for themselves, they can 'feel' the tissues and can develop manual skills which are useful clinically.

We have been unable to reach a consensus on the justification for using animals in experiments in first degree courses. Some members of the Working Party believe that there is now no justification for such use of animals, arguing that an undergraduate can acquire all the relevant skills and knowledge by using the various alternative systems described above. (Although in some cases, the use of animals in individual student projects might be justified.) On the other hand, some members believe that there should be continued, but carefully controlled and limited, use of animals in biomedical degree courses. It is argued that the use of non-animal alternatives cannot achieve all of the desired educational objectives. In particular, it is said that the alternatives lack immediacy and do not introduce students to the problems of setting up and carrying out experiments using living animals.

9.2.5 Manner in which animals are used

In the end (as in schools) the decision about whether or not to use animals in a particular degree course rests with the teachers. If it is decided to include the use of animals in such a course, then very careful consideration should be given to the manner in which the animals are to be used. It is possible that in some cases animals are used because of tradition rather than because such use is necessary. Teachers should therefore think carefully about the use of animals in their courses and consider whether it might be possible to use alternatives to whole animals. Where work using animals is proposed, as in research, the likely harm caused to the animals must be justified in terms of the perceived educational benefits of the work. Such justification may not be easy, however, since there have been few objective analyses or scientific studies of the effectiveness of the various teaching methods used in undergraduate biomedical education.

In only a few cases will it be possible to justify the use of animals in projects other than those which involve observation of the animals (with or without minor manipulations, such as changing the animals' environment or altering the animals' diet and so on) or in which the animals are terminally anaesthetized. Many may well find it even more difficult to imagine cases where surgery with recovery could be justified. We would therefore urge that the use of animals in survival surgery in medical and veterinary schools in the USA and in a veterinary school in the Netherlands should be stopped, unless and until its necessity can be demonstrated.

The context in which demonstrations or class practicals using animals are carried out is also very important. Classwork using living animals should be carefully introduced and supervised, so that the animals are not subjected to any unnecessary distress, and attention should be paid to the training of students, in order to ensure good animal welfare whilst the animals are alive in the laboratory. The experiments should be carried out in a sensitive manner and the students should be given the opportunity to discuss the ethical issues before the work is started.

There may be students who do not wish to work with animals at all. There are numerous anecdotal cases of students refusing to work with living or dead animals in classes or even to attend such practicals. In recent years student opposition to the use of animals in education and training, as well as in research, has been growing. In 1986 the National Anti-vivisection Society began a campaign for 'Violence Free Science', and a Students' Charter was drawn up (Table 9.4). The charter aims to give students the right to opt out of using animals in their educational studies, without prejudice to their marks or prospects.

The first International Student Meeting to discuss the use of animals in education was held in Bergen, Holland in May 1988

Table 9.4 The Students' Charter (National Anti-vivisection Society 1986)

1. As a student, the right and opportunity to practise violence-free science will be recognized as mine.
2. Such choice on my part is dictated by material, intellectual, and moral considerations.
3. I will be free to follow my conscience in refusing to perform violent practices in experimentation (required of me) which are not in accordance with the Universal Declaration of the Rights of Animals.
4. Within an educational institution, no disciplinary or administrative action shall be imposed on me simply because I have exercised such a right of conscience.
5. Similarly, I shall have the right to object to violent practices in applied science in every case which would tend to involve me.
6. I will act with dignity at all times when exercising my right to study and put into practise non-violent science.
7. I will invoke this Charter whenever violent practices in human or animal experimentation are imposed upon me during my studies or professional activities.
8. I will defend and propagate the spirit of the Charter to ensure that science deals with mankind, animals and nature with comprehension, sympathy and peace.
9. To give full support and back up to any fellow student who wishes to invoke the Charter.
10. To seek, through appropriate channels, a huge reduction in the number of animal experiments in this university and if necessary actively campaign to alleviate animal suffering inflicted in the 'interests of education'.

and, arising from this meeting, a student group, the European Network for Independent Campaigns for Humane Education (EURONICHE), has been formed. EURONICHE aims to help students in 'conscientious objection' to the use of animals in education and to provide information and resources on animal welfare and the development of techniques which can be used to replace the use of animals in education.

In view of these feelings amongst students, it would seem important to make information about animal use in undergraduate courses available to applicants for those courses, and to discuss such use at interviews with potential students. Even then, if after the course has started, it becomes apparent that a student has a reasonable objection to the use of animals, which could not have been anticipated when entering the course, the student should be allowed to opt out of this use (perhaps by studying an alternative system), without prejudice to marks or prospects.

9.3 USE OF ANIMALS IN RESEARCH FOR POSTGRADUATE DEGREES

In addition to their use in undergraduate courses, animals may be used in postgraduate research leading to a higher degree, such as MSc, MPhil, PhD, DPhil, MD or MS. Again, in the UK, if vertebrate animals are to be used in such work in ways that are likely to cause the animals pain, distress or lasting harm, then the appopriate Home Office licences must be obtained.

The main cause for concern lies in the adequacy of the supervision of such students. It has, for example, been suggested that inadequate supervision may be a major reason for failure in an MD degree: 25 doctors who had failed or were failing to obtain an MD were invited to list the problems they had encountered; 36 per cent of these doctors reported that lack of supervision had contributed to their failure (Wood and Catford 1986). Postgraduate students, particularly those who have followed the 'professional' courses in medicine, dentistry or veterinary science, may embark on higher degrees with little or no experience in designing scientific experiments for themselves.* The student's supervisor will therefore

* Many undergraduate courses in the basic biological sciences include project work and this can provide a valuable background in experimental design. In the UK, some medical students may add a year to their formal medical training in order to follow an intercalated degree course leading to a BSc in basic medical science and this may also provide 'an invaluable early introduction to scientific method' (House of Lords Select Committee on Science and Technology 1988). At present, however, it is difficult to obtain funding for this

play an important role in ensuring that the student's work constitutes 'good science' and that animals are not wasted in badly designed experiments.

Students also require supervision when learning experimental techniques. Where UK Home Office licences are required for the work, the student's personal licence, like that of all new licensees, will contain a supervision condition which will normally remain in force for a year. This condition may only be lifted when the student has attained sufficient competence to perform the techniques specified in the licence without supervision (Home Office 1986). Students therefore require careful instruction from their supervisors in order to ensure that the animals used are treated as humanely as possible and that the techniques are carried out efficiently and reproducibly, in order to ensure 'good science'.

9.4 TRAINING FOR LICENSEES

As discussed in section 9.3, all licensees must show 'competence' in the techniques they wish to use before they can be issued with a personal licence under the Animals (Scientific Procedures) Act. Until competence is gained the licence issued will bear a supervision condition. New licensees will therefore require training on two levels:

(a) some form of training in their proposed techniques; and,
(b) more general training concerning the use of animals in research.

These two kinds of training are important for anyone starting research using animals, not only UK licensees. In the UK, the latter form of training is becoming increasingly well-covered, since many establishments now run short in-house training courses for new licensees. There are also a few generally available courses, such as those run by the Royal College of Surgeons and the North-East Surrey College of Technology, although it seems that attendance on these is falling, probably because more provision is being made in-house. The Institute of Biology has launched a

---

extra year – the MRC has been pressed to stop funding the degrees because they constitute training and not research and the LEAs will not support many students whilst the grants remain discretionary (Smith 1986). The House of Lords Select Committee on Science and Technology (1988) has recommended that 'an intercalated honours degree should now be available for all medical students who opt for it' and that 'funding for intercalated degrees should be provided by LEAs as a mandatory part of their funding for undergraduate degrees, of which intercalated degrees should be treated as a necessary part'.

scheme for the accreditation of such courses and intends to award certificates to those satisfactorily attending accredited courses.

The training of licensees in the procedures to be used on the animals is problematic. As discussed earlier, the UK Act expressly forbids the use of living animals in training for manual skills, with the exception of microsurgery. Animals may not be used for training in any other kind of manual skill, such as inoculation, gavage, venepuncture and so on. Hence, a paradox: competence is required before animals may be used and yet animals may need to be used in gaining that competence. The supervision condition does, however, provide for this competence to be gained on-the-job, during the course of real experiments carried out under supervision. What, therefore, is the moral difference between learning a technique during the course of an experiment, which may well fail because competence has not yet been achieved, and using animals for the sole purpose of learning that technique (admitting that it will take some practice to become competent and so achieve useful results)? In fact, it might be better to have such standard techniques taught officially and properly by expert teachers rather than by supervisors on a part-time basis.

In addition, some skills, such as techniques for handling laboratory animals and Schedule 1 methods of euthanasia do not come within the Animals (Scientific Procedures) Act and so may be practised freely. What is the difference between practising these techniques and learning others, such as oral dosing with saline, administration of anaesthesia using a chamber technique or learning surgical techniques under terminal anaesthesia, which (except for microsurgery) are forbidden under the Act? There would not seem to be much difference in the likely adverse effects on the animals concerned.

Although, in some instances, it might be desirable for licensees to be able to practise manual skills and so to gain competence, it is not desirable that animals should be used to gain manual skills in any other form of education (such as in schools or in tertiary education). Since licensees can learn on-the-job, it would seem best to avoid further complicating the law by allowing an exception for licensees but not for other kinds of student to practise manual skills. Furthermore, the political dimension has to be taken into account: if the restriction to microsurgical skills is lifted it is likely that animal welfare organisations will lose confidence in the legislation. It is important that the legislation is *seen* to be effective.

9.5 CONCLUSIONS

In education and training, as we have seen, animals are used to teach *known* facts, ideas and techniques. We have reviewed the ways in which animals are used in such teaching and have considered the justification for such uses. Our conclusions are presented as answers to the three morally relevant questions set out in the introduction to this chapter.

Is it acceptable to use animals to teach facts and ideas which are already known?

We are agreed that some use of animals in teaching known facts and ideas can be justified, but that such use should be limited, carefully considered and controlled.

In *primary and secondary schools* living animals might be observed and, in secondary schools, dead animals might be dissected. Keeping and observing animals in schools can help pupils to develop an interest in and respect for animals in general. The pupils can learn about the animals' husbandry and, through non-invasive experiments, can gain some understanding of the animals' behaviour and physiology. At secondary level, dissection may help pupils to understand how the body is organized. Both of these kinds of use require animals to be maintained in schools. If such use is to be justified, the animals should be kept in the best possible conditions and should not be subjected to any unnecessary stress. Schools keeping animals should be required to register with a local veterinary surgeon.

When animals are used in dissection they are usually killed specifically for this purpose. Thus some pupils (as well as teachers) might find such use of animals unjustified, and wish to opt out of dissection. The view has been expressed in the Working Party that there should now be no animal dissection in schools. However, some members agree that dissection should be permitted, and that students who object should be able to opt out, in favour of studying non-animal alternatives, without prejudice to their marks. Most are agreed that it is unnecessary, and therefore unjustified, to kill and dissect animals simply in order to gain the manual skills needed to pass a practical examination. Invasive studies of living vertebrates in schools would also seem unjustifiable, since it is very unlikely that the knowledge gained by the pupils will outweigh the costs of the studies, in terms of animal suffering. For this reason, we are very concerned to learn of the use of animals in invasive studies in US High Schools and we urge that this practice be abolished.

With regard to *tertiary (undergraduate) education*, there is no common view on whether invasive uses of animals are justified. Some members believe that there is now no justification for such use of animals, whilst others believe that there should be continued use of animals in biomedical degree courses. In the final analysis, however, the choice of whether or not to use animals rests with the teacher concerned. We are agreed that if animals are to be used, this use should be controlled and limited (both in the number of animals used and in the amount of harm caused to them). Furthermore, the use should be explicitly justified, carried out in a sensitive manner and well supervised. Where surgery is involved, this should be under terminal anaesthesia, and animals should not be used in 'survival' surgery, as has been reported in some US and Dutch veterinary schools and in US medical schools.

There is a wide range of alternative systems available for use in tertiary education, so that the possibility of not using animals, in favour of studying a non-animal alternative, should be carefully considered. In this context, there is a need for evaluation of the effectiveness of the various (animal and non-animal) teaching methods. There is increasing student objection to the use of animals in undergraduate studies and these views should be considered. We suggest that, provided they could not have been expected to entertain their objections before starting the course, students should be allowed to opt out of using animals (studying non-animal systems instead), without prejudice to their marks or prospects. It is important that course organizers should be willing to provide genuine prospective students with information about any proposed use of animals in their courses.

In pursuit of *higher degrees*, students embark on a research training, so that invasive uses of animals may be justified. However, this use of animals should be carefully supervised; there is a need for review of the checks and balances which operate to ensure that such students receive adequate supervision.

Is it acceptable to use animals in order to gain manual skills which may be used to benefit humans or other animals?

We are agreed that there is no need for pre-university students to use living animals in order to develop or practise manual skills. The same is true in most university courses. In professional training in veterinary and medical courses, where the development of surgical skills is an important aim of the course, such skills may be learnt by dissection of dead animals or humans and by 'apprenticeship' with practising veterinary or medical surgeons.

The question of the use of animals in training for those wishing to carry out experiments on animals (prospective 'licensees' in the UK) is more problematic, since these people must gain competence in the techniques they propose to use. Such competence is important for the welfare of the animals involved. However, provided new licensees are carefully supervised, they should, like doctors and veterinarians in training, be able to gain competence on the job (in the context of real experiments), without compromising the well-being of the animals involved. Thus in the present political climate in the UK, we accept that it is reasonable to restrict the use of animals in gaining manual skills to the development of microsurgical skills by practising surgeons.

How can educational experiences with animals be used to encourage the development of balanced, humane attitudes towards animals?

The manner in which animals are treated in schools and in undergraduate and research training can play an important role in shaping students' attitudes towards the animals. It is therefore essential that students see that any animals used in their courses are treated in a sensitive and humane manner and that the teacher justifies the use, so that animal life is not seen as being taken lightly. In this context, there is a need for the issues surrounding the use of animals in education (as well as human uses of animals in general) to be discussed in a sensitive, accurate and well-balanced manner.

REFERENCES

Association for Science Education (1988). *Issues surrounding the use of animals in science lessons: a resource pack.* Association for Science Education, Hatfield, Hertfordshire.

Association for Science Education, Institute of Biology, and Universities Federation for Animal Welfare (1986). The place of animals in education. *Biologist* **33,** 275–8.

Clarke, K. A. (1987). The use of microcomputer simulations in undergraduate neurophysiology experiments. *Alternatives to Laboratory Animals* **14,** 134–40.

Clarke, K. A. (1988). Microcomputer simulations of mechanical properties of skeletal muscle for undergraduate classes. *Alternatives to Laboratory Animals* **15,** 183–7.

Cochrane, W. and Dockerty, A. (1984). The role of dissection in schools. *Biologist* **31,** 250–4.

Cooper, M. E. (1987). *An introduction to animal law*. Academic Press, London.

Computers in Teaching Initiative Centre for Medicine (1989). The interactive video club. *CTICM Update* **1,** 3–4. (Available from Department of Pathology, University of Bristol.)

Dickinson, C. J., Ingram, D., and Ahmed, K. (1985). The 'Mac' family of physiological models. *Alternatives to Laboratory Animals* **13,** 107–16.

EURONICHE (1989). *Newsletter number 2, September 1989*. Published by Royal Society for the Prevention of Cruelty to Animals, Research Animals Department, Horsham, West Sussex.

Home Office (1986). *Notes on the completion of the personal licence application form*. HMSO, London.

Home Office (1990). *Guidance on the Operation of the Animals (Scientific Procedures) Act 1986*. HMSO, London.

House of Lords Select Committee on Science and Technology (1988). *Priorities in medical research*. HMSO, London.

Morton, D. B. (1987). Animals in education. *Alternatives to Laboratory Animals,* **14,** 334–43.

National Anti-vivisection Society (1986). *The Students' Charter*. National Anti-vivisection Society, London.

O'Donoghue, P. N. (1988). The law and animals in schools. *J. Biol. Educ.* **22,** 13–15.

Office of Technology Assessment, US Congress (1986). *Alternatives to animal use in research, testing and education*. US Government Printing Office, Washington DC.

Orlans, F. B. (1988). Debating dissection. *The Science Teacher,* November 1988, 36–40.

Remfry (1987). Euthanasia of animals in schools – trends and methods. In *Euthanasia of animals* pp. 3–7. Universities Federation for Animal Welfare, Potters Bar.

RSPCA (1986*a*). *Small mammals in schools*. Royal Society for the Prevention of Cruelty to Animals, Education Department, Horsham, West Sussex.

RSPCA (1986*b*). *Dissection*. Royal Society for the Prevention of Cruelty to Animals, Education Department, Horsham, West Sussex.

Sharpe, R. (1988). Programs of the body works. *The Guardian,* 26 April 1988.

Smith, R. (1986). A senseless sacrifice: the fate of intercalated degrees. *Br. Med. J.* **292,** 1619–20.

Tricker, B. J. K. and Dowdeswell, W. H. (1970). *Projects in biological science*. Nuffield Advanced Science. Penguin, Harmondsworth (for the Nuffield Foundation).

Wood, G. M. and Catford, J. C. (1986). Research post-MRCP. *J. Roy. Coll. Physicians Lond.* **20,** 89–94.

Zahn-Waxler, C., Hollenbeck, B. and Radke-Yarrow, M. (1985). The origins of empathy and altruism. In *Advances in animal welfare science 1984* (ed. M. W. Fox and L. D. Mickley), pp. 21–41. Martinus Nijhoff, Dordrecht.

10

From theory to practice

Control of biomedical research involving animal subjects

So far in this report, we have been concerned mainly with considering how ethical decisions regarding the use of animals in biomedical research might be made. To this end, we have identified, and reflected on, a number of morally-relevant questions. In particular, we have asked questions about the harm caused to animals in research, the likely benefits of that research, the possibilities for achieving these same benefits without the use of animals, and how harm and benefit might be weighed against one another in order to decide whether or not animals should be used in the manner proposed in particular pieces of research. Now, in this chapter, we move on to examine where, in practice, such ethical review and decision might occur. In practice, where should responsibility for controlling and safeguarding standards in animal research lie?

In general terms, such controls and safeguards can be divided into two kinds: those which have legal authority and those which are set up voluntarily. Legal, and legalistic, controls are considered in the first part of this chapter, and a variety of other controls are examined in the second part.

Part 1: Legal control of research involving animal subjects

10.1 NEED FOR LEGISLATIVE CONTROL

In the UK, research involving animal subjects has been subject to special legal control for over 110 years, yet, even today, there is no equivalent legislation controlling the ethical conduct of research involving human subjects. Such research is controlled by voluntary adherence to guidelines or codes of practice. The situation was, until very recently, the same in most European countries. In

Denmark, for example, the first legislation to protect laboratory animals was passed in 1891, and a second, more comprehensive statute was put in place in 1953 and amended in 1977 (see Søndergaard 1986); in contrast, although in Denmark there is a nationwide ethics committee system for the review of research on human subjects, this system does not yet rest on legislation. Moves towards legal control have taken place only recently: a law, passed in 1987, established a National Ethical Council, and a commission has been appointed to advise the government on how to give a statutory backing to the regional ethics committee system (Riis 1988). Similarly, most other Western European countries have well-established animal protection laws, but are only just beginning to implement legislation relating to research involving human subjects (see IME Bulletin 1988).

Why, it might be asked, has it been thought fit (in Europe at least) to implement legislation controlling the use of animals but not the use of human subjects in research? The reasons are complex, but two general considerations may have contributed. First, consent is required for humans to take part in research procedures. In ethical terms, there is a respect for the principle of human autonomy. In research on human subjects obtaining true (or informed) consent is seen as 'central to the ethical conduct of clinical investigation'.* The requirement that consent be true, means that prospective subjects should be provided with all the information necessary to enable them properly to judge whether they wish to take part in the research or not. Such information should include 'the objectives and consequences of their involvement' in the research, including particularly any 'identifiable risks and inconvenience' (World Health Organization 1982). Once they have consented to take part in the research, the human subjects have to trust the researchers to safeguard their interests. Animals, however, are obviously incapable of giving such consent. In this context, animals lack the capacity even to begin to look after their own interests, so that the potential for abuse is much greater than with humans. There would therefore seem to be a greater need for legal control over the use of animals when compared with the use of humans in research. Since animals cannot look after their own interests, the trust obligation of researchers engaged in research using animals has to be reinforced by law. Second, common law

* There are obvious practical difficulties where, for example, children or mentally incompetent subjects are involved: in such cases consent is usually obtained from the subject's immediate family or guardian (see for example, *Guidelines on the practice of ethics committees in medical research involving human subjects*, second edition, Royal College of Physicians of London, 1990)

can act paternalistically with respect to humans killing or harming others or indeed themselves. This will prevent research procedures likely to harm the interests of the human subject involved, even if the subject has consented to such procedures. (This principle of 'not causing harm' applies in therapeutic medicine as well.) The common law, however, would not readily intervene in this way in animal research. Indeed, animal research is fundamentally different from research on humans, in that much of this research involves procedures which are not at all in the particular animal subject's interests. Much of what is done routinely to animals would infringe general statute and common law if done to humans. Again, therefore, there would seem to be a greater need for special legal control of the use of animals in research.

10.2 TYPES AND EXTENT OF LEGAL CONTROLS: INTERNATIONAL COMPARISONS

Although most of the developed countries have established special legal control over the use of animals in scientific procedures, worldwide legal provisions governing such animal use show considerable variation. Almost all countries have some kind of legislation which can be applied to the use of animals in laboratories, but this legislation varies from the most general and broad to the most detailed and specific of provisions. Often, in addition to this legal framework, there is some system of voluntary control, but, once again the extent and nature of such voluntary control varies considerably from country to country. The following account presents an overview of some of these legislative frameworks, setting descriptions of selected international laws against the perspective offered by an historical analysis of the development of legislation in the UK.

10.2.1 An historical perspective: development of legislation in the UK

General anti-cruelty legislation and its application to animal experiments

The first British laws relating to the treatment of animals came at the time of the late eighteenth century rise in humanitarian concern for animals. In 1781, a law requiring scrutiny of the treatment of cattle in Smithfield market was passed and in 1786 legislation requiring the licensing of slaughterhouses was enacted (see Thomas 1983). These laws, however, were concerned with protecting animals as human property, rather than for their own sakes. The first measure specifically concerned with preventing cruelty to animals was passed in 1822. This Act, known as Martin's Act after its sponsor Richard Martin, related mainly to horses and cattle. In 1835, amendments resulted in an Act which protected all domestic animals, including dogs and cats, against 'wanton' cruelty (see Harrison 1982). After 1835, a splinter group of the RSPCA,† the Animals' Friend Society, tried several times to obtain convictions under Martin's Act for cruelty in animal experiments. However, the group was unsuccessful, and there was no prosecution for cruelty in such experiments until 1874 (see French 1975). This prosecution followed the BMA's annual meeting held in August 1874 in Norwich, at which a French physiologist, Eugene Magnan, demonstrated the induction of epilepsy in two dogs, using intravenous injections of absinthe and alcohol. There were protests at the experiments and the RSPCA later instigated proceedings against Magnan and three Norwich doctors who had arranged the meeting, charging them, under Martin's Act, with 'wanton' cruelty. Although the prosecution proved unsuccessful (since Magnan had returned to France and the Norwich men had not been involved in the actual demonstration), the magistrates considered that the RSPCA had been justified in bringing the action and declined to award defence costs (French 1975). This was a rare success. Even though, in 1849, Martin's Act was amended so as to omit the word 'wanton' (removing the need to prove that the cruelty was intentional), it became clear that it would be difficult ever to achieve a successful prosecution under this Act for cruelty in animal experiments. This was because cruelty under the statute referred to the '*unnecessary* abuse of the animal', and hence, in order to prove cruelty, the experiments would have had to be shown to be useless (Hampson 1978).‡

Attempts at self-regulation

British physiologists themselves, before the Norwich prosecution, had drawn up and attempted to apply guidelines for carrying out animal experiments. In 1831, Marshall Hall, one of the few scientists practising 'experimental medicine' in Britain at that time, suggested five principles for animal experiments (Table 10.1). Hall

† The Royal Society for the Prevention of Cruelty to Animals was founded in 1824, although it was not granted its prefix 'Royal' until 1840.

‡ Hampson (1978) reports that the case of Budge vs. Parsons (1863) under Martin's Act established the *locus classicus* definition of cruelty, that is, 'the cruelty intended by the statute is the *unnecessary* abuse of the animal'. In a later case (1889) it was argued that what was simply convenient or profitable could not be regarded as necessary.

Table 10.1 Marshall Hall's 'Principles'. 'We are greatly in need of a code of medical ethics. I have ventured to sketch in what appears to me to be ethical in regard to experiments in physiology' (p. 58).

First principle
'We should never have recourse to experiment in cases in which observation can afford us the information required.'

Second principle
'No experiment should be performed without a distinct and definite object, and without the persuasion, after the maturest consideration, that that object will be attained by that experiment, in the form of a real and uncomplicated result.'

Third principle
'We should not needlessly repeat experiments which have already been performed by physiologists of reputation.'

Fourth principle
An experiment 'should be instituted with the least possible suffering' . . . 'in all cases the subject of the experiment should be of the lowest order of animals appropriate to our purpose, as the least sentient; whilst every device should be employed, compatible with the success of the experiment, for avoiding the infliction of pain'.

Fifth principle
'Every physiological experiment should be performed under such circumstances as will secure due observation and attestation of its results, and so obviate, as much as possible, the necessity for its repetition.'

Drawn from an article published in 1847 in *The Lancet*, **i**, 58–60 (reprinted from Hall's work, *On the circulation of the blood* (1831)).

was sensitive to the popular disapproval of such work and appealed to experimental physiologists to form a Society, which could discuss proposed experiments and use the principles to decide whether the proposals were acceptable or not. Such a Society might have been the first ever 'animal ethics committee', but it was not forthcoming (see Manuel 1987).

During the 1860s experimental physiology gradually became institutionalized in Britain and this same period saw the growth of an organized movement opposed to animal experiments. In this decade, there were repeated calls in the medical press for a physiologists' 'moral code' of practice for experimental work on animals. Such a code was eventually drawn up by a Committee of the British Association for the Advancement of Science (BAAS).

At the annual meeting of the BAAS in 1870 a representative of the RSPCA had suggested that BA grants should be withheld from members of the Association who caused pain to animals. This led the BA to set up a committee to prepare a statement of physiologists' views on how to minimize animal suffering in experiments. The committee reported at the BAAS annual meeting in 1871, giving four main guidelines for animal experiments (Table 10.2).

Table 10.2 Guidelines produced by the British Association for the Advancement of Science (1871), as quoted in Hampson (1978)

1. No experiment which can be performed under the influence of an anaesthetic ought to be done without it.
2. No painful experiment is justifiable for the mere purpose of illustrating a law or fact already demonstrated; in other words, experimentation without the employment of anaesthetics is not a fitting exhibition for teaching purposes.
3. Whenever, for the investigation of new truth, it is necessary to make a painful experiment, every effort should be made to ensure success, in order that the sufferings inflicted may not be wasted. For this reason, no painful experiment ought to be performed by an unskilled person, with inefficient instruments and assistants, or in places not suitable to the purpose; that is to say, anywhere except in physiological laboratories, under proper regulations.
4. In the scientific preparation for veterinary practice, operations ought not to be performed upon living animals for the mere purpose of obtaining greater operative dexterity.

However, the BA's hope that this code would reassure those who criticized their experiments was disappointed, both by the Norwich trial and also by the publication in 1873 of a two volume '*Handbook for the Physiological Laboratory*'. This handbook, written by some of the eminent physiologists of the time, contained step-by-step instructions for classical physiological experiments and stated that it was 'intended for beginners in physiological work' (see Richards 1987). The book was, to the anti-vivisectionists, proof that the physiologists could not be trusted to work to their own ethical code, since it omitted reference to the use of anaesthesia in the majority of invasive experiments.

Cruelty to Animals Act, 1876

In 1875 two Bills for the regulation of animal experiments were put before Parliament, one emanating from the anti-vivisectionist campaigners, and the other as a response from the scientists. Neither Bill was adopted and, following calls in the popular and medical press, a Royal Commission was set up to examine the whole area. The Report of the Royal Commission recommended legislation and, in May 1876, a Bill was presented to parliament by the Colonial Secretary, Lord Carnarvon.

According to French (1975), 'the medical and scientific press expressed itself as appalled at the Government Bill'. There was scientific opposition to two sections of the Bill in particular, sections which it was thought would unduly restrict scientific work involving animals. One such section would have restricted the purpose of experiments to medical advance, rather than allowing the use of animals to further scientific knowledge in general; and the other would have promulgated a complete ban on the use of dogs and cats in experiments. The scientists' lobby was very active and took advantage of Carnarvon being called away from London, to press for amendments to the Bill. By the time Carnarvon returned to London, the scientists' position was very strong, and he had to concede several amendments, including one which broadened the permitted purposes of experiments to include scientific advance, and another which allowed scientists to perform experiments on cats and dogs provided they obtained a special certificate. The amended Bill was then 'rushed through in the dying hours of a very crowded [parliamentary] session' (French 1975), and there was little time to debate the proposed measure. The Bill received its second reading in the Commons on 9 August 1876, its second reading in the Lords on 12 August, and obtained Royal assent on 15 August 1876, becoming 'An Act to Amend the Law Relating to Cruelty to Animals', short title the 'Cruelty to Animals Act'. This Act was the first of its kind in the world. Despite its rushed passage, it stood in place for 110 years, being repealed only in 1986, to make way for new and improved legislation.

The Act related to 'experiments calculated to cause pain' to 'living animals' (that is, vertebrates, since the Act did not apply to invertebrates). The term 'experiment' was not defined in the statute, but was taken to refer to 'any procedure performed upon a living animal that is designed to find the answer to a problem or test a hypothesis' (House of Lords, 1980). All such experiments were subject to certain restrictions, which are shown in Appendix 1. Any person contravening the Act was liable to prosecution, although proceedings could only be instituted with the written assent of the Secretary of State. Inspectors, appointed by the Secretary of State, visited registered places so as to ensure compliance with the Act and assessed licence applications, deciding whether licences should or should not be awarded. On the recommendation of a Second Royal Commission (1906–12), a non-statutory Advisory Committee was set up in 1913, to advise the inspectors and the Secretary of State on individual cases of special difficulty and, in later years, on more general issues—such as the use of the LD50 test (Home Office 1979).

When the Act was passed, animals were used mainly in surgical experiments and the legislation was aimed primarily at ensuring that such procedures were carried out using appropriate anaesthesia. The term 'pain' was not defined, but in later years, as animal experimentation included a more diverse array of techniques, it was interpreted in a broad sense, so as to cover other adverse effects, including disease, discomfort or disturbance of normal health, suffered by the animals (House of Lords 1980). In very general terms, the Act could be seen as licensing people to perform specific techniques. The legislation exerted little control over the purposes for which such techniques were used. Once a licence had been granted, the techniques permitted could, in theory, be used by the licensee in almost any kind of experiment. No explicit justification was required for the use of particular techniques in particular experiments, so that, in theory, there was no limit to the amount of suffering that could be caused to animals, even for trivial purposes. In addition, since licensees did not need authorization to perform specific experiments, there was no control over scientific design. Hence the legislation did not explicitly control factors such as the number of animals used, the choice of the most suitable species, the choice of techniques which would cause the least possible suffering to the animals, the competence of licensees, or proper consideration of replacement alternatives in experiments. Strictly speaking, the Act applied to animals only when they were under experiment. There were no provisions for the care and welfare of animals outside this time. Furthermore, the Act exerted no control over the breeding and supply of animals for use in experiments.

In spite of its shortcomings, the Act was generally effective for most of its life. In recent times this was helped by the Home Office inspectors' broad and flexible interpretation of the legislation. The inspectors extended their concern to the care and welfare of animals both on and off experiment, considered the purpose and scientific validity of proposed work when deciding whether or not to grant licences, and, as discussed above, defined pain in a broad sense. From the 1960s onwards, however, there was seen to be a need to give statutory backing to this good practice and to provide a licensing framework that would require *all* work to have specific authorization, requiring explicit justification for the use of particular techniques for particular purposes and taking in regard the scientific design of proposed work.

A review of the operation of the 1876 Act was carried out by the Littlewood Committee in 1963–5. The committee's report, published in 1965, concluded that, although generally effective, the provisions of the 1876 Act had 'not matched up with modern scientific and technological requirements' and new legislation was required. There was no new law, but some of the Littlewood Committee's recommendations were implemented administratively. The pressure for change built up in the 1970s. The RSPCA and the St Andrew Animal Fund designated 1976 'Animal Welfare Year', coinciding with the 100th anniversary of the Cruelty to Animals Act. This year saw the formation of the Committee for the Reform of Animal Experimentation (CRAE), following the presentation to the Home Secretary of what became known as the 'Houghton–Platt Memorandum': a document under the signatories of Lord Houghton of Sowerby, the late Lord Platt of Grindleford and others, urging that action be taken to revise the law relating to animal experimentation. In 1978 there was the start of a campaign to 'put animals into politics' in order to achieve this aim. At the 1979 General Election, all the major parties declared their intention to reform the nineteenth century Act (Balls 1985).

Moves towards a change in the law began in 1980, with, amongst other events, the publication in April of a House of Lords Select Committee report on Lord Halsbury's Laboratory Animal Protection Bill (House of Lords 1980). In May 1980, the Home Secretary asked his Advisory Committee, which had been set up under the 1876 Act, to consider a framework for new legislation.

The Advisory Committee, in its report published in 1981, made 23 main recommendations for legislative reform (see Balls 1985). Also around this time, CRAE began discussions with the British Veterinary Association (BVA) and the Fund for the Replacment of Animals in Medical Experiments (FRAME), and, in March 1983, a joint set of proposals was submitted to the Home Secretary. The Government, anxious to meet its 1979 commitment to reform the 1876 Act, drew greatly on the BVA/CRAE/FRAME proposals in drafting a White Paper (Command 8883), which was published in May 1983. Following submission of written comments and discussion between the responsible Minister, David Mellor, and representatives of organizations spanning a wide range of opinions, a Supplementary White Paper (Command 9521) was published in

1985. The two White Papers, together, formed the basis for new legislation, which was passed in 1986.

10.2.2 Current UK legislation: Animals (Scientific Procedures) Act, 1986

Implemented in January 1987, the Animals (Scientific Procedures) Act replaced the outdated Cruelty to Animals Act. The Act is consistent with the principles of the 1985 Council of Europe Convention and the 1986 European Community Directive (see section 10.2.3), which both set minimum standards for the protection of animals in European member states.

The new Act has a broadened scope, relating not to 'experiments', but to 'scientific procedures'. Hence certain procedures which were not covered by the 1876 Act now fall within the legislation. Such procedures include the use of animals in the production of antisera or other blood products, the maintenance of disease-causing organisms and in the passaging of tumours. These uses previously were not covered since their outcome was 'known' and hence they could not be classified as 'experiments'. The Act also makes explicit that all possible adverse effects on animals are to be considered. It regulates scientific procedures which may cause vertebrate animals 'pain, suffering, distress or lasting harm' and controls the overall severity of scientific procedures, with severity being taken to include all such adverse effects.

The Act extends previous statutory control to cover, in explicit terms, the care and welfare of animals both on and off experiment. Places where scientific procedures are performed on animals must obtain 'certificates of designation', issued by the Secretary of State. Such certificates must specify a person responsible for the day to day care of the animals kept in the establishment and a veterinary surgeon, who has a statutory duty to provide advice on the animals' health and welfare. These 'named persons', if they have cause for concern over the welfare of any animal in their charge, must notify the relevant personal licensee or, if the licensee is unavailable or none exists, care for the animal themselves, killing it if necessary. The Act also provides for the issue of Codes of Practice and the first such Home Office Code lays down guidelines for the housing and care of animals. Establishments breeding and supplying animals for laboratories are also now brought under legislative control. They too must obtain certificates of designation, which must name a person in day to day care and a veterinary surgeon. Mice, rats, guinea-pigs, hamsters, rabbits and primates used under the Act must either have been bred at a designated breeding establishment or obtained from a designated supplying establishment. Cats and dogs must be obtained from designated breeders. In addition, the Advisory Committee is now replaced by a statutory committee, the Animal Procedures Committee (APC). The APC has wide powers to advise the Secretary of State on matters of policy and practice. It consists of a chairman and at least 12 members, of whom no more than half may be licensees.

Perhaps the most important new controls relate both to the purpose and scientific design of proposed work using animals. First, the purposes for which scientific procedures may be performed on animals are drawn up in more detail than in the previous legislation (see Table 10.3), with the aim of achieving 'a proper balance between two public interests; on the one hand avoiding prejudice to work which is essential in the prevention or cure of human or animal disease, or important in extending knowledge, and on the other responding to public concern for the welfare of experimental animals' (Home Office 1983).

Table 10.3 Animals (Scientific Procedures) Act, Section 5(3). Permissible purposes for scientific procedures performed on animals

The prevention (whether by testing of any product or otherwise) or the diagnosis or treatment of disease, ill-health or abnormality, or their effects, in man, animals, or plants.
The assessment, detection, regulation or modification of physiological conditions in man, animals, or plants.
The protection of the natural environment in the interests of the health or welfare of man or animals.
The advancement of knowledge in biological or behavioural sciences.
Education or training otherwise than in primary or secondary schools.
Forensic enquiries.
The breeding of animals for experimental or other scientific use.

Second, and most importantly, each individual programme of work now requires specific authorization, whereas previously an experimenter, provided he used only the techniques covered by his licence, could move on from project to project without having to obtain any further authority. The 1986 Act uses a two-tier licensing system, retaining individual (now called 'personal') licences, which authorize experimenters to perform certain scientific procedures on animals, and adding a new kind of licence, the 'project licence', which authorizes the use of such scientific procedures in a specific programme of work. Personal licensees may work on living animals only if that work has been authorized by a project licence. Such project licences are held only by individuals who have overall responsibility for particular projects, but cover all the other personal licensees engaged on that work. Personal licensees must demonstrate competence in the techniques they propose to use, and they must work under supervision until such competence has been achieved.

The requirement to obtain authorization for each and every project in which animals are used means that the legislation, as administered by the Home Office inspectors, can now exert greater control over both the purpose and scientific design of such work. As described in previous chapters, under section 5.4 of the Act applicants for project licences must now justify the severity of the procedures which they propose to use in terms of the likely benefits of the proposed work; and, furthermore, such applicants must testify that the aims and objectives of the project cannot be achieved without using protected animals: they are required to sign a declaration which states that '. . . in my opinion, no such alternatives will achieve the objectives of the project'.

The proposed work must be justified not only in these general terms but also in terms of its precise scientific design. The proposed experimental protocol must be spelt out in detail in the project licence application, including information about the number and species of animal, the specific techniques (including an assessment of their likely severity) and, where appropriate, the anaesthesia and analgesia which will be used. The Home Office inspectors can therefore scrutinize the details of the work before it is carried out and can make sure that the proposed techniques are the most humane available, that the minimum number and least sentient species of animals possible will be used and that the experimental design is likely to produce satisfactory results. The inspectors have the power to stop any project immediately, if any of the licence conditions are breached, and can also order any animal which is undergoing 'excessive suffering' to be humanely killed.

Overall, UK legislation is now explicit in its ethical control of what is done to animals in scientific research, testing and education.

10.2.3 International comparisons

The above historical analysis has illustrated one particular pattern of change in legislation controlling animal experiments. UK law has developed from simple anti-cruelty law to, in the first instance, legislation licensing people and techniques and then, very recently, to legislation controlling the precise way in which animals are used in scientific experiments (and other procedures)—with attempts at self-regulation by scientists along the way. Bearing this development in mind, we can now take a selective look at various extant laws controlling animal experiments around the world, and examine some of the advantages and disadvantages of the different systems of control.

Anti-cruelty legislation

Many countries, particularly those in the developing world (such as Egypt, Nigeria, and Malaysia) and also, for example, Bulgaria, South Africa and most provinces in Canada, have no specific legislation dealing with the use of animals in scientific experiments or other procedures (Universities Federation for Animal Welfare 1986). In some of these countries the broad animal protection laws have basic provisions concerning the use of animals in experiments, whilst in other countries only very general anti-cruelty statutes apply (Abdussalam 1984). However, as was noted in the historical context, it is very difficult to apply such laws effectively to the use of animals in the laboratory. An illustration of this difficulty was seen recently in the USA. In January 1982, the Maryland State attorney brought 17 charges of cruelty against the researcher Edward Taub, who was using monkeys in stroke research funded by the National Institutes of Health. Taub was charged under the Maryland state anti-cruelty law and not under the main US Federal law relating to the use of laboratory animals, which specifically excludes the actual research practices.§ Although initially found guilty of failing to provide adequate veterinary care for six animals, Taub was eventually acquitted, not because his cruelty was disproved, but because it was ruled that the Maryland statute was not intended to apply to the use of animals in research. The Court of Appeal interpreted the state law as referring only to 'unnecessary' or 'unjustifiable' pain to animals (just as Martin's Act of 1835 was interpreted).

The Maryland law's stated intention was to protect all animals from 'intentional cruelty', but not to make any person 'liable for

§ This law (the Animal Welfare Act, which was passed in 1966 and amended in 1970, 1976 and 1985) can only 'establish and enforce standards for care and treatment of experimental animals outside the laboratory door' and 'require covered research facilities to certify that professional standards of care, treatment and use are being followed in the laboratory, including "appropriate" use of anaesthetics and pain relievers, except when their use would interfere with the experimental objectives' (Office of Technology Assessment 1986).

normal human activities to which the infliction of pain to an animal is purely incidental and unavoidable' and there was, as in the mid-nineteenth century in Britain, a 'judicial reluctance to find cruelty in an activity of some recognized social utility' (Office of Technology Assessment [OTA] 1986). This kind of case gives weight to the claim that general anti-cruelty laws are insufficient for the adequate regulation of animal experiments. Such laws can only take effect after the event and it is clear that it may be difficult to apply them to research even in the 'worst cases', when apparently obvious and unnecessary cruelty has occurred.

Further controls: self-regulation and legislation

The use of animals in biomedical studies in countries which rely on general anti-cruelty laws to regulate such work is not insignificant: India, for example, which has only a few general regulations governing animal experiments within more general animal welfare legislation, uses upwards of one million animals a year (Abdussalam 1984). In view of the difficulties in applying the more general laws, as described above, there would seem to be a need for some further form of control.

In some countries, this control takes the form of self-regulation. Canada, for instance, has adopted a comprehensive system of voluntary self-regulation (see section 10.3); and many other countries have *ad hoc* systems of self-regulation, in which universities and other research institutions have developed their own in-house rules and principles for the conduct of research using animals (Abdussalam 1984). There is, however, great variation in the extent and rigour of this type of self-regulation.

In 1985, in response to the variation in controls governing the use of laboratory animals worldwide, the Council for International Organizations of Medical Sciences (CIOMS), a body established jointly by the World Health Organization (WHO) and the United Nations Educational, Scientific and Cultural Organization (UNESCO), published a set of International Guiding Principles for Biomedical Research Involving Animals. The eleven basic principles are summarized in Table 10.4. Aimed at countries which have no legislation relating to the use of laboratory animals, the Guiding Principles are intended to provide a 'conceptual and ethical framework' within which any country or institution can draw up more formal and specific regulatory measures (CIOMS 1985).

Within the narrower European sphere, two initiatives have been instrumental in improving and standardizing regulations concerning the use of laboratory animals. In 1985 the Council of Europe adopted a Convention for the Protection of Vertebrate Animals

Table 10.4 International Guiding Principles for Biomedical Research Involving Animals (CIOMS 1985)—Summary of the Basic Principles

I The advancement of biological knowledge and the development of improved means for the protection of the health and well-being both of man and of animals require recourse to experimentation on intact live animals of a wide variety of species.

II Methods such as mathematical models, computer simulation and *in vitro* biological systems should be used wherever appropriate.

III Animal experiments should be undertaken only after due consideration of their relevance for human or animal health and the advancement of biological knowledge.

IV The animals selected for an experiment should be of an appropriate species and quality, and the minimum number required to obtain scientifically valid results.

V The proper care and use and the minimization of discomfort, distress and pain in animals should be regarded as ethical imperatives.

VI It should be assumed that procedures that would cause pain in human beings cause pain in other vertebrate species, although more needs to be known about the perception of pain in animals.

VII Procedures that may cause more than momentary or minimal pain and distress should be performed with appropriate sedation, analgesia or anaesthesia in accordance with accepted veterinary practice.

VIII Any waiver of article VII should be made by a suitably constituted review body, with due regard to the provisions of articles IV, V and VI, and not solely for the purposes of teaching or demonstration.

IX Animals suffering severe or chronic pain, distress, discomfort or disablement which cannot be relieved should be painlessly killed, either after or, where appropriate, during an experiment.

X The best possible living conditions should be maintained for animals kept for biomedical purposes. Normally, their care should be under veterinary supervision and in any case veterinary care should be available as required.

XI Investigators and other personnel should have appropriate qualifications or experience for conducting procedures on animals and there should be adequate opportunities for in-service training.

Used for Experimental and Other Scientific Research Purposes, which lays down minimum standards for the care and use of laboratory animals (Table 10.5). Countries ratifying the convention are required to give legal force to its provisions. In addition, all EEC countries had, by November 1989, to bring their legislation into line with the provisions of a 1986 Council Directive 'on the approximation

Table 10.5 General principles underlying the Council of Europe Convention, 1985.

1. The purpose for which experiments should be permitted should be clearly defined and limited. [Use in education and training should be allowed only for the 'professions or other occupations'.]
2. Non-sentient alternatives to animals should be used wherever practicable.
3. The minimum number of animals should be used consistent with the objective.
4. Animals bred, supplied and used for experimental procedures should be cared for in accordance with the best standards of modern animal husbandry. [Breeding, supplying and user establishments to be registered; an Appendix to the Convention gives detailed guidelines on accommodation and care of animals.]
5. In the application of the controls the concept of pain should be applied in a wide sense.
6. The infliction of unnecessary pain should be avoided.
7. Appropriate measures should be taken to reduce pain and suffering.
8. There should be control over the severity and duration of any pain which is unavoidable if the object of the procedure is to be achieved [animals may be kept alive at the end of a procedure only if there is no lasting pain or distress, otherwise they must be killed humanely].

As described in Home Office (1983), p. 3, with some additional notes.

of laws, regulations and administrative provisions of the Member States regarding the protection of animals used for experimental and scientific purposes' (Directive 85/609/EEC, see Appendix 2). Hence, Spain and Portugal, which previously had no specific laboratory animal laws, have had to draft legislation in order to meet the requirements of the Directive.

Laboratory animal laws

We have seen that, in the UK, legislation developed from a system of control which licensed people to perform certain techniques for very broad purposes, to a system which now more directly controls the design of studies using animals and requires specific justification for each and every project. Countries which have specific legislation relating to the use of laboratory animals seem to have moved, or be moving, to the latter type of control. A notable exception to this trend is the United States, where legislation still specifically excludes the research practices themselves.

10.2.4 West European legislative controls

The last few years have seen a large number of changes in Western European laws concerning animal experiments. As was mentioned in section 10.2.3, all European Community (EC) members have had to bring their national legislation into line with the 1986 EC Directive, whose main provisions are shown in Appendix 2.

In its own words, the 1986 EC Directive aims to 'ensure that the number of animals used for experimental and other specific purposes is reduced to a minimum, that such animals are adequately cared for, that no pain, suffering, distress or lasting harm are inflicted unnecessarily and ... that, where unavoidable, these shall be kept to a minimum', also that 'unnecessary duplication of experiments should be avoided'. The Directive therefore requires adherence to the 'Three Rs' principles of reduction, refinement and replacement of animal use where possible. Standards are laid down for the housing and care of laboratory animals and for reducing and controlling any adverse effects suffered by animals during experiments. The Directive's provisions are the minimum acceptable requirements, and the legislation of some EC countries further restricts the use of laboratory animals. In particular, the laws of some countries are more strict about the re-use of animals (either prohibiting re-use, or allowing it only if the adverse effects of the first experiment were slight); several countries require that the purpose of proposed work (its likely benefit) be taken into account when deciding whether or not the work should be authorized; and, furthermore, some countries have made specific provisions regulating the use of animals in education and training—a use which is not covered by the EC Directive (see Chapter 9). By now, all EC countries should have enacted legislation which is in line with the provisions of the Directive. Since 1986, most EC countries have amended their laws, and some have drafted completely new statutes, in response to the Directive.

The principles underlying the second major European initiative, the 1985 Council of Europe Convention, are shown in Table 10.5. The European Community, several EC Member States and non-EC countries have signed the Convention, although, at the time of writing, only Norway and Sweden have ratified the Convention. Ratification by four Council of Europe member states is required before the Convention can enter into force. A European Communities Council Decision (COM[89]302 final-SYN 198) issued on 10 July 1989, requires all EC member states to ratify the Convention by 1 August 1990, so as to 'clear the way' for the Convention to enter into force.

Overall, Western European countries have enacted, or are moving towards enacting laws which are broadly similar in their provisions. The effectiveness of the legislation, however, will depend on the manner in which the laws are administered, and this is where the European legal provisions differ most significantly. Because the picture has been changing rapidly, and still is changing, and because each country's administrative provision is different in its details, the following notes simply give an idea of some of the main kinds of system adopted by the different countries, for reviewing and authorizing experimental work involving animals.

In the UK, as we have said, each project involving animals has to be authorized by the Home Office before it can be carried out. The Home Secretary, via the Home Office Inspectors, issues a project licence for each project, and a personal licence to each individual working under a particular project licence. Most Western European countries, similarly, require statutory review and authorization of each and every project, although the methods of review and authorization vary from country to country. In West Germany, for instance, 1987 amendments to the 1972 Law for the Protection of Animals, require that all proposals for experiments involving animals are reviewed by State Advisory Commissions appointed by State Welfare Officers. According to *New Scientist* (MacKenzie 1986) the Commissions 'are instructed only to approve work which is "indispensable", taking into account the state of scientific understanding, and the availability of alternative methods. Procedures expected to cause "pain and distress" to vertebrates are permitted only if the scientific goal is judged sufficiently important.' The Commissions will be composed mainly of scientists, doctors and veterinarians, but up to one third of members may be nominated by animal protection organizations (Neffe 1986). In Switzerland, similarly, each experiment or series of experiments must be reviewed and authorized by Cantonal Commissions, whose work is overseen by a Federal Commission on Animal Experiments—a committee of experts, including animal welfare specialists (Hampson 1987).

In Sweden, 1979 amendments to the 1944 Protection of Animals Act (amended again in 1982 and 1988) require that ethical committees be established in each of Sweden's six university regions, for the review of proposed experiments involving animals within that region. A National Board for Laboratory Animals (NBLA) advises the committees and oversees their work. Both the regional ethical committees and the NBLA are responsible to the government's National Board of Agriculture (NBA). (For more details, see Thelestam 1983 and Britt 1985). Licences are issued to individual experimenters by the NBA, following ethical review of their proposed work by the regional committees. Under 1988 amendments, all applications to perform experiments on living vertebrates are subject to ethical examination by the committees, and experiments cannot be started until this ethical review has taken place. The committees' decisions are advisory in nature: they are not binding on the researcher. Where the committee considers the proposed experiments to be of 'fundamental importance from an ethical point of view', or to involve procedures in which animals might suffer severe pain, the decisions must be reported to the NBLA (Statute Book of the NBA: *LSFS* **41**, 10 (1989)). In Denmark, until recently, there were few enough experimenters for all experiments involving animals to be reviewed and authorized by a single committee, the Animal Experiments Committee (AEC) of the Ministry of Justice, consisting of eight members, four from research institutions, three from animal protection groups, and a chairman (OTA 1986). However, the number of experimenters in Denmark has increased in the last few years and now amendments to the law require the setting up of local ethics committees (Hampson 1987).

In several other Western European countries, licences are issued to institutions rather than individuals, and responsibility for reviewing and authorizing particular projects is delegated to the heads of the institutions. In Norway, for example, most experiments are carried out under establishment licences, issued to a designated Responsible Person (RP) in the establishment. These establishment licences are issued by an Experimental Animals Board (EAB), appointed by the Ministry of Agriculture. The RP must authorize all projects involving animal experiments which are to be carried out within the establishment. Applications for RP approval must explain the purpose of the work and give details of the animals to be used, of the proposed experimental design and the likely effects on the animals involved. Copies of all applications are sent to the EAB, who can ask for further details or, if necessary, can stop particular experiments (see *Alternatives to Laboratory Animals* **15**, 260 (1988)). In the Netherlands, similarly, most of the responsibility for administering the 1977 Law for Animal Experiments is entrusted to the Heads of Institutions, to whom licences are issued by the Ministry of Public Health. In addition, regulations imposed in January 1986 require each licensed institution to establish an ethics committee composed of persons of several disciplines (including ethics) to oversee all experiments (OTA 1986). There is also a central advisory committee to the Ministry of Public Health, composed of 'persons skilled in animal experimentation, laboratory-animal science and/or animal welfare science'.

In all of the above countries, compliance with the legislation can be monitored and sanctions applied in the case of breaches of the law, since there is provision for inspection of premises where animal experiments are performed. However, not all European countries have such well-established administrative provisions. Spain and Portugal have had to draft legislation for the first time, and Greece and Italy, for example, have had to revise their laws considerably, in order to comply with the Directive. Hampson (1987) reports that, in 1987, in Greece there was 'no inspection' and in Italy, 'services were undermanned' with 'no prospected change'. How the new regulations will be administered and whether they will prove effective is still unclear.

10.2.5 USA: Problems in implementing effective legislative control

The US Animal Welfare Act (passed as the Laboratory Animals Welfare Act in 1966 and amended in 1970, 1976 and 1985) is the main Federal law relating to laboratory animals. The Act is administered by the US Department of Agriculture (USDA), and responsibility for monitoring compliance with the Act lies with USDA's Animal and Plant Inspection Service (APHIS). Under 1970 amendments, the Act covers 'any live or dead dog, cat, monkey (nonhuman primate animal), guinea pig, hamster, rabbit or other such warm-blooded animal as the Secretary of State [of USDA] may determine . . .'.

'Cold-blooded' animals (invertebrates, fish, amphibia and reptiles) are, by definition, not covered by the legislation, and 1977 regulations further restricted the coverage of the Act, by excluding rats, mice, birds, horses and other farm animals from the definition of 'animal'. The 1970 amendments provide for the humane care and treatment of animals, in that they 'should be accorded basic creature comforts of adequate housing, ample food and water, reasonable handling, decent sanitation, sufficient ventilation, shelter from extremes of weather and temperature, and adequate veterinary care' (OTA 1986). The provision of 'adequate veterinary care' includes 'appropriate use of anaesthetic, analgesic and tranquillizing drugs'.

Research facilities are required to register with APHIS. APHIS should inspect all registered facilities at least once a year and carry out follow-up inspections if deficiencies are identified. The Office of Technology Assessment (OTA 1986) reports that the 'most important standard' to be established during inspections is that of 'professionally acceptable standards of relief of pain and distress ... during and after experimentation, except where administration of anaesthetics or pain relievers would interfere with the purpose of the experiment'. Research facilities are also required to submit to APHIS annual reports to show that professional standards of care and treatment of animals and appropriate control over pain and distress during research practices have been followed.

This, however, is the limit of the Act's statutory control of research practices. The Secretary of State is not allowed to establish 'rules, regulations or orders with regard to the design, outlines, guidelines or performance of actual research or experimentation by a research facility'. It is intended that the Secretary shall 'neither directly nor indirectly in any manner interfere with or harass research facilities during the conduct of actual research or experimentation. The important determination of when an animal is in actual research is left to the research facility itself' (see OTA 1986).

Amendments sought in 1985 to strengthen the provisions of the Act. The amendments required, amongst other things, that minimum standards for the care of animals be issued (including provision for the exercise of dogs and the psychological well-being of non-human primates); that training on aspects of humane practice be provided for personnel working with animals; that reporting and record-keeping requirements be strengthened (including a requirement to provide written assurance that no suitable alternative exists in experiments which might involve causing pain to the animals); and that each research facility appoint an Institutional Animal Care and Use Committee (IACUC), which must include at least one member not affiliated with the facility, to represent the general community's interest (OTA 1986; *Nature*, 13 July 1989, p. 88).

There has since been great debate on how best to implement these requirements and APHIS has spent four years writing new regulations to incorporate the amendments. When the new regulations were first published, for consultation, in March 1989, they were strongly resisted by the scientific community. In particular, researchers argued that meeting the new standards of animal care would cost around $1 billion, and so 'would render much research unaffordable' (Holden 1989). When a revised version of the regulations was published, early in September 1989, USDA had made a 'dramatic about-turn' on these regulations, bringing the provisions on animal care into line with the current, less rigorous, Public Health Service guidelines, and conceding that 'research institutions should themselves decide how to meet the required standards' (McGourty 1989). However, the provision that IACUC's, which have the power to suspend research immediately, should include a member unaffiliated with the research facility, has been retained and this has been welcomed by animal welfare groups (McGourty 1989). As noted in Chapter 5, early in 1990 the US Office of Management and Budget rejected the amended regulations because of White House opposition (Anderson 1990); and they are being rewritten in 'performance' rather than 'engineering'-based language.

In addition to the Animal Welfare Act, the 1985 Health Service Extension Act applies to a large number of US research institutions. The Act makes two features of Public Health Service (PHS) Policy statutory requirements for PHS-funded research. Each institution funded by the PHS must establish an IACUC and must submit a written assurance to the National Institutes of Health (NIH) that it will comply with the NIH *Guide for the care and use of laboratory animals*. The Health Service Extension Act also contains provisions for the development of alternative research methods—the first explicit description in Federal law of the concept of alternatives to animal use. According to the OTA (1986), 'taken together, the . . . PHS Policy and Federal statutes bring the overwhelming majority of animal users in the United States under the oversight of Institutional Animal Care and Use Committees'. Under PHS Policy, the IACUC's have the functions shown in Table 10.6.

The US system of Federal regulation of animal research has been criticized since its inception. The OTA (1986) describes USDA's approach as 'cautious and literal with regard to research facilities', tracing this position to two influences: first, '. . . the [Animal Welfare] Act and its legislative history make clear Congress' desire to avoid any entanglement in the actual conduct of research' and, second, 'both the legislative and executive commitments of funds and personnel for enforcement have never lived up to the expectations of those who believe the primary mission of the existing law to be the prevention or alleviation of animal suffering'. USDA itself has reported that failure of research facilities to register and report is a significant problem and that there is 'no effective system for detecting research facilities that use laboratory animals without being registered' (OTA 1986). The choice of APHIS, whose main concern is disease control in domestic plants and animals, as the enforcement agent is also open to criticism. APHIS, it is argued, has 'neither the staff nor the budget to secure compliance even with the limited provisions of the Act relating to

Table 10.6 PHS Policy. Functions of IACUCs

With respect to PHS-funded activities to:
1. review at least annually the institution's program for humane care and use of animals;
2. inspect at least annually all of the institution's animal facilities;
3. review concerns involving the care and use of animals at the institution;
4. make recommendations regarding any aspect of the institution's animal program, facilities or personnel training;
5. review (and approve, withhold approval or require modifications to) sections of PHS applications or proposals or on-going activities—checking accordance with the Animal Welfare Act and NIH *Guide* as well as assuring:
   - minimization of pain and distress caused to the animals;
   - use of appropriate sedation, analgesia and anaesthesia;
   - humane euthanasia for animals in severe or chronic pain or distress which cannot be relieved;
   - appropriate housing and care for animals;
   - provision of veterinary care as necessary; and
   - appropriate qualifications and training for personnel carrying out procedure.
6. be authorized to suspend an activity involving animals.

such a massive research activity' (OTA 1986). Furthermore, the exclusion of rats and mice from the Animal Welfare Act's provision means that the majority of animals used in research in the US are not afforded Federal legislative protection, and, even for the species which are protected, there are at present no rules, regulations or standards concerning the actual practice of research involving the animals.

In addition to the Federal law, State or other, more local, laws may regulate the use of animals in research. According to the OTA (1986), 21 State jurisdictions regulate research specifically. These laws mostly concern the procurement of animals (especially dogs and cats) for research, and individual States seem 'as reluctant as Congress to go behind the laboratory door' (OTA 1986). However, a bold step was taken recently by the Cambridge (Massachusetts) City Council, in establishing the United States' first local law relating to the care of laboratory animals. This local ordinance, passed in June 1989, requires all research institutions in the city to conform to Federal laws and regulations concerning the care of laboratory animals (at present most of the city's laboratories follow Federal guidelines on a voluntary basis—see Shulman

1989*a*). The law also *extends* the Federal provisions to include rats and mice, birds, fish, reptiles and amphibians and establishes a 'Commissioner of Laboratory Animals', who has the right to conduct surprise inspections of research facilities within the city. Whether local laws will follow in other parts of the USA remains to be seen. However, the law appears to have satisfied the Cambridge community since, according to *Nature* (Shulman 1989*b*) animal-rights supporters see 'a victory in the precedent set by the new law and . . . the size of the animal population affected' and researchers in the city also 'say they can live with the city council's decision'.

10.3 VOLUNTARY SELF REGULATION

At present, Canada is the only country in the world to have adopted a national system of voluntary self-regulation of laboratory animal use. This system is not required by any legislation: in all provinces except Ontario the only legal control which can be applied to the use of animals in laboratories is Section 402 of the Canadian Criminal Code, which relates only to general cruelty to animals (see Rowswell 1988). Although it is not a system of legal control, the Canadian system attempts to regulate research using animals in a legalistic manner and it is interesting to examine this method of control in the light of the more formal legal controls already described.

The self-regulatory system is run by the Canadian Council on Animal Care (CCAC)—a national advisory body which establishes guiding principles, oversees their application and advises the provincial governments. The Council has 21 members: eight from national associations of higher education; five from departments of Federal Government; four from national agencies providing research grants; two from the Canadian Federation of Humane Societies; one from the Pharmaceutical Manufacturers' Association of Canada and one from the Canadian Association for Laboratory Animal Science (this last member was added to the list in 1988).

The CCAC's activities are organized around its major publication, *Guide to the care and use of experimental animals*, which consists of two volumes, first published in 1980 and 1984, and frequently updated. The CCAC's programme covers 'any vertebrate animal that is separated from its natural environment and utilized in research, teaching and testing'. It requires the formation of an Animal Care Committee (ACC) in every institution (that is, university, government department or pharmaceutical laboratory) in which experimental animals are used. The committees should be informed of all research procedures in which animals are used, and no work should be carried out without their prior approval. The total complement of each ACC is 'determined by the needs of the institution', but a member or consultant should be a veterinarian, biologist or animal scientist experienced in animal care and use. The Committees may, if they wish, co-opt lay members and 'over 100' such members serve on the ACCs (see March 1988 update to CCAC *Guide*).

The ACCs review protocols 'from the viewpoint of the ethical aspects of the procedures and the acceptability of the methodologies proposed, particularly if they may appear to involve pain or distress'. The committees' remit includes responsibility for co-ordinating and reviewing the activities and procedures that involve the use of animals; the standards of facilities, equipment and care for animals; the procedures for ensuring the health and welfare of experimental animals; the training and qualifications of personnel that are engaged in the care of animals; procedures for prevention of unnecessary pain including the use of anaesthetics and analgesics and procedures for euthanasia of animals. They are not, however, concerned with the scientific merits of the proposed work, on the basis that this has already been assessed by 'government departments and granting agencies'. An ACC is empowered both to 'stop any objectionable procedure' and to destroy humanely any animal which is experiencing pain, distress, etc. which cannot be alleviated. The CCAC *Guide* stresses the need for proper consideration to be given to alternatives, stating that 'animals should be used only if the researcher's best efforts to find an alternative have failed', and 'those using animals should employ the most humane methods on the smallest number of appropriate animals required to obtain valid information' (see October 1989 update to CCAC *Guide*).

Compliance with the *Guide*, and the functioning of the ACCs, is overseen by assessment panels chosen by the CCAC. These panels should visit institutions at least every three years in order to assess the effectiveness of the institutional ACC, the facilities and animal care practices and procedures. Each panel usually consists of three scientists and one member of the Canadian Federation of Humane Societies. The scientific members are drawn from a pool of some 95 individuals with experience in laboratory animal experimentation and care. All the panelists serve voluntarily and do not receive payment, except expenses.

It was noted that the development of legislation in the UK

included at least two attempts by scientists to establish voluntary systems of self-regulation—neither of which met with much success. Indeed, it has been argued that the lesson from history is that such self-regulation can never work, since science is essentially 'self-serving'. Rupke (1987) has written that, 'today, as much as a century or so ago, scientific curiosity and professional ambition easily over-ride concern for animal welfare, if legal enforcement of the latter is not assured'.

However, according to the CCAC, the Canadian system is highly successful. A 1988 update to the *Guide* states the belief that the CCAC programme

is successful in that Canadian research and teaching institutions enjoy

1 Animal care facilities as good as, or better than most countries in the world.
2 A system of local animal care committees, that are only now becoming a requirement in other countries (e.g. USA).
3 An increased level of awareness and sensitivity to the ethics of animal experimentation amongst scientists and investigators.
4 Effective liaison with the animal welfare movement as exemplified by the Canadian Federation of Humane Societies.

The preface of the 1980 CCAC Guide (Part 1) gives reasons for preferring voluntary over statutory control:

It is believed that the achievement of optimal conditions for animal care and use is more likely to be approached through the CCAC program which is considered preferable, under Canadian constitutional conditions, to attempting national legislative control. Legislation has always tended to enforce only minimal acceptable standards. Historically, it has failed to involve or satisfy the concerns of the animal welfare movement. Furthermore, the implementation of legislation through a national inspectorate would be difficult where institutions are sparsely scattered and frequently isolated over many thousands of miles. Additionally, no government inspectorate could expect to enjoy the benefit of the scientific expertise presently provided, without honorarium, to the CCAC assessment program.

Although the regulation is voluntary,

institutions whose practices or facilities are in continued non-compliance with CCAC requirements face withdrawal of all funding by Canada's major granting agencies ... Since the inception of non-compliance status [in 1984] ... seven institutions have been so designated. After notification, the CCAC carefully monitors their progress towards implementing its panels' recommendations. Construction of new facilities or the refurbishing of those in use commonly result.

The system, however, has been criticized recently (see, for example, Clark 1987; Tatum 1988). Critics have claimed that the voluntary controls are effective in protecting the scientists, giving them the freedom to do as they wish, but ineffective in protecting the animals. There seems to be variation in the effectiveness of the ACCs, with some meeting very infrequently, or even not at all, and in some cases it has been alleged that the chairman routinely signs all research proposals without reference to the committee. Furthermore, it is suggested, the peer review system makes it very difficult for the ACCs to block objectionable research—it is argued that it is 'impossible' for a local ACC, composed almost entirely of researchers, to stop the research of one of their colleagues (Tatum 1988). Even though the voluntary nature of the control is reinforced by the threat of sanctions in the form of removal of all grant funding, several authors have argued that the system is now in need of some legislative back-up (see also Hampson 1987). Clark (1987), for instance, has suggested that the lack of legal control makes the protection of animals, via the work of the ACCs, seem unimportant:

> If we really wanted to be sure the animals were protected, we would not leave their protection to the conscience of individual investigators ... If we want to protect laboratory animals from neglect or abuse, we must insist on strong laws that can be enforced. In collecting taxes, the government does not rely on morals or attitudes or an education or gentle persuasion; it has forceful means—because collecting taxes is important.

All of the criticisms listed above might be, and indeed have been, applied to legislative systems of control. Nevertheless, it seems that implementation of legislation might help to improve public confidence in the system of control. There appears to some pressure for legislation in Canada (notably from the Canadian Federation of Humane Societies), though whether a law will result remains to be seen. If, however, Canada eventually gives statutory backing to the system of voluntary control then the country will be following what seems to be a common historical pattern.

10.4 RESEARCH REVIEW COMMITTEES

10.4.1 Types of research review committee

In reviewing legal (and legalistic) control of experiments using animals, we have seen several examples of the use of research review (or ethics) committees to discuss the ethical aspects of proposed work involving animals, and in some cases to decide whether or not the work is acceptable. These committees are, in the main, established within scientific institutions. They usually consist of several members of the institution and, in some cases, one or two members from outside the institution. The internal members might include working scientists, animal technicians, the institutional veterinary surgeon, administrators and perhaps a student representative. The external members might include, for example, a practising veterinary surgeon, an animal welfarist, a philosopher, and a member of the general community in which the institution is based.

We have encountered such committees in two different situations:

(1) in countries which have no legal control over the use of laboratory animals. Some of these countries have *ad hoc* systems of self regulation, which may include institutional ethical committees, and one country, Canada, has a national system of voluntary self regulation based on institutional Animal Care Committees;
(2) in countries where there is a legal requirement for research proposals to be reviewed by a committee, as, for example, in Denmark, the Netherlands, Sweden, and the USA.

It is also possible that such committees might be established in a third situation:

(3) in countries where there is legal control and where a government authority takes decisions about what should and should not be allowed, as in the UK, for example. Although in such countries there is no statutory requirement for institutional research review committees, some institutions may, voluntarily, establish such committees. These committees may carry out ethical review of research proposals, or of research in progress, and may establish ethical guidelines or give general guidance to the researchers within an institution.

In addition to these local committees, two further types of review committee are common: journal committees, which advise the editor of the journal whether or not on ethical grounds a particular paper should be published; and learned society committees, which may issue guidelines and give advice to workers contemplating particular experiments in the society's field of interest, and also take ethical decisions about what might and might not be published in the society's journal. Journals and learned societies may play an important non-statutory role in the discussion and taking of decisions on ethical issues involved in the use of laboratory animals. This role is considered further in the second part of this chapter.

10.4.2 Possible roles for local research review committees

In some countries, such as Canada, the Netherlands, Denmark and Sweden, local committees are responsible for deciding whether or not, on ethical grounds, particular proposals for scientific work using animals should be allowed to proceed. In other countries local ethics committees may not play such a decisive role, either because the law requires that such decisions are made by the relevant government authority (for example, in the UK the Home Office inspectors make these decisions, except in marginal cases) or because they are *ad hoc* committees in countries with no special legal or legalistic control of laboratory animal use. In both situations local committees would have no powers to impose their decisions on the researchers, so that, whilst they may advise, they cannot formally decide whether or not particular pieces of work should take place. In these cases (and also in cases where the committees can take binding decisions) local research review committees might play a number of other important roles.

1. Such committees could provide clearly identified forums for ethical discussion of proposed research (or research in progress), which might otherwise be lacking within an institution. The committees might also engage in post-study review and so built up valuable case history experience which would help in future discussions and decisions.

2. In the context of ethical discussion, the committees might also be seen as educative bodies: they might play a part in informing the consciences of researchers, by providing a variety of independent opinions on a particular protocol or aspect of animal use, and so helping investigators to develop a balanced and humane attitude towards their use of animals. Although it might be argued that scientists already have well-informed consciences, it should be noted that it has sometimes been found necessary for ethics committees to fill this role in the field of medical research involving humans. Care should be taken, however, to ensure that the committee does not have an opposite effect, providing a refuge from conscience for the researcher ('if the committee has said it's all right, then it must be all right!').

3. Lay representation on institutional committees could provide reassurance about what is going on in that institution within that community. One example of the way in which review committees might help to reassure the public concerns the vetting of all applications for project licences for cosmetics testing by the UK Animal Procedures Committee (a kind of national research review com-

mittee). All such applications are referred to the APC because the Government has taken note of public anxiety about cosmetics testing. Lay representation can, however, be a difficult business—choosing the right person is not always easy. It is not possible to describe the ideal lay member or to give general criteria which would assist in a suitable choice: it is clear, however, that individual committees would need to think very carefully about any choice of lay representative. If, for instance, an ardent anti-vivisectionist was elected to the committee this could hold up research (a problem which has been encountered in Sweden, where there is a strong anti-vivisection movement, see Britt 1985).

4. Research review committees could be concerned with all forms of animal use within an institution, not only those covered by local laws and codes of practice. In the UK, for instance, ethics committees could consider the use of invertebrates in scientific procedures, or the killing of animals (by Schedule 1 methods) in order to obtain parts of their bodies for *in vitro* work, neither of which uses requires a licence under the Animals (Scientific Procedures) Act. The committees might also have the option of setting standards higher than those of the local laws or codes.

5. Such committees could also concern themselves with the wider aspects of animal use within the institution—making sure, for instance, that the animal care and breeding facilities are meeting the needs of both the researchers and animals; co-ordinating animal use between different research groups within an organization (so that, for instance, organs and/or tissues for *in vitro* work can be taken from animals being used in terminal studies); providing a focal point for the exchange of information on local facilities and expertise, so helping to ensure that researchers are able to carry out procedures using the best available equipment and techniques; and acting as a resource centre on animal welfare issues.

10.4.3 Some conclusions regarding research review committees

In countries where there is no legal control, local research review committees could be established as a first way forward, providing an opportunity for discussion and review of research proposals, so as to help sort out a coherent ethical position, delineating what is and is not thought to be acceptable practice in that country. Learned society and journal ethics committees could also play a valuable part in this process. As discussed above, in some countries research review, or ethics, committees are already a part of the system of statutory control, where they play a central role in deciding whether particular pieces of research should or should not be allowed. In countries, such as the UK, where such decisions are taken by representatives of the government, local committees could have an important complementary role. They could, for instance, act as a preliminary screen, by reviewing research proposals before application to the Home Office, and could consider the uses of animals which are not covered by current legislation, as well as giving some local accountability and public relations coverage.

10.5 AN OVERVIEW

10.5.1 Ethics and the law

Laboratory animal protection laws are all based on the premise that, in certain circumstances, it is morally acceptable to use animals in scientific procedures for the benefit of humans or other animals. As we have seen, in the past such legislation was mainly concerned with improving the treatment of animals in laboratories: stopping cruelty, improving animal husbandry, ensuring the use of anaesthetics where appropriate, and promoting the refinement of procedures. Decisions about what ought and ought not to be allowed were taken mainly on the basis of the likely cost to the animals used. In recent years, however, there has been a general move towards a broadening of the moral perspective, so as to consider not only what is to be done to the animals, but also the purpose for which the animals are to be used. The legislation of some, but not all countries, now requires an ethical weighing of the likely benefits against the likely costs to the animals when deciding whether work should or should not be allowed. Thus the laws of some European countries, including the UK, incorporate this requirement. The US Animal Welfare Act, however, has no such provision—and explicitly refrains from 'interfering' in the actual research. Similarly, the Canadian ACCs are mainly concerned with the costs side of such ethical decisions, and are not required to consider the scientific merits of proposed research.

In terms of the likely costs of proposed research, most laws now incorporate a further requirement: it must be shown that the chosen methods of achieving the aims of the research are those which will cause least animal suffering. EC laws, for instance, require that the minimum number of animals needed to attain the scientific objective, and the least sentient species possible, be used. Furthermore, it must be shown that there is no valid replacement alternative method available, and hence that the only way of achieving the aims of the research is by the use of sentient animals. Thus, laws regulating the use of animals in experiments are making explicit the factors which ought to be taken into account when coming to a decision about what is and is not to be allowed. In some cases these factors refer only to the likely harm caused to the animals, and in others, they refer to both the costs and likely benefits of the proposed research.

Some countries' laws, however, go further than this and specifically prohibit certain kinds of research. Here, the laws regard certain practices as morally unacceptable. Some laws prohibit particular scientific procedures, because the suffering which they impose (or are likely to impose) on the animals is thought too great ever to justify their use. In the UK, as in most other countries, the use of neuromuscular blocking agents, such as curare, instead of, or apart from, anaesthetics is prohibited. In Canada, the CCAC has made a list of 'categories of invasiveness in animal experiments' which includes a category (E) for experiments which are considered 'highly questionable or unacceptable, irrespective of the significance of the anticipated results'. Thus, the 'use of muscle relaxants or paralytic drugs without the use of anaesthetics', or 'burn or trauma infliction on unanaesthetized animals' might for example, be included in this category (Update to CCAC *Guide*, October 1989). Some countries have legal provisions which prohibit the use of animals for certain scientific purposes. In the UK the use of living animals to gain manual skills, other than in microsurgery, as well as the use of animals in scientific procedures in primary and secondary schools, is prohibited. It is thought that manual skills can be acquired without the use of living vertebrates, and the use of animals in scientific procedures in schools is considered both unnecessary and undesirable. (Neither of these uses of animals is prohibited in the USA). Other countries have prohibited whole areas of research and testing, presumably because the likely benefits of the work are thought to be insufficient ever to justify the use of animals. Thus, in response to public pressure, 1987 amendments to West Germany's 1972 Law for the Protection of Animals have prohibited the use of animals in the development and testing of weapons, and in the testing of tobacco products, washing powders and decorative cosmetics (Neffe 1986).

There are, however, dangers inherent in this kind of blanket ban. First, such areas of research present complex moral and practical issues (see Chapter 8, on Toxicology and Toxicity Testing), so that certain, but perhaps not other, projects within each area might be justified. A complete ban does not allow for any flexibility of approach in deciding what should and should not be allowed. Second, and perhaps of greater concern, is that investigators may still want to carry out such research and may therefore either disguise the real purpose of their work (perhaps by reclassifying it into another allowed area), or export the work to another country, which has no ban, and which might have lower standards of animal care. In either case the national law loses control over this use of animals. In the UK, the 1986 Act does not impose a blanket prohibition on any particular type of work but rather seeks 'rigorously to control all experimental and scientific work involving living animals, each proposal being considered on its merits' (Home Office 1988, p. 8). However, the Home Secretary, in recognition of public concern, has decided routinely to refer all project licence applications for microsurgery training, the testing or use of tobacco and tobacco products and cosmetics testing to the Animal Procedures Committee. This is

> ... in recognition of the particular controversy such work arouses; in the case of cosmetics and tobacco products because the nature of the products is not fully accepted by public opinion as sufficiently important to warrant the sacrifice of animal life in production or safety testing; and in the case of microsurgery techniques, because any use of animals for the acquisition of a manual skill had previously been prohibited under the Cruelty to Animals Act 1876 controls (Home Office 1988, p. 8).

The Home Secretary is empowered to apply this referral system to other kinds of research as he thinks appropriate, and in 1987 the list was extended to include work involving procedures of substantial severity carried out on non-human primates. This way of responding to public concern ensures that the law retains control of such areas of research, and at the same time provides an opportunity for ethical deliberation aimed at establishing firm criteria for deciding what, if any, of this kind of research, should be allowed. The Animal Procedures Committee's reports (Home Office, 1988, 1989, 1990) give an indication of its approach to such casework and of the way it handles the issues involved.

10.5.2 Law, education, and society

In some areas of life there are laws which are aimed at ensuring certain precise things: on building sites, for example, Health and Safety legislation ensures, amongst other things, that workmen wear safety helmets. In other areas the law has a different, more educational, impact: the laws against racial and sexual discrimination, for example, aim both to stop discrimination and, at the same time, to educate people to set aside any prejudices they may have.

Laws regulating the use of laboratory animals may perhaps be considered to fall into the latter category. Such laws aim both to restrict the use of animals according to a set of defined criteria and to educate people so as to help them to make their own decisions about what is and is not acceptable. The legislation is helping to change society's attitudes towards the use of animals in scientific procedures. It seems important, however, that, in framing such legislation, a balance is struck between the leading and following roles of the law. Whilst the law can play a vital role in moulding social values, its provisions should also take into account the current attitudes and values of the society (including both the professionals and the public) to which it is to apply (see Brown 1989, for further discussion).

In the UK, one hundred years of experience with the 1876 Cruelty to Animals Act set the scene for the more searching provisions of the 1986 Act. Scientists had already accepted the constraints of the 1876 Act, and had become used to thinking (perhaps without realising it) about some of the moral aspects of their use of animals. The 1876 Act had generated a particular moral atmosphere, and the climate was right for the implementation of more explicit ethical control, under the terms of the 1986 Act. In the USA, on the other hand, Federal legislation (the Animal Welfare Act) seems to have met with only limited success. As the vigorous debate concerning the implementation of 1985 amendments to the Act has shown, there is enormous difficulty in gaining agreement on the terms of legislation pertaining to the use of laboratory animals in the USA. This lack of consensus about what is thought proper is perhaps not surprising, given the vast and diverse population of the USA, and given the fact that animal research continued, until relatively recently, with very little legal or other control.

If legislation is to take the lead, there would seem to be a need for a programme of education to help in developing a moral consensus. Such education, indeed, seems to be taking place in the USA, where efforts are focused particularly on training the members of IACUCs, including the lay members, in ethical aspects of animal experiments.

10.5.3 Public reassurance

The law can be viewed as attempting to ensure that researchers are both responsible and accountable for their actions. We have seen that the educational impact of the law can help to promote responsibility; but what of accountability? A common criticism of laboratory animal protection laws is that they protect the scientists as much as, or even more than, they protect the animals. This view may arise because the regulatory process is seen as being highly secretive, and it may be reinforced by the lack of prosecutions under the laws. Whilst the law must be flexible—so as not unnecessarily to fetter people operating within its framework—it must also be enforceable, and, from the public's point of view, must be *seen to be* enforceable.

There are several ways in which the public might be assured that the law really is protecting the animals. One way might be to pursue more prosecutions under the law. This would help to assure people that the legislation has 'teeth'. In the UK, however, (and probably elsewhere in the world) there seems to be a reluctance to bring transgressors before the courts. Although, over the years, a number of infringements of the 1876 Act and now of the 1986 Act have been brought to the attention of the Director of Public Prosecutions, there have been no prosecutions under either Act.¶ The number of infringements and action taken are recorded in the annual Home Office Statistics, which are available to the public. Examination of these records over the period 1978–89 (latest available Statistics) shows that all those infringing the Act were 'admonished' or 'warned about their future conduct'. In each year except 1987 and 1989, some cases were referred to the DPP, who, after consideration, chose not to prosecute. In some cases licences were revoked and in one case the DPP instructed the police to issue a Caution (see Table 10.7).

As criticism of the Canadian system illustrates, the public may have little faith in a system that is not seen to be enforced. In a similar way, the lack of prosecutions under UK law has caused some public concern. Indeed the UK statutory advisory committee, the APC, has stated in its report for 1987 that it is 'concerned ... that, where appropriate, people who offend against the Act should be prosecuted' (Home Office 1988, p. 13). The Committee endorsed a 1986 letter from the Advisory Committee under the 1876 Act to the Secretary of State, which included the following points:

Administrative sanctions can no doubt be effective, but discretion to enforce the criminal law should not mean that it is never enforced in the

¶ In 1978, a licensee was successfully prosecuted for causing unnecessary suffering to laboratory animals by carrying out unlicensed work, in which canaries, mice, goldfish and rats were used as live prey in a study of the predatory behaviour of cats and kittens (Scottish Society for the Prevention of Vivisection 1978). However, for technical reasons, this prosecution was brought under the Protection of Animals Act (Scotland), 1912, rather than under the 1876 Act (Home Office 1988).

courts. . . . The Animals (Scientific Procedures) Bill rightly makes provision for substantial penalties for the most serious offences involving animals used in scientific procedures. We are concerned that the new legislation will fail to command the respect it deserves, unless there is a willingness to bring its full force to bear when the circumstances warrant it. . . . It is the hope of the Committee that the official procedures for dealing with serious infringements may be improved so that the possibility of prosecution is more vigorously explored. We believe this will be of particular importance for both the public credibility and the effectiveness of the legislation now before Parliament, which has your Committee's full confidence and support (Home Office 1988, pp. 13–14).

It is clear that there may be a gradation in the culpability of offences under laboratory animal protection laws, from the most minor, inadvertent infringements, to more substantial neglect or irresponsibility and even deliberate evasion of the law (see Home Office 1988). It would seem important that this gradation of offence is matched rigorously with a gradation in penalty, including prosecution for the most serious offences. For other, less serious, offences, the removal of licences may be a sufficiently severe sanction, since this can mean that offenders may never again be allowed to perform experiments involving animals.

Another way of increasing public confidence in the legislation might be to increase public participation in the regulatory process, by having lay members on ethics committees. Although, in view of the threats of extremist anti-vivisectionists, it is understandable that many research institutions are wary of providing public information about their use of animals, such lay representation might, if carefully organized, help to allay public fears and suspicions about what goes on behind the 'closed doors' of such establishments. This kind of reassurance is likely to work best on a local basis, provided that there is good communication between an institutional ethics committee and the community in which it is based.

A further possibility would be to provide some possibility for external, impartial review of the way in which the legislation is administered. Looking at the UK system as an example, the Animals (Scientific Procedures) Act 1986 gives researchers a statutory right of appeal to the Secretary of State against refusal or revocation of permission to carry out scientific work using animals. The possibility of appealing against a decision to allow certain scientific work using animals, however, is more limited.

On a day-to-day basis, those caring for the animals involved in scientific procedures (the technicians, the Named Veterinary Surgeon and the Named Person in Day-to-Day Care), as well as

Table 10.7 Infringements and prosecutions under the UK Cruelty to Animals Act 1876 and the Animals (Scientific Procedures) Act 1986. Infringements of the Acts or licence conditions usually fell into one of several broad groups: 1. Carrying out unlicensed work. 2. Continuing to work after the expiry of a licence. 3. Exceeding the authority of a licence by using techniques or species of animals without permission, or carrying out work at a place other than that authorized by the licence. 4. Failing to take proper care of an experimental animal.

Year	No. of infringements	No. referred to DPP	Outcome[a]
1978	22	5	1 prosecution, under Protection of Animals (Scotland) Act;[b] 1 other case: licence revoked[c]
1979	28	13	1 case: licence revoked
1980	25	8	1 case: licence revoked 1 case: certificate disallowed
1981	16	4	—
1982	17	2	1 case DPP instructed Police to issue a caution
1983	55	32	1 case: licence revoked
1984	23	5	2 cases: licences revoked 1 case: 'authority curtailed'
1985	32	5	—
1986	9	2	—
1987	9	0	1 case: licence revoked
1988	20	2	2 cases: licences revoked
1989	14	0	1 case: licence revoked

[a] People infringing the Act were usually 'admonished' or 'warned about their future conduct'. 'Outcome' here refers to other positive action taken.

[b] See footnote to p. 279.

[c] This case is not included in the 22 infringements. The licence was revoked because, according to the 1978 Statistics, 'the Secretary of State decided that the licensee was no longer a fit person to perform experiments on living animals ... as a consequence of his conviction under the Protection of Animals Act 1911 for offences which were unrelated to work under the 1876 Act'.

other members of the particular institution, can take any complaint to the Home Office Inspector, who will deal with it as he or she thinks fit. On the other hand, members of the public who wish to complain about a particular use of animals, can do so through a parliamentary route. They can write to their Member of Parliament, who may then put the complaint as a formal parliamentary question, to which the Minister for the Home Office will reply. This procedure, however, is lengthy and cumbersome and the ministerial replies are usually non-committal. The issues arising from such questions, and from other public expressions of concern can be taken up by the statutory advisory body under the Act, the Animal Procedures Committee (APC). The APC can investigate and advise the Secretary of State on any aspect of the legislation and administration, and it can initiate its own lines of enquiry. The APC, therefore, is in a position to take up appeals, on behalf of the animals, from those outside the legislative system. Thus, as was noted earlier, the APC advises the Home Secretary on all applications for licences for microsurgery training, the testing or use of tobacco and tobacco products, cosmetics testing and work involving procedures of substantial severity carried out on non-human primates. The APC has also considered other matters of public concern (including the use of animals in behavioural and psychological studies, the use of non-human primates in research, the use of animals to obtain manual skill and the handling of infringements under the Act), and has published reports on these matters (Home Office 1988, 1989, 1990).

Nevertheless, although its right of enquiry is absolute under the law, the APC may be regarded as being 'within the system', since its members are appointed by the Secretary of State. There might therefore be a need for some further means of review, which is seen to be totally independent. What form a further system of review might take, and whether or not such a system would be practical, is not clear. However, it seems likely that the public might be reassured by a system which would provide a formal route for complaints about the implementation and administration of the Act. One possible model for such review might be the ombudsman principle, already successfully applied in central and local government, in the Health Service and in a variety of other areas, including banking and insurance.

'Ombudsman' is a Scandinavian word which, in its special sense, means a commissioner who has the duty of investigating and reporting to Parliament on citizens' complaints against the Government. He has no legal powers except powers of inquiry. His effectiveness derives entirely from his power to focus public and parliamentary attention on citizens' grievances. This, however, is a powerful lever—where a complaint is justified, an ombudsman can often persuade the government department to modify its decision (or, where relevant, to pay compensation). Such an ombudsman was first instituted in the UK by the Parliamentary Commissioner Act of 1967. The 1967 Act says that the ombudsman is to investigate maladministration, which is not defined in the statute, but was described in a 1966 parliamentary debate as including 'bias, neglect, inattention, delay, incompetence, ineptitude, arbitrariness and so on'. It is now generally accepted that the ombudsman's remit is to investigate 'any action or inaction by government which he feels ought to be criticized, including anything which is unreasonable, unjust or oppressive'. Complaints to the ombudsman may be made by any member of the public. The ombudsman receives the complaint informally, and, on deciding that the complaint should be investigated, enters the relevant government department, speaks to officials, reads the files and finds out exactly who did what and why. The findings are then published in a public report. There is no formal procedure, nor any possibility of legal sanction, involved. The ombudsman exerts pressure for change, or the reversal of a decision, where deemed justified, simply through the publication of the impartial report (see Wade 1988).

The possibility of instituting an 'animal ombudsman' for complaints in relation to the legal control of the use of laboratory animals might be worthy of further consideration. Such a review procedure might provide concerned members of the public with a mechanism for making complaints, on the animal's behalf, against decisions which they consider to be unjust, or ethically unacceptable. The publishing of publicly available reports on cases which the ombudsman investigates might help to strengthen confidence in the legislation.

10.5.4 When things go wrong

In practice, the effectiveness of the various legislative controls, whatever they say on paper, depends on how well they are administered and enforced. Sadly, things can go wrong even when stringent controls exist.

Section 10.5.3 gave details of the infringements under UK controls, and the action taken by the Home Office, between 1978 and 1989. The Animal Procedures Committee's report for 1989 notes that for offences other than technical infringements, 'the Home Office has increasingly adopted the practice of revoking licences or imposing additional and more specific or stringent licence requirements on infringers' (Home Office 1990).

In May 1990, two leading British newspapers disclosed evidence that rabbits were operated on under inadequate anaesthesia in a London laboratory (see *Alternatives to Laboratory Animals* **17**, 284–6 (1990) for further details). A Medical Research Council enquiry found that those involved (the project and personal licensees, the Certificate Holder, the Named Veterinary Surgeon, the Home Secretary, and the Home Office Inspectors) on this occasion failed to fulfil their statutory duties under the 1986 Act. Such an event does not imply that the statutory provisions themselves are inadequate. The event, although likely to be an isolated incident, does, however, give pause for all concerned with animal experiments to reflect on how well they are fulfilling their various statutory and moral responsibilities.

No system of control, no matter how effective generally, can be perfect. However as the editor of ATLA has noted, whenever and wherever failures occur, 'honest enquiries can reveal lessons to be learned', so that the implementation of the controls can be strengthened and 'such incidents avoided in the future'.

Part 2: Other controls on research involving animal subjects

In addition to the legal and legalistic national controls described in the previous sections, there are other less formal, less direct, ways in which the use of animals in scientific procedures may be regulated. All of these means of control come from within the scientific community. They all involve some form of peer review, in which the scientists themselves consider the ethical dimensions of the scientific use of animals, and decide what might and might not be acceptable practice.

10.6 PEER REVIEW WITHIN INSTITUTIONS

As noted in section 10.4, scientists may engage in informal or formal discussion and review of research in progress or planned. Informal discussion takes place routinely, as part of the scientific process, and there may also be discussion within a more formally constituted forum, such as a Research Review Committee. This kind of peer review may lead to the refinement of research protocols (helping to improve scientific design and reducing the harm caused to animals), and may help in deciding whether or not the protocol constitutes good, or the best, practice. In this context, discussion with those experienced in laboratory animal procedures (laboratory animal scientists, the technicians and others responsible for the care of the laboratory animals, the institution's veterinary surgeon and, in the UK, the Home Office Inspector) can be particularly useful in promoting refinement of research involving animals.

10.7 REVIEW BY GRANT-GIVING BODIES

Applications for financial support for proposed research may also be subjected to peer review. It has been argued that, in the UK at least, the system of allocating grants for research plays a large part in determining the quality of the science (Bateson 1986). There is great competition for limited funds and this plays a part in ensuring that only the best research receives funding. In each of the five years 1983 to 1987, fewer than 50 per cent of applicants for Medical Research Council grants were successful (MRC Annual Report 1986/7). Indeed, the Trustees of the Nuffield Foundation (a charitable organization which awards grants for Medical Research and has a policy of supporting 'speculative or preliminary investigations of a kind that are not yet sufficiently developed to warrant support from the Research Councils') were forced to take the decision, in 1982, 'no longer to consider applications which are within the remit of the Research Councils'. The Foundation lacked the capacity to respond to the number of applications which would in better days have been strong candidates for Research Council support. Some of these were even alpha-rated (the highest possible grading), but had been rejected by the Research Councils through lack of funds (Nuffield Foundation 37th Report 1986).

However, although such peer review committees for funding bodies are obviously concerned with the scientific aspects of research proposals, and cost to the animals (amongst many other factors) may be taken into consideration in the review, the questions asked of referees often do not specifically address the 'animal' issues involved. Furthermore, it is not always the case that funds for research involving animals are intensely competed for and that applications for funds are subjected to rigorous peer review. Sometimes money is obtained from institutional funds, with little or no review of the work, some local charities may have very little scientific peer review of grant applications, and some clinicians may be able to afford to run their own studies, with no outside funding being sought. Therefore, whilst peer review by grant-giving bodies plays an important role in ensuring the quality of scientific aspects of research involving animal subjects, it should not be assumed, simply because research is funded, that the quality of that research is high and that issues concerning the use of animals have been addressed.

It might be desirable for review by funding bodies to include more explicit consideration of the ethical aspects of proposals for research involving animal subjects. This might be especially important in countries where there is an absence of effective legal provision governing such use of animals. Funding bodies might be in a good position to implement humane policies on animal use, since the threat of refusal or removal of funding would help to ensure that investigators attended to these aspects of their projects. Such a role for funding agencies has been recommended in the USA (Orlans 1982), and in Canada, where the threat of removal of funding is used as a sanction to try to ensure compliance with the voluntary system of self-regulation. Whilst it might be difficult to build such questions into routine procedures, since funding bodies are already faced with difficult, multi-dimensional, decisions, it would seem important to include animal welfare considerations in the process of review. The inclusion of a specialist laboratory animal scientist on each review panel might be one way of ensuring such consideration.

10.8 ROLE OF LEARNED SOCIETIES

Another way in which scientists can regulate their own practices is through the various learned societies (societies for the study of particular areas of medicine). These societies may issue ethical guidelines for research in the particular specialist field. The International Association for the Study of Pain, for example, has drawn up a set of ethical guidelines for the investigation of pain in conscious animals (Zimmerman 1983—see Table 10.8). In the field of behaviour research, the British Psychological Society (1985) and the Association for the Study of Animal Behaviour (1986) have produced Guidelines for the Use of Animals in Research, which give welfare and ethical advice on carrying out specific kinds of technique, including the use of aversive stimulation, deprivation of food and water; social deprivation and the use of surgery in behaviour experiments. In addition, some societies (such as the Association for the Study of Animal Behaviour) have ethical committees which may discuss and advise on ethical questions posed by particular research practices and may also review research in progress as well as manuscripts submitted for publication in the society's journal.

The role of learned societies in considering ethical aspects of

Table 10.8 Ethical guidelines for the investigation of experimental pain in conscious animals (Zimmerman 1983)

1. It is essential that the intended experiments on pain in conscious animals are reviewed beforehand by scientists and lay persons. The investigator should be aware of the ethical need for a continuing justification of his investigations.
2. If possible, the investigator should try the pain stimulus on himself; this principle applies for most non-invasive stimuli causing pain.
3. To make possible the evaluation of levels of pain, the investigator should give careful assessment of the animal's deviation from normal behaviour. To this end, physiological and behavioural parameters should be measured. The outcome of this assessment should be included in the manuscript.
4. In studies of acute or chronic pain in animals measures should be taken to provide reasonable assurance that the animal is exposed to minimal pain necessary for the purposes of the experiment.
5. An animal presumably experiencing chronic pain should be treated for relief of pain, or should be allowed to self-administer analgesic agents or procedures as long as this will not interfere with the aim of the investigation.
6. Studies of pain in animals paralysed with a neuromuscular blocking agent should not be performed without a general anaesthetic or an appropriate surgical procedure that eliminates sensory awareness.
7. The duration of the experiment should be kept as short as possible and the number of animals involved kept to a minimum.

research involving animals is particularly important, since the societies can give expert advice and formulate guidelines based on a detailed knowledge of the particular field of work, which might involve the use of special techniques or procedures.

10.9 EDITORIAL POLICIES

Many journals in biological science now include some reference to the ethics of animal use in their Instructions to Authors. The journal *Medical Science Research*, in its Manual for Authors, gives the following general instruction:

Ethical principles: Only those papers reporting work which conforms to high ethical standards will be considered for publication. In general, these standards are those which are governed by statute in the United

Kingdom, but the Editor reserves the right to apply more stringent criteria where these are felt to be appropriate. The fact that the country of origin of a particular study does not have legislation governing animal or human experimentation which is as stringent as those operating in the UK, will not be relevant to the editorial decision. A discussion of some of the ethical principles governing experimentation using animal subjects is given in Appendix A. [Appendix A is a three-page document, covering six general principles: the responsibility of the investigator; the value of the proposed experiment; the choice of an appropriate animal species; the minimization of distress; limitation of animal numbers; and the general welfare of animals.]

In some journals, referees reviewing manuscripts may be asked to comment on the ethical acceptability of the work described and on any animal welfare issues. Several other journals (including *Animal Behaviour*, *British Journal of Pharmacology*, *Journal of Physiology*, *Journal of Small Animal Practice*, and *Veterinary Record*) also adopt the policy of *Medical Science Research* in requiring that, to be acceptable for publication, research using animals, regardless of its country of origin, should be permissible under UK law.

Such editorial policies may discourage researchers from performing ethically unacceptable experiments using animals, since the results of the research would then be very difficult to publish. However, this is not the view taken by one important international journal, *Nature*. Following correspondence (Hollands 1989) questioning the licitness, under UK law, of research carried out in France and described in one of the journal's published papers, *Nature* published an editorial which stated:

> This journal's rule of thumb for at least a quarter of a century has been that the results of animal experiments that would not easily win general regulatory approval had better be of exceptional interest if they are to be published. This position is admittedly pragmatic (and requires a subjective judgement of what may be exceptionally interesting). The obvious counter-argument, that a refusal to publish will ensure that contentious experiments are not undertaken in the first place, might have more force if there were not, for other good reasons, such a variety of journals (*Nature* 1 June 1989, p. 324).

Whilst, as *Nature* has expressed, there might be some scepticism about the ability of editorial boards to influence research practices, editorial guidelines would seem, at the very least, to have an important educational value. Even if some researchers, perhaps from countries with different standards for research involving animals, disagree with the guidelines, the very existence of guidelines might serve to make researchers think about the issues raised by the use of animals. In the longer term, such guidelines might, through their educational role, lead to changes in practice.

Guidelines on preparing manuscripts for submission might also play a role in improving or maintaining standards for the use of animals in biomedical research. It is often very difficult to establish information about morally important aspects of such research from published papers. For example, the effects of the scientific procedures on the animals and steps taken to prevent or alleviate any adverse effects may be reported in insufficient detail, so making it difficult or impossible to judge the likely costs of the experiments to the animals. Furthermore, the broader purpose of the experiments, and how the work relates to other published work, may not be mentioned in a form which would allow someone not intimately associated with the particular scientific field to judge the likely benefits of the research. There may be difficulty in including this extra information, since journals often limit the length of the articles they will publish. Nevertheless, some simple guidelines for reporting the results of experiments involving animals might help to remove these difficulties and hence allow those outside a particular specialist field to understand the justification for the work which is being reported. Indeed, some journals already give guidance on certain aspects of the reporting of work involving animals. The *Journal of Physiology*'s Notice to Contributors (1989) states that:

> For experiments on living animals authors must provide a full description of their anaesthetic and surgical procedures, and provide evidence that adequate steps were taken to ensure that animals did not suffer unnecessarily at any stage of the experiment. For experiments on isolated tissues authors must indicate whether the donor animal was killed or anaesthetized, and how this was accomplished.

Such an approach might be extended, so as to ensure reporting of any loss of animals due to the effects of scientific procedures and also any steps taken to refine the procedures so that they cause less harm to the animals involved. Reporting of these aspects of the work would help those wishing to use similar procedures to ensure that they employ the best possible practice. In general, better reporting of the costs and benefits aspects of such work could help critics to pinpoint research which is of ethical concern and help to reduce criticism, through misunderstanding, of work which might otherwise be considered ethically acceptable.

REFERENCES

Abdussalam, M. (1984). Animals for biomedical research: perspective of developing countries. In *Biomedical research involving animals* (ed. Z. Bankowski and N. Howard-Jones), pp. 34–43. Proceedings of the XVIIth CIOMS Round Table Conference, Council for International Organizations of Medical Sciences, Geneva.

Anderson, G. C. (1990). White House says no. *Nature* **344,** 604.

Association for the Study of Animal Behaviour (1986). Guidelines for the use of animals in research. *Anim. Behav.* **34,** 315–18.

Balls, M. (1985). Scientific procedures on living animals: proposals for reform of the 1876 Cruelty to Animals Act. *Alternatives to Laboratory Animals* **12,** 225–42.

Bateson, P. (1986). When to experiment on animals. *New Scientist* **109,** 30–2.

British Psychological Society, Scientific Affairs Board (1985). Guidelines for the use of animals in research. *Bull. Br. Psych. Soc.* **38,** 289–91.

Britt, D. P. (1985). Local committees for the review of experiments involving animals. *Alternatives to Laboratory Animals* **12,** 171–4.

Brown, L. (1989). *Cruelty to animals: the moral debt.* Macmillan, London.

Canadian Council on Animal Care (1980, 1984). *Guide to the care and use of experimental animals.* 2 volumes. Canadian Council on Animal Care, Ottawa, Ontario.

CIOMS (1985). *International guiding principles for biomedical research involving animals.* Council for International Organizations of Medical Sciences, Geneva.

Clark, J. (1987). Public concerns for animals in research. *Lab. Anim. Sci.* Special Issue, January 1987, pp. 120–1.

French, R. D. (1975). *Antivivisection and medical science in Victorian society.* Princeton University Press, Princeton.

Hampson, J. E. (1978). *Animal experimentation 1876–1976: historical and contemporary perspectives.* Unpublished PhD thesis, University of Leicester.

Hampson, J. E. (1987). Legislation: A practical solution to the vivisection dilemma? In *Vivisection in historical perspective* (ed. N. A. Rupke), pp. 314–39. Croom Helm, London.

Harrison, B. (1982). *Peaceable kingdom.* Clarendon Press, Oxford.

Holden, C. (1989). Compromise in sight on animal regulations. *Science* **245,** 124–5.

Hollands, C. (1989) *Nature* **339,** 248.

Home Office (1979). *Report on the LD50 Test.* Home Office, London.

Home Office (1983). White Paper: *Scientific procedures on living animals.* Command 8883, HMSO, London.

Home Office (1988). *Report of the Animal Procedures Committee for 1987.* HMSO, London.

Home Office (1989). *Report of the Animal Procedures Committee for 1988.* HMSO, London.
Home Office (1990). *Report of the Animal Procedures Committee for 1989.* HMSO, London.
House of Lords (1980). *Report of the Select Committee on the Laboratory Animals Protection Bill [H. L.]*, Vol. II, *Minutes of evidence and appendices.* HMSO, London.
IME Bulletin (1988). *Ethics committees flourish in Europe,* IME Bulletin No. **39,** pp. 13–21. IME Publications, London.
McGourty, C. (1989). About-turn on regulations. *Nature* **341,** 6.
MacKenzie, D. (1986). German scientists rail at animal welfare law. *New Scientist* **110,** (1506), 20.
Manuel, D. (1987). Marshall Hall (1790–1857): Vivisection and the development of experimental physiology. In *Vivisection in historical perspective* (ed. N. A. Rupke), pp. 78–104. Croom Helm, London.
Neffe, J. (1986). Critics assail West German law. *Nature* **321,** 5.
Orlans, F. B. (1982). Summary of workshop on funding agency responsibilities. In *Scientific perspectives on animal welfare* (ed. W. J. Dodds and F. B. Orlans), pp. 85–91. Academic Press, Orlando, Florida.
Office of Technology Assessment [OTA], US Congress (1986). *Alternatives to animal use in research, testing and education.* US Government Printing Office, Washington DC.
Richards, S. (1987). Vicarious suffering, necessary pain: physiological method in late nineteenth century Britain. In *Vivisection in historical perspective* (ed. N. A. Rupke), pp. 125–48. Croom Helm, London.
Riis, P. (1988). Denmark. In *Ethics committees flourish in Europe,* IME Bulletin No. **39,** pp. 15–6. IME Publications, London.
Rowswell, H. C. (1988). How Canada protects its experimental animals. *PsyETA Bulletin* **VIII,** 1–2.
Rupke, N. A. (1987). Introduction. In *Vivisection in historical perspective* (ed. N. A. Rupke), pp. 1–13. Croom Helm, London.
Scottish Society for the Prevention of Vivisection (1978). *Annual pictorial review, 1978.* SSPV, Edinburgh.
Shulman, S. (1989*a*). Cambridge first with local law. *Nature* **339,** 496.
Shulman, S. (1989*b*). Cambridge law passes unanimously. *Nature* **340,** 88.
Søndergraad, E. (1986). Danish legislation on laboratory animal experiments. *Alternatives to Laboratory Animals* **13,** 206.
Tatum, J. B. (1988). On the ineffectiveness of Canadian Animal Care Committees. *PsyETA Bulletin* **VIII,** 3–5.
Thelestam, M. (1983). Swedish National Board for Laboratory Animals (NBLA). *Alternatives to Laboratory Animals* **11,** 89–90.
Thomas, K. (1983). *Man and the natural world, changing attitudes in England 1500–1800.* Penguin, Harmondsworth.
Universities Federation for Animal Welfare (1986). *Laboratory animal legislation.* Universities Federation for Animal Welfare, Potters Bar.
Wade, H. W. R. (1988). *Administrative law* (6th edn). Clarendon Press, Oxford.

World Health Organization (1982). *Proposed international guidelines for biomedical research involving human subjects.* Council for International Organizations of Medical Sciences, Geneva.
Zimmerman, M. (1983). Ethical guidelines for the investigation of pain in conscious animals (guest editorial). *Pain* **16,** 109–10.

Appendix 1 Main 'restrictions as to the performance of painful experiments on animals', imposed by the Cruelty to Animals Act, 1876

1. Experiments were to be performed only 'with a view to the advancement by new discovery of physiological knowledge or of knowledge which will be useful for saving or prolonging life or alleviating suffering'.*
2. Persons performing experiments had to be licensed by the Secretary of State and places where experiments were performed were registered.
3. Experiments had to be performed wholly under anaesthesia and the animal killed before recovery, unless certain 'certificates' were obtained:
   Certificate A allowed experiments to be performed without anaesthesia and with no requirement to kill the animal. A condition attached to such a licence allowed 'no operative procedure more severe than simple innoculation or superficial venesection'.
   Certificate B permitted the animal to recover from anaesthesia, provided the animal was killed as soon as the object of the experiment had been attained.
4. Curare was not to be used as an anaesthetic.
5. Following the recommendations of a second Royal Commission a 'pain condition' was attached to all licences which carried Certificates A or B.
   This condition stated:
   (a) that if an animal was found to be in severe *or* enduring pain, and if the main result of the experiment had been attained, then the animal had to be killed and
   (b) that if an animal was found to be in pain which was severe *and* likely to endure, then it had to be killed, regardless of whether the object of the experiment had been attained.
6. Experiments on cats, dogs and equidae were allowed only when it could be shown that no other species were suitable. Experiments on dogs and cats performed using terminal anaesthesia were

* Certificate D, which could be attached to licences held under the Act, was intended to allow repetition of experiments, 'for the purpose of testing a particular former discovery alleged to have been made for the advancement of such knowledge' if 'such testing is absolutely necessary for the effectual advancement of such knowledge'. Certificate D, however, was never used.

Appendix 1 *(contd)*

allowed under licence alone, but other experiments required certificates:
Certificate E, in conjunction with certificate A, allowed the use of cats and dogs in non-surgical experiments without anaesthesia and *Certificate EE*, in conjunction with certificate B, allowed the use of cats and dogs in recovery experiments.
Certificate F allowed the use of equidae in any experiment.

7. Experiments were not to be performed to illustrate lectures at medical schools, hospitals, colleges or elsewhere unless it could be shown that such experiments were 'absolutely necessary' for such instruction—in which case *certificate C* was attached to the licence.
8. Experiments were not to be performed for the purpose of attaining manual skill.
9. Public demonstrations of experiments on animals were prohibited.

Appendix 2 Main provisions of the 1986 EC Directive 'On the approximation of laws, regulations and administrative provisions of the Member States regarding the protection of animals used for experimental and other scientific purposes' (Directive 86/609/EEC)

Animals covered

Live non-human vertebrates [including free-living larval and/or reproducing forms but excluding embryonic forms], used in product development and testing, or in research to benefit the health and welfare of man or animal.

Definition of 'experiment'

'Any use of an animal for experimental or other scientific purposes which may cause it pain, suffering, distress or lasting harm, including any course of action intended, or liable, to result in the birth of an animal in any such condition, but excluding the least painful methods accepted in modern practice (that is, 'humane' methods of killing or marking an animal).'

Distinction between different species of animal

Animals with the 'lowest degree of neurophysiological sensitivity' possible to achieve the aims of the experiment should be used. Animals taken from the wild may not be used, unless this is the only way to achieve the aims of the experiment.

Source of animals

Mice, rats, guinea-pigs, golden hamsters, rabbits, dogs, cats, quails and non-human primates must be purpose-bred, unless special exemption is given. Stray animals of domestic species are not to be used. Breeding and supplying establishments are to be approved and registered.

Appendix 2 *(contd)*

Number of animals used
The minimum possible number of animals must be used. Unnecessary duplication of experiments must be avoided. Where experiments are carried out in order to satisfy 'national or Community health and safety legislation, Member States shall as far as possible recognize the validity of data generated by experiments carried out in the territory of another Member State'.

Use of alternatives
'An experiment shall not be performed if another scientifically satisfactory method of obtaining the result sought, not entailing the use of an animal, is reasonably and practically available.' The Commission and Member States are to encourage research into the development and validation of alternative techniques (covering all 'Three Rs').

Animal pain and distress
All experiments must be designed to avoid unnecessary pain and suffering to animals. Where unavoidable, these must be kept to a minimum. Experiments likely to cause severe or prolonged pain require specific authorization and justification.

Anaesthesia, analgesia and euthanasia
All experiments must be carried out under general or local anaesthesia except when (i) anaesthesia would be more traumatic than the experiment itself, or (ii) anaesthesia is incompatible with the object of the experiment. Anaesthesia is to be used in the case of severe injuries which may cause pain. If anaesthesia is not possible analgesics should be used to ensure that pain, etc. is kept to a minimum consistent with the aims of the experiment and, in any event, should not be allowed to become severe. At the end of the experiment no animal may be kept alive if it is likely to remain in lasting pain or distress – such animals must be humanely killed.

Re-use of animals
No animal may be used more than once in experiments entailing severe pain, distress or equivalent suffering.

Housing and care of animals
Standards of housing and care must accord with guidelines issued by the EC. All animals must be provided with housing, an environment, at least some freedom of movement, food, water and care which are appropriate to their health and well-being; any restriction on the extent to which an experimental animal can satisfy its physiological and ethological needs must be limited to an absolute minimum; environmental conditions must be controlled; the well-being and state of health of the animals must be monitored and action must be taken to prevent pain, avoidable distress or lasting harm.

Appendix 2 *(contd)*

Competence of personnel
Experiments are to 'be performed solely by competent persons, or under direct responsibility of such a person, or if the experimental or other procedure is authorized in accordance with provisions of national legislation'.

Training of personnel
Those carrying out or taking part in experiments and those taking care of the animals used for experiments must have appropriate education and training.

Establishments using animals
These must be registered or approved; identify person(s) responsible for the care of animals and functioning of equipment; provide sufficient trained staff; make adequate arrangements for provision of veterinary advice and treatment; and charge a veterinarian or other competent person with supervisory duties in relation to the well-being of the animals.

Notification of experiments
Procedures must be established for the notification, to the appropriate authority, of experiments themselves, or details of persons conducting them. Member states may require that work is authorized before it is carried out.

Statistics of animal use
Member States must collect, and as far as possible periodically make available, statistical information on the use of animals in experiments (including the number and kinds of animals used; the purposes of the work; and the number of animals used in experiments carried out for legislative reasons).

11

The philosophical and moral debate on the use of animals in biomedical research

11.1 PHILOSOPHICAL ARGUMENTS AND MORAL CONCERNS: AN INTRODUCTION

Ethical, or moral, arguments about the use of animals in biomedical research are keenly disputed. Were this not so, the present study might not have been undertaken. These arguments reflect the moral concerns of scientists, animal welfarists, and other members of the public. In recent years, philosophers also have become increasingly involved in this area of public debate. The philosopher's particular contribution is careful analysis of the arguments used to express moral concern. At this point, it may be helpful to say a little more about this contribution, both in general and in terms of what some recent philosophers have written about the use of animals in biomedical research. Something should also be said here about the relation between philosophical analysis and our everyday concerns, and how this is reflected in the Working Party's deliberations and conclusions. The first part of this chapter provides a general introduction to these questions. The second part then sets out a more detailed philosophical analysis and defence of the Working Party's agreed conclusions.

Most expressions of moral concern arise from reflection on experience. In this sense, most people *have* a reason for their moral views. When these views are challenged, moreover, we often attempt to *give* reasons, either to defend our own, or to criticize the challenger's views. It is at this point that the philosopher's contribution may be helpful. The reasons we give for our moral views sometimes are based on inadequate reflection on limited experience; and few of us are sufficiently impartial for our views never to be tainted by self interest. Philosophy provides a means of counteracting this, by, among other things, examining the consistency and coherence of the reasons we give, in the light of more disinterested reflection on a wider range of experience.

In making this contribution, it should be added, philosophers do not claim to be morally wiser than the rest of us. Their contribution depends rather on their knowledge of the canons of logic and rational argument, and on being kept up to scratch in this by continually fencing with their fellow philosophers. The experience they bring to public debate is that of a long tradition, in which their intellectual tools have been finely honed. But their moral experience, of life generally (reflected in their fundamental presuppositions and, often, in the common sense or intuitive examples they give to illustrate their argument) is as variable as that of other people.

Because their moral experience varies, one philosopher's analysis of a moral argument may differ quite markedly from another's in terms of the kind of reasoning involved, the conclusions reached, and even to some extent the rules of the philosophical game. An English philosopher, for example, may scarcely recognize what his French counterpart writes as philosophy at all; and this is the result of a long history, including not only the different schools of Continental Rationalism and British Empiricism in the seventeenth and eighteenth centuries, but also the differentiation of the natural sciences from natural philosophy and of psychology from mental philosophy, as well as that of philosophy itself from theology.

Developments of this kind mean that philosophers today often express their disagreements with one another in what sounds to the outsider like highly technical language. This is one reason why their potentially helpful contribution is not always initially perceived as such, for example by scientists who have agreed criteria for establishing the facts, or by people who have to resolve moral problems as part of the practical business of their professional or private lives. A related reason also has to do with a contrast between factual and moral questions. Our view of the facts normally can be revised if sufficient new information persuades us that it should be. But no amount of new information may be sufficient to persuade us to revise our moral views.

This difference seems to arise because our moral views are an aspect of how we each view the world, and view it differently. To change our moral views, in some sense, is to change ourselves. This is not necessarily the case when we revise our view of the facts, since unless they are facts about ourselves (an example would be discovering the fact that we have a terminal illness), they seem part of the world we view rather than of our view of the world. Thus when a philosopher points to inconsistency, incoherence, or the bias of self interest in the reasons we give for our moral views, we are likely to take it, if not personally, then at least as a challenge to find better reasons for our views, which are less vulnerable to the philosopher's criticism. But if the philosopher has been sufficiently skilful, this may prove difficult. It is not surprising therefore, that some of the most skilful philosophers, as Socrates remarked, are regarded by practical people rather as a gadfly is regarded by a horse. Unlike the gadfly, however, the skilful philosopher, if we attend to his criticism, may sting us into thinking both more critically and more creatively about our own arguments. In this way, he may perform the service (again using a metaphor of Socrates) of an intellectual midwife, bringing deeper understanding to birth.

The aim of philosophy then, is to goad us into thinking, by revealing not only the incoherence and inconsistency of our moral views, but also the extent to which they are influenced by individual or collective self-interest. Thus to paraphrase Socrates again, the beginning of philosophical understanding is to realize that we are ignorant, and that our moral views often reflect mere opinion rather than wisdom. Where to go beyond this realization of our own ignorance, however, philosophy cannot tell us. For while there will always be plenty of individual philosophers, or schools of philosophy, willing to chart the way forward, there will always be others, with no less cogent counter-arguments to block the proposed path.

Whether we regard philosophy's contribution to public debate as helpful then, will depend on whether or not we think it helpful to learn how difficult and well-nigh impossible it is to be consistent, coherent and disinterested in our moral thinking. In political and inter-personal contexts, for example, we may well not regard this as helpful, particularly if other people do not share this assumption. This observation is particularly pertinent to debate about the ethics of using animals in biomedical research, and we shall return to it after the following brief discussion of what some recent philosophers have written on the subject.

11.1.1 Philosphers on animals

Modern philosophical debate about the use of animals in biomedical research has roots in ancient Hebrew and Classical culture. In the *Genesis* myths (*Genesis* 1 and 2), man is given dominion over the animals, and the opportunity to name each species. Man, in other words, is expected to exercise his dominion wisely. But man does not always do this, as *Genesis*' accounts of his expulsion from Eden and of the Flood illustrate. After the Flood, not just green plants, but also the animals are given to man for food. At the same time, however, God's covenant is made not only with Noah and mankind, but also with every living creature (*Genesis* 9). The implications of this are drawn out, both in the injunctions of *Deuteronomy* 22 and 25, and *Proverbs* 12, against cruelty to farm animals, and in *Isaiah's* vision of the future, when 'the wolf shall lie down with the lamb' (*Isaiah* 11). What these Old Testament references suggest then, is that man has been given a 'limited dominion' over other creatures, to be exercised responsibly, in the awareness of man's power to harm, and in the future hope of more harmonious relations between man and the animals (see Dunstan 1980).

New Testament sources add little that is explicit to this. Jesus reassures people that they are 'of more value than many sparrows', not one of whom 'is forgotten by God' (*Luke* 12). St Paul, renewing Isaiah's vision, writes that 'the creation itself will be set free from its bondage to decay' (*Romans* 8). But the development of Western Christendom's attitudes to animals also owes much to Greek philosophy, mediated by theologians such as St Augustine and St Thomas Aquinas.

Plato and Aristotle, the dominant figures of ancient Greek philosophy, referred to animals mainly in order to draw a contrast between them and man, the 'rational animal' as Aristotle described him. Aristotle was also interested in animals as a biologist, and in this capacity restated the human 'naming' of animals in more abstract terms. The principle of life, or 'soul', making plants, animals and humans different from one another was, in plants, the capacity for nourishment and reproduction, in animals, the additional capacity for sensation and movement, and in humans the further capacity to reason. Man alone had this capacity, although a limited capacity of memory was present in some higher animals.

Like Aristotle after him, Plato also believed that man alone possessed the faculty of reason. In the *Republic*, it is true, he did remark that a watchdog 'shows a fine instinct, which is philosophic in the true sense' (p. 65, Cornford trs. 1941), when he distinguishes between a friend and a stranger; but the remark was not entirely serious. Plato's high view of reason, as opposed to the experience of the senses, and his interest in being, as opposed to becoming, indeed, were pervasive influences present at the birth of modern philosophical debate on man's use of animals.

The modern debate began with the seventeenth century French philosopher Descartes, who shared Plato's distrust of sense experience, and set out to rethink philosophy by doubting everything that could not be established by reason alone. This led Descartes, influenced by the mechanistic science of his time, to doubt that the bodies of humans and animals were anything more than machines. Descartes, however, did not doubt the traditional theological view, which built on Plato, that the immortal part of man was his rational soul. This theological development of Plato's philosophy represented the soul, less as a principle of becoming, as in Aristotle's thinking, than as an entity. An immortal soul, however, was not possessed by animals. The consequence of this was Descartes' view that animals lacked not only reason, but also consciousness and any capacity to suffer pain.

This view of the leading Continental Rationalist was not accepted by British Empiricists who, in the seventeenth and eighteenth centuries, were also reshaping philosophy. David Hume (1739), for example, remarked that 'no truth appears to me more evident than that beasts are endow'd with thought and reason as well as men' (p. 176, Selby-Bigge (ed.) 1888). Nevertheless, Descartes' denial that animals (despite all appearances to the contrary) were able to suffer, appears to have been widely used as a justification for experimenting on live animals, at a time when that practice was becoming more common.

Although the implications of Descartes' and, in particular, Hume's thinking set his agenda, the eighteenth century German philosopher Kant reverted to something closer to an older tradition on the subject of animals. He quite clearly opposed wanton cruelty to them, but mainly because 'He who is cruel to animals becomes hard also in his dealings with men'. Kant did not believe that animals were self conscious, and argued that they were 'there merely as a means to an end. That end is man' (Infield trs. 1963). In the same year that Kant said this, however, the British Utilitarian Bentham, was to ask his famous question about animals: 'The question is not, Can they *reason*? nor Can they *talk*? but Can they *suffer*?' (Bentham 1789).

Bentham's question was answered, in practice, by those, of all religious and philosophical persuasions, who brought about nineteenth century legislative and other improvements in terms of animal welfare and against cruelty. Major philosophical debate about the morality of using animals for human purposes, including biomedical research, however, did not begin until the second half of the twentieth century, when later Utilitarian and other philosophers began to claim, on behalf of animals, what Bentham (1789) had called 'those rights which never could have been withholden from them but by the hand of tyranny'.

The question of animal 'rights' is central to the present philosophical debate about the use of animals in biomedical research, though not determinative of that debate. (Some philosophers feel bound to set aside the language of 'rights' applied to animals in order to mount what they see as more logically consistent arguments.) Few philosophers now are willing to defend Descartes' view that animals lack sentience and cannot feel pain. (The arguments of those who do are discussed in the second part of this chapter.) Most, at the very least, would accept some version of Kant's view that wanton cruelty to animals demeans human dignity. The question, rather, is whether humans may use animals for purposes they deem good, but for which they would not use other humans, either because of the suffering involved, or because human rights preclude this.

The philosophers Peter Singer and Tom Regan are among those who make the strongest claims for animal rights, and against their continued use in biomedical research (see Singer 1983 and Regan 1983). They, and other philosophers, generally ask the following kind of question. Is it not precisely because animals are like us that they are useful for biomedical research, and the more like, the more useful? But if they are so like us, why do we treat them so differently? These philosophers reply by arguing that animals (or at least the higher mammals) do not differ sufficiently from humans to justify our treating them in these ways. To do so, Singer argues, is to disregard their interests, while Regan argues that it is to violate their rights.

Although Singer and Regan use different arguments, the crucial question which each raises is whether there is any morally significant feature which all humans, but no animals, possess, and which hence could justify treating them differently. The morally significant features under discussion are variously referred to in terms of self consciousness, rationality, language and so forth. Singer and Regan do not claim that any other animal possesses these features in as developed a form as do adult human beings. But some of these features are possessed, to a greater or lesser extent, by some animals; and they are absent in some human beings, for example infants, the severely mentally retarded and the senile. If we believe that the interests or rights of these humans should be respected, these philosophers argue, then it is inconsistent not to respect analogous interests and rights in animals.

There is no good reason, Singer argues, not to respect or give equal consideration to the interests of animals alongside those of humans. Our reason for not respecting those interests, when we use animals in ways in which we would not use humans, appears to be a rationalization, which Singer terms 'speciesism'. Speciesism, he argues, is a form of human prejudice, not unlike the racist prejudice of whites or the sexist prejudice of males; if we reject these forms of prejudice on moral grounds, it is only consistent to reject speciesism. (The nature of the moral challenge represented by talk of speciesism is more fully discussed in the second part of this chapter.) Singer here is following the Utilitarian tradition and is prepared to accept that animals may be used for purposes which serve the common good, but only if we are prepared to use humans in the same way. Regan, by contrast, is not prepared to concede even this. Writing in the tradition of respect for the rights of the individual, Regan argues that the value of an individual (whether human or animal) cannot be measured by its usefulness to others, but only by its experience of the importance of its own life to itself.

The arguments of Singer and Regan have been challenged or criticized by other philosophers on a variety of grounds. Some argue, for example, that rights can be attributed only to persons who claim them, and that this can be done only by those who understand, in some sense, that they have rights. Regan's attempt to claim rights, for animals, analogous to those of humans, seems particularly vulnerable to this criticism. Critics who take this view, however, in turn have been criticized because it appears to imply denying human rights also to human beings, such as infants, the severely mentally retarded and the senile, who lack the capacity to understand and claim them, or (in other versions of this argument) to accept duties as correlative to rights.

The claim that animals have rights, other philosophers again argue, need not be advanced in Regan's terms for it to be more than rhetoric. The British philosopher Stephen Clark, for example, is not willing to deny rights to any human beings, regardless of their capacity to understand or claim them (Clark 1984). But using the philosophical as opposed to the legal distinction between positive rights (involving the claim that someone should do something for one) and negative rights (involving the claim that someone should refrain from doing something to one), he argues that some such negative rights (for example, a dog's right not to be treated cruelly) are enshrined in law. It is not necessary, therefore, he concludes, to establish that animals have 'abstract metaphysical rights', before recognizing their entitlement to these negative ones.

A related argument is advanced by another British philosopher, Timothy Sprigge, who suggests that affirming a human being's right not to be tortured, for example, adds something to simply condemning this (Sprigge 1990). What it adds, he argues, is that it 'highlights the fact that there is something about people, namely their capacity to suffer, that makes it wrong to treat them in this way'. But if this is the reason why it is wrong to ill-treat humans in this way, Sprigge continues, surely it is also the reason why it is wrong to treat animals similarly.

Arguments of this kind, it might be added, are no longer confined to exchanges between philosophers. A recent article in a British daily newspaper examined the claim that;

Self-consciousness and reason and language give humans a dimension of suffering that mere animals lack; because we can anticipate pain and death; and because we know that death will represent the end of our consciousness forever, and because we recognize that threats to one citizen may represent a threat to us all—because of all this, the protection of human rights is essential to everyone's peace of mind. So a robust conception of individual rights is essential for a human society in a way that it isn't for, say, the welfare of a chicken society.

The problem with this argument however, the journalist (Wright 1990) continues, is that;

it amounts to philosophical surrender. To rely completely on this argument is to concede that language, reason and self-consciousness are morally important only to the extent that they magnify suffering or happiness. Pain and pleasure, in other words, are the currency of moral assessment.

The journalist agrees, however, that this conclusion might be avoided if a 'non-arbitrary reason' can be advanced to distinguish the moral status of humans from that of animals.

The task of making such distinctions is an essential part of the contemporary philosophical effort to refute claims of the kind advanced by such philosophers as Singer and Regan. For a more detailed examination of the subject the reader is referred to the second part of this chapter. There the reader will find discussion on the important questions of whether the notion of rights is the best means of ordering our thoughts about the wrongness of invasive research, of the precise relevance of rationality and self-consciousness to these debates, and of what might serve as a non-arbitrary basis for making distinctions of moral status between creatures.

A further question, which may be asked here however, is why these philosophical arguments are now rehearsed, in some detail, in the arena of public discussion. The American philosopher James Rachels has recently argued that this owes less to the influence of philosophers than to that of Darwin (Rachels 1990). Some philosophers, Rachels notes, wrote about Darwin's theories in the immediate aftermath of their publication. But subsequently they had little to say on the subject, to some extent because the idea of 'the survival of the fittest' was used to support a variety of morally unjustifiable social and political claims. Moral philosophers in particular, Rachels suggests, also suspected that to draw any ethical conclusions from the theory of evolution was to make the logical mistake of deriving an 'ought' from an 'is'. Rachels agrees that this disallows the claim that Darwinism disproves the traditional view of the unique value of human life. But he argues that Darwinism undermines the support given to this view by 'the idea that man is made in the image of God and the idea that man is a uniquely rational being'. Gradual undermining of these ideas in the century since Darwin wrote, in other words, is an important part of the reason why 'speciesist' arguments now find a more hospitable hearing in public discussion.

This shift in public sentiment, Rachels claims, is well founded. He supports this with a variety of empirical evidence, particularly from Darwin's own writings, which leads him to conclude that, in terms of rationality, language and intelligent behaviour, and also a capacity for moral behaviour, animals differ from humans in degree but not in kind. To discuss this evidence again would go beyond the scope of the present chapter, as would discussion of Rachels' arguments against the idea that man is made in the image of God. Rachels' general claim however is that the pre-Darwinian view, 'that membership in a species was determined by whether the organism possessed the qualities that defined the essence of the species' and that this 'essence was something real and determinate, fixed by nature itself', has been so undermined as to be no longer scientifically plausible. 'Evolutionary biology', following Darwin, sees 'no fixed essences', but 'only a multitude of organisms that resemble one another in some ways but differ in others' (p. 195).

This post-Darwinian way of looking at humans and animals, Rachels goes on to argue, 'fits', in a way that is 'weaker than logical entailment' but 'stronger than mere consistency' the ethical approach which he calls 'moral individualism'. This approach is based on the Aristotelian principle that 'individuals are to be treated in the same way, unless there is a difference between them that justifies a difference in treatment'. The evidence he has offered, Rachels continues, shows that 'humans and animals are not radically different in kind: they are similar in some ways, and different in others, and these differences are often merely matters of degree.' He concludes:

> Therefore, our treatment of humans and other animals should be sensitive to the pattern of similarities and differences that exist between them. Where there is a difference that justifies treating them differently, we may; but where there is no such difference, we may not' (p. 197).

Rachels' arguments then, not only help to explain the more hospitable hearing now being given to critics of 'speciesism', but also defend and refine the case being made by these philosophers. In doing this, however, Rachels relies, crucially, on controversial arguments about a) species nature and the implications of neo-Darwinism and b) rationality, language and intelligent behaviour, which deny a difference in kind between humans and animals sufficiently radical to justify treating them differently. Whether this way of arguing is philosophically acceptable, thus again must await the more detailed discussion in the second part of this chapter. Particular attention will be given there to discussing how far species-membership can be considered (*pacé* Rachels) to be a real fact about moral subjects that may influence our treatment of them.

One further aspect of Rachels' discussion merits attention at this point, since it concerns a moral factor, not so far mentioned, which also has to be weighed in the balance when examining the ethics of using animals in biomedical research. This factor is the obvious one, that such research may result in great practical benefits and the reduction of human and animal suffering. On the subject of using animals in biomedical research, Rachels reaches conclusions not dissimilar to those of other defenders of animal rights or interests. He does this, however, by choosing to focus attention on behavioural and psychological research. Illustrating this with the familiar example of Harlow and Suomi's work on maternal deprivation in monkeys, he argues that 'if the animal subjects are not sufficiently like us to provide a model, the experiment may be pointless', but 'if the animals are enough like us to provide a model, it may not be possible to justify using them in ways we would not treat humans' (p. 220). This conclusion, however, would not be accepted by all behavioural scientists: the utility of a particular animal model for the study of particular forms of human behaviour is not in direct and uncomplicated proportion to what Rachels calls the animal 'being like us'. The conclusion fits even less well when the model is used for physiological rather than psychological comparison.

With special reference to psychological models, moreover, Rachels goes on to point out the irony that, while his own moral objection to such work rests on Darwinian foundations, as he interprets them, so too does the 'whole idea of using animals as psychological models for humans' (p. 221). Since his argument has frequently cited Darwin's own writings, moreover, Rachels here also quotes Darwin's views on the subject. Although the subject of 'vivisection', Darwin wrote, was one which made him 'sick with horror', he also had

> long thought physiology one of the greatest of sciences, sure soon, or more probably later, greatly to benefit mankind; but, judging from all other

sciences, the benefits will accrue only indirectly in the search for abstract truth. It is certain that physiology can progress only by experiments on living animals (see Rachels 1990, p. 216).

In the light of the conclusion he has drawn from 'moral individualism', illustrated by the example of behavioural research mentioned above, Rachels comments that 'it is fair to say that Darwin himself did not fully appreciate the implications of his own work'. But the moral implication of this comment itself seems less than fair, both to physiologists, whose science Darwin admired, and to those behavioural scientists whose methods of observation are either minimally or non-invasive.

11.1.2 Moral concerns and conflicts

From these introductory remarks about the philosophical debate, two conclusions may be drawn. One is that, from the point of view of rigorous philosophical argument, a relevant difference between human and animal moral status must be alleged and morally supported, if present discriminatory practices between human and animal research subjects are to be justified. The second is that rigorous philosophical argument has not yet reached agreement, among philosophers, on whether any such relevant difference can be morally supported. In the light of these conclusions, it is not surprising that the multidisciplinary Working Party responsible for this study is unable to offer arguments for or against the use of animals in biomedical research, to which philosophers will not take exception. What it does in this book, rather, is to record (in Chapter 3) a general conclusion, based mainly on an argument from temporary (but by no means perpetual) necessity, with which members of the Working Party, most of whom are not philosophers, felt able to agree. It then explores (in the second part of this chapter), with particular help from its members who are philosophers, one view of the moral status of animals, resting on a particular basis, which illustrates how its general conclusion might be defended by rigorous philosophical argument. Before turning to this, however, and in concluding the introductory part of this chapter, some moral concerns and responses expressed in the Working Party's deliberations by members who are not philosophers, should be recorded.

In the first place, the Working Party believed, it is important to record the variety of reasons why people may be morally concerned about the use of animals in biomedical research. The following may be suggested:

(a) the moral concern may be about pain and suffering, and may arise because human feelings are projected into animals;
(b) the moral concern may be about an animal's awareness of what is being done to it, and may arise because human self-consciousness is projected into it;
(c) the moral concern may be about the mutilation or destruction of any living organism, and may derive from views about the connectedness and wholeness of all forms of life;
(d) the moral concern may arise because the organism is valuable, decorative or much loved. Exactly the same kind of concern might be prompted on behalf of an inanimate work of art such as a Leonardo;
(e) the moral concern may arise because attitudes to animals may affect attitudes to human beings. Many people remain sceptical about consciousness in animals rather than humans, but take the view that human readiness to emphathize with other animals means that if we deny our feelings in this direction we are also likely to deny them in relation to fellow humans;
(f) the moral concern may arise from respect for the views of those who care deeply about animals and from an unwillingness to cause offence to such people.

These concerns it may be noted, are not mutually exclusive, and each may be held with or without the others. The concern may be greater for animals of some species than for those of others: commonly it is greater in the case of animals whose biological similarities to human beings make them more likely to be of use in biomedical research. There may also be greater concern for those which interact with humans in ways reminiscent of interaction between infants and adults, for example, dogs, cats, horses, primates or parrots.

Each or all of these moral concerns may persuade people that animals should not be used in biomedical research. The strongest objections however are likely to arise from the first two concerns, when human feelings or self-consciousness are projected into animals; and these objections in turn are most likely to be advanced in the case of animals with the greatest biological similarity to humans; and these as has just been noted, are normally the animals likely to be of most use in biomedical research. Leaving aside now the question of self-consciousness as a matter of philosophical and scientific debate, serious concern still may be expressed on the grounds that most animals used in biomedical research are conscious creatures who can suffer pain (see Chapters 4 and 5). Even if we cannot be sure what this means to them, it may be argued, as conscious creatures who ourselves can suffer pain, we can only assume that it is sufficiently similar in their case to make it wrong to inflict pain on any conscious creature, which is not necessary for its own good. How then, it may be asked, can this be morally justified?

One response to this question, which appealed to some in the Working Party but not to others, may be mentioned here. This took a broadly historical and moral form. In history, it suggested, human beings normally have tended to become aware of the moral claims of their own kin and kind before those of others; and awareness of responsibility for those outside their immediate circle has tended to depend on how far humans have secured the means for their own survival. For the practice of virtue, St Thomas Aquinas remarked, a modicum of comfort is necessary. For much of history, it was argued, human beings have not enjoyed sufficient material comfort to spare many thoughts for the morality of animal–human relations. The practice of inflicting pain on conscious creatures in biomedical research thus was adopted and accepted before any moral debate which could raise objections to it arrived on the ethical agenda. Any moral doubts which might have arisen were dealt with rather as Darwin dealt with his—by pointing to how the practice served humanity's and individual's survival.

According to this response then, the practice of inflicting pain on animals in biomedical research arose because stronger creatures largely unreflectingly used weaker ones in the quest to secure means for their own survival. Having done this, however, the stronger creatures stopped to reflect on what they had done, and concluded that their only logical way forward was either to find a moral justification for continuing to use the weaker in this way, or, if they could not, to cease doing so. The reasons for continuing, at least until some alternative to their practice could be found, were very strong—the great benefits to animals as well as humans, resulting from this. But the moral offence caused by their practice to many people also was great; and until it could be justified in terms other than simply those of its great potential benefits, many, perhaps the majority, who realized what was involved were likely to regard it with sentiments stronger than regret, if weaker than remorse.

The development of this line of thinking may help to explain why the Working Party, having concluded that the 'necessary evil' of animal use in biomedical research should continue until a better alternative was found, felt it necessary to offer some further philosophical justification for this. Had the Working Party reached the opposite conclusion, philosophical justification of this also would have been no less necessary. In the event of the conclusion going either way, moreover, it is unlikely that everyone who agreed to the general conclusion would agree to every link in the chain of the philosophical argument supporting it.

One link in the chain which some members of the Working Party, again not philosophers, were uneasy about was its denial that a 'tragic conflict' was involved, and the reasons for their unease may be mentioned here. In terms of strict philosophical entailment, they agree, the term might be inappropriate. But it accorded more closely with the 'sentiments stronger than regret, if weaker than remorse' with which they regarded the practice of inflicting pain on animals used in biomedical research. This suggests, of course, an inconsistency—that while being willing to accept the conclusion based on temporary necessity, they were not willing to accept the philosophical implication that there was a relevant difference between human and animal moral status. It suggests also, that on another occasion, after further thought, they might not be willing to accept the conclusion based on necessity.

The criticism of inconsistency thus is one which some members of the Working Party may have to accept. The idea of a 'tragic conflict' was originally introduced to the Working Party's deliberations from Calabresi and Bobbitt's (1978) study of the 'allocation of tragically scarce resources'. In that context, 'tragic' is defined in terms of social decisions in which, whatever method is chosen, 'the distribution of some goods entails great suffering or death' (p. 23). No available method, Calabresi and Bobbitt argue, can accommodate all the values which society wishes to be accommodated. Thus:

> Since the values endangered by any given approach vary, a society which wishes to reject none of them can, by moving, with desperate grace, from one approach to another, reaffirm the most threatened basic value and thereby seek to assure that its function as an underpinning of the society is not permanently lost (p. 198).

This understanding of tragic choices also seems relevant to the moral problems raised by the values which the critics and the defenders of using animals in biomedical research each wish to accommodate, and also to the 'great suffering or death' entailed in accepting, and acting upon the logic of either position, and indeed of most if not all compromises between them. In this respect, the ethics of animal use may be appropriately regarded as related to the ethics of scarcity of which Calabresi and Bobbitt write:

> We doubt whether there could be an open society whose values were sufficiently consistent to obviate the possibility that scarcity would bring

about tragic choices. Morality—since the terms in which it is stated and by which it is understood must be grounded in culture and tradition—is not simply the aggregate of individuals atomistically wishing to do right. And therefore a moral society must depend on moral conflict as the basis for determining morality (p. 198).

The idea that a tragic conflict is involved in the use of animals in biomedical research, it must be emphasized, was accepted only by some members of the Working Party; and even those who accepted it accepted also the Working Party's general conclusion and decision, that this was best supported philosophically by the arguments in the second part of this chapter. The discussion of this idea is included here simply to illustrate some aspects of the Working Party's deliberations. It may also serve as an illustration of the point made earlier in this section, that philosophy demonstrates how difficult it is to be consistent, coherent and disinterested in our moral thinking, and also a final observation of Calabresi and Bobbitt—that when tragic decisions need to be made, 'they are not the easier for the understanding' (p. 199).

11.2 THE WORKING PARTY'S CONCLUSIONS: A PHILOSOPHICAL ANALYSIS

11.2.1 The Working Party's general approach

The Working Party contains a variety of views on the moral status of animals but is prepared to accept for the present that biomedical research using animal subjects is justified as an undesirable but unavoidable necessity. This working agreement extends to the following points:

1. In the absence of any scientifically and morally acceptable alternative, some use of animals in biomedical research can be justified as necessary to safeguard and improve the health and alleviate the suffering of human beings and animals.
2. The benefits, in turn, depend on the advancement of fundamental scientific knowledge. But even when no therapeutic or other practical benefit can yet be derived from it, any significant advance in scientific knowledge is a good, and *may* serve as a justification for using animals to that end.
3. However, not every projected improvement to human health or addition to scientific knowledge is sufficiently significant to justify every use of animals. Some uses of animals may have adverse effects too serious to justify them at all, while in other cases the adverse effects may be considered disproportionately serious in relation to the significance of the results gained.

4. In the latter case especially, both the potential benefits of a particular research project and the likelihood of the project achieving those benefits need to be assessed carefully before they, in turn, are weighed against the likely adverse effects to the animals.

All the members of the Working Party are prepared to accept these statements, though there are differences between members on the detailed comparative judgements implied in them.

The conclusion offered here makes sense on the dual assumptions a) that the moral status of at least many laboratory animals has to be treated for practical purposes as less than that of human beings, but yet b) that animals do enjoy a moral status which is not merely dependent on their use by humanity, a status which requires the checking and limiting of the licit uses human beings can put them to. Or in other words: we owe separate duties to promote human welfare (including advancing scientific knowledge) and to care for and protect animals. We must perforce act as if the former obligation generally outweighs the latter in cases of conflict; but the latter is never annulled and genuinely limits what we may do to pursue our obligations to human welfare. The Working Party affirms the necessity of a process of moral weighing between conflicting duties, one in general to be treated as stronger than the other, with the genuine possibility that the circumstances of particular cases may force the weaker to triumph. There are then three component parts to the Working Party's overall conclusion: the obligation to human welfare (including here the advancement of science), the obligation to animal welfare, and the analysis of such conflicts as arise between these obligations.

On the first: no one in the Working Party would question the strength of the general imperative for the pursuit of genuine medical advance and the duties to humanity that lie behind it. In general there is no future in arguing for enhanced concern for animal welfare by lowering the moral status of human beings and the strength of our obligations to humanity.

On the second: the facts about animal awareness and sensitivity assembled by the Working Party convince us that all vertebrates and possibly some invertebrates are capable of experiencing pain, distress, and the like, and therefore fall within the scope of virtues such as benevolence, and vices such as cruelty. The Working Party acknowledges a more general respect owed to all living things and an obligation to foster and protect the existence and life-styles of animals. We are aware of the objection that the difference in the range of self-consciousness between normal human beings and even the highest mammals makes it impossible to predict pain, distress, etc. in animals (see Harrison 1989). This conclusion draws on the claims of the seventeenth century French philosopher Descartes, who taught that animals were little more than complex machines: lacking souls they could not be credited with sentience. However, this Cartesian tradition of thought about animal sensitivity faces two major difficulties. It forces us to deny what everyday observation and scientific research both confirm: the apparent manifestation of sensitivity in animal behaviour. It also generates the paradox that pain, distress, etc. are not really present in the behaviour of human infants (for they lack self-consciousness too). It seems best to cope with such differences as clearly exist between human and animal sensitivity by concluding that the range of psychological concepts that can be used of animals may be limited and their sense perhaps altered analogically. We may use descriptions entailing sensitivity and awareness in extended but related senses of animals, but only someone determined to defend a thesis at all costs would question as mere sentimentality our reaction to higher animals as creatures capable of suffering. Our obligation to avoid animal suffering is therefore real, and varies in strength in line with the degree of sensitivity we can attribute to different species of vertebrates and invertebrates.

On the third issue we note that the clash of genuine obligations owed to humanity and animal welfare is generated by empirical facts. There are differences among us on how temporary or easily alterable are these facts. This is a matter of judgement on how far a programme of alternatives to work on whole animals can be developed and perhaps on the scope available to modern medicine to alter its goals and methods in the light of these alternatives. We shall comment below on the ethics and the duties that arise out of the recognition of conflict.

So far we have outlined the general conclusions about the moral issues that the Working Party is prepared to accept as a group. However, there are differences in the Working Party as to how best to articulate these very general conclusions and on how they should be defended and applied in detail. These disagreements reflect the difficulty in producing an agreed statement of the ethics of human use of animals in the light of the differing philosophical and theological backgrounds represented in this study. Accordingly, the defence of the general conclusions about the weight to be given to our obligations to human welfare, on the one hand, and our obligations to animal welfare on the other, offered in this part of the chapter is limited in its scope. It offers a defence, of the balance struck between these obligations by the Working Party in its deliberations on the details of biomedical research, which draws upon the minimum of metaphysical or religious construction and which allows considerable room for differences over specific judgements on the licitness of particular procedures.

Within the Working Party, just as in society at large, there is no shared religious or philosophical world-view which might dictate an agreed answer to the question of how we can adjudicate on conflicts between obligations owed to humanity and to animals. If there is to be any generally acceptable defence of the use of animals in biomedical research it must seek a basis of moral consensus other than at the level of agreed world view or metaphysic. The following discussion seeks to find such a basis at the level of reflective practice, through the underlying belief that the best practices in the conduct of biomedical research reveal a wisdom about the relative status of animals and mankind. Major forms of objection to present practices will be considered as the argument proceeds.

11.2.2 The ethics of use and moral consensus

Research ethics and justice

The essence of the approach to the moral issue of man's treatment of animals outlined here seeks acceptable arguments which do not appeal to shared religious or philosophical world outlooks, and is based on the thought that intelligent human choices constitute in themselves a form of knowledge about how to act. Moral knowledge does not arise solely or mainly through the application of grand theories to the particularities of human conduct. It arises equally through the education of judgement in decisions about particular cases. This idea goes back to Aristotle. We might expect, if this is true, to see a wisdom about the relative moral status of man and animals behind the best reflective practices in the treatment of man and animals in biomedical research—a wisdom that can be articulated without drawing upon a grand ethical or metaphysical theory. According to this approach, the way a moral argument on this subject (or any) should best proceed is by appealing to an agreed aspect of reflective, intelligent practice, and reasoning from the values implicit in that to the desired conclusion. The Working Party's deliberations on specific areas of biomedical research recorded in the previous chapters are in fact an exemplification of this method in practice. They represent, not the effort to apply a grand, agreed philosophy of human and animal status to particular issues, but a reflective attempt to examine practice and sift out the best and worst in it.

The argument of this section therefore begins with the differences that exist in the best research practice using human subjects and the best practice using animal subjects (though it is important for the argument that there are likenesses too).

For humans there is widespread agreement that research is governed by the jointly sufficient but separately necessary conditions of consent and the intent to do no harm (hereafter, non-maleficence). Real or presumed or proxy consent must be present, for the human subject must be capable of being seen as a willing partner in the medical enterprise. No undue harm or risk of harm should come from research procedures, even if the subject has consented to them (for simplicity's sake we ignore the possibility of justified self-sacrifice in the cause of important medical research). Non-maleficence has to be judged in relation to the circumstances of the individual patient/subject. A research procedure that is also a therapy may justifiably risk great harm to the patient if it offers a reasonable, or the only, hope of remedy for a desperate condition (as in normal medicine: we may mutilate through amputation where this is the only means of warding off yet greater harm).

How can we make sense of these different research ethics? Research ethics for humans reflects a feature of general ethics: it rests on the presumed inviolability of the human subject. The inviolability of the human person links with the Kantian dictum that humanity is always to be treated as an end and never merely as a means. It does not forbid us to risk or bring harm to human beings, but to let such evils be governed by the criteria of justice, understood as the overarching duty of giving to each his due. Following John Lucas (1980, pp. 6–19) we see justice as governing the pursuit of general good by the conditions that just treatment must be backed by inescapable and individualizing reasons: justice '. . . enables the individual to identify with the actions of society, even those that are adverse to him or to someone else, because such actions are taken only if they are required by individualised reasons into which the individual can enter' (p. 18). If human beings are to be justly treated, harms inflicted on them must be justified by impersonal reasons into which even the object of harm can enter and they must be reasons which justify infliction of them on *this* man. Thus they must relate to what he has done, is, or how he relates to others.

Many moral philosophers see the need for a developed ethics of biomedical research as arising from the perceived tension between the desire to pursue utility (the long term good of the community) and the imperative to do no wrong to the subjects used in research (that is, respect the demands of justice in the treatment of others).

The two aspects of human research ethics apply constraints arising from justice on the pursuit of general good. Non-maleficence forbids the researcher from inflicting serious harm or undue risk of harm on the subject. Consent entails using subjects of research who are true volunteers, willingly and knowingly accepting whatever risks are involved in pursuing the general good. It also means that there is opportunity to consult the subject's own conception of his interests in considering whether use of him would threaten those interests too much. Hence, consent and non-maleficence in medical research on the human subject are the means of translating and applying constraints from justice on the pursuit of the general goods aimed at by medicine.

The contrast between the ethics of animal and human research is based on the violability of the animal subject and the absence, for the most part, of the peculiar restraints of justice in this case. Analogies with human research ethics exist, for the researcher must consider the cost to the animal, in terms of pain, distress, discomfort, etc., of his proposed experiments. Yet the cost to a particular animal subject may be justified if counterbalanced by sufficient ensuing overall good either to human beings or other animals. The cost to this particular research subject remains as an important check to be weighed in the balance against good achievable for others; as such it behoves the good researcher to keep it to the minimum necessary and it may prevent doubtful procedures from ever being undertaken. But the cost to the animal subject is not reckoned in the light of the requirements of justice. We do not need to cite facts about this particular animal to justify intended research procedures in which it figures (unless they be facts which bear on the utility of using this particular animal to achieve the research goal). We seem to accept that even if a procedure inflicts substantial extra harm on the animal subject (to the point of bringing about its death in some cases), it may still be open to us to argue that this is justified if sufficient general good can be expected to arise from it. There is no clear sense in which animals can consent to the treatment meted out to them in research. We cannot ask them to consider how a proposed use of them in research matches up to their own sense of their interests. They cannot therefore be considered as volunteers. Only reflections of concerns about justice can therefore restrain our use of animals in this pursuit of the general good. (A reflection of justice might be the requirement, on grounds of fairness, that no one animal should be made to undergo repeated procedures which involve major harm or mutilation. It is fairer, we feel, to humanely kill it and repeat the procedure on a new subject.)

On what general grounds might these contrasts rest? Animals evidently cannot directly enter into a community governed by the restraints of justice, being unable to appreciate the reasons behind the distribution of benefits and burdens in such a community. But we do allow human beings who cannot directly enter into such a community to be protected by the constraints of justice (infants, the senile, etc) and recognize or appoint proxies on their behalf capable of appreciating the inescapable and individualized reasons on which justice rests. The decision not to do so on behalf of animals must reflect a prior decision on the moral status of animals, one to the effect that they do not have the kind of good or interests that deserve protection by the requirements of justice. It is altogether too simple to argue that because animals cannot recognize the claims of justice, therefore they cannot be protected by them.

The moral status of animals

There is an almost endless variety of metaphysical views about the relations between humanity and animals on which opposing sides on these questions might draw. The moral view we have summarized above can claim, however, to be implicit in common morality, that is in intelligent, reflective choices we make about the treatment of animals. It does indeed have a particular metaphysical origin, reflecting a Christian vision of man as God's viceroy in creation, to whom is delegated a rule over other creatures that is not absolute and cannot be exercised tyrannically. This led to the following ethical stance, as described by Keith Thomas (1983):

> Man, it was said, was fully entitled to domesticate animals and to kill them for food and for clothing. But he was not to tyrannize or to cause unnecessary suffering. Domestic animals should be allowed food and rest and their deaths should be as painless as possible. Wild animals could be killed if they were needed for food or thought to be harmful. But although game could be shot and vermin hunted, it was wrong to kill for mere pleasure (p. 153).

This practical stance, minus its metaphysical, religious background, can properly be said to reflect common morality's perspective on the status of animals. Some may be sceptical of the idea of there being a common morality in twentieth century Western society which might provide the basis for reflection on the ethics of any practice. Yet such practical views about animals underlie widely supported legislation governing what may be done to animals in ordinary life, sport, husbandry and of course scientific research. They undergird the work of and public support for many charities and voluntary bodies. There is every reason to think that at the level of intelligent practice there is a common morality on the treatment of animals, one which of course allows for widespread disagreements and which is not beyond challenge and change. If successful arguments are to be mounted to overturn the contrast in research ethics expounded here they will have to come from that common morality and not from the more private metaphysical visions which inform the personal lifestyles of some radical animal welfarists. It is notable that contemporary philosophical challenges to customary treatment of animals tacitly acknowledge this point by appealing first to some item thought to be at the heart of common morality (that is, a value or values implicit in other reflective, intelligent choices we make) before then condemning accepted treatment of animals as wrong (thus: natural rights, or equal respect for interests are appealed to as the basis for 'animal liberation').

The radical philosophical challenge to the ethics outlined here asks for the relevant moral difference between animal and human subjects which might justify a contrast between the one as violable and the other as inviolable. Its basic weapon is the charge of 'speciesism'. This alleges that the mere fact of membership/non-membership of the human race cannot in itself be the criterion for judging some creatures to have an enhanced moral status. This argument is analogical. In rejecting racism we accept that mere membership of an ethnic group does not in itself justify greater burdens or less benefits in one's social lot. Discrimination in benefits and burdens must relate to specific facts about the subject discriminated against (or for). Here we can see an appeal to the same thoughts about justice we found in Lucas: justice in the treatment of others must rest on individualizing reasons.

It is customary for philosophers pressing the evils of speciesism to claim that its evident errors demonstrate that animals have rights. This connects with our remarks about justice. Talk of 'rights' may be seen as a way of speaking of the demands of justice from the point of view of those who would benefit from justice's demands (Finnis 1980, pp. 204–8). The notorious vagaries in the language of rights make an approach to these issues through the category of justice preferable in our view (see Byrne 1987, pp. 23–5).

Note what is at the heart of this argument for the conclusion that, if human beings are protected by justice, animals must be (or: if humans have rights, animals must have them too). It appeals to no religious or metaphysical view about the status of non-human creation. It uses a principle of fairness, consistency and equality which is implicit in intelligent and reflective moral thought:

namely, that if we treat individual creatures differently in conduct we must find relevant differences between them as individuals. To this is added the thought that membership of a biological species does not of itself individualize and is morally unimportant. If we are to discriminate between creatures in our conduct, such discrimination must be based on traits they possess as individuals. These traits must provide in their own right a basis for just discriminations in treatment.

The challenge of animal liberation

It is well to define more clearly the objection from animal liberationists to the use of animals as research subjects before proceeding to respond to it.

One popular line of argument against the use of whole animals in biomedical research appeals to the alleged fact that such research could continue without serious detriment even if work with animal subjects were abandoned. It would question the usefulness of, and therefore the need for, research on animal subjects to further worthwhile medical goals. The earlier chapters in this study attempt to answer this kind of argument.

Philosophical objections to animal research start at a more fundamental level: granted that human welfare (broadly defined to include the advancement of science) can genuinely be promoted in this way, it is wrong because it is inconsistent and unjust to invade the lives of animal research subjects in ways we would never dream of doing for non-consenting human beings. Simple consistency demands that if we cannot separate the moral status of animals and human beings in a relevant way, then our conduct cannot discriminate between them. Merely citing the fact that animals are not human does not discriminate for these purposes.

The power of this argument to extend modes of reasoning accepted without question in other moral contexts to the case of animals cannot be underestimated. It is a fundamental principle of practical reason (and hence of justice) that like cases are to be treated alike, and that differences in treatment must rest on morally significant discriminations. We rely on this principle when we challenge the political racist to show why some ethnic groups are not to be given the full respect of citizenship in a human community while others are. We would never accept the reply 'But they are different: they don't belong to my (favoured) group.' Without a basis in morally relevant, distinguishing traits, harsh treatment of groups different from one's own stands as the expression of unjustified power without legitimacy. As such it can continue only if those who are behind it are morally and rationally blind or just plain wicked.

While this report carefully shows ways in which the lot of animal research subjects can be improved and points to methods for reducing the number of animals used in biomedical research, it continues to support the idea that some invasive research using animals can continue in the absence of better alternatives. Someone who can allege no morally relevant differences between animals and human beings must object to this and contend: either that human research ethics should be extended to animals or that animal research ethics should be extended to humans. The first of these conclusions would spell the end of invasive research on animals, since the animal subject would then have to be considered inviolable. The second conclusion would in practice lead to the same result. No one would accept the price of consistency if it led to overturning the whole thrust of the development of human research ethics in the modern age. Moreover, if we ever accepted that the liberty, good and lives of human subjects could be sacrificed for medical and scientific goals in the way the liberty, good and lives of animals are sacrificed, we would find it hard to justify continuing using mere models for human biological processes when we could get at the real thing without the troubling ethical constraints operating at present.

The negative critique of the philosophical animal liberationist appears to leave open only one line of reply; finding a morally relevant difference between humans and animals that might justify discriminatory treatment between them as research subjects. It must be a reply which is framed in the light of the fact that the liberationist will allege relevant moral similarities between animals and humans. As this report acknowledges, animals, like human beings, have interests and are sentient subjects of experience. The challenge is to show why these are not enough to demand that humans and animals are to be treated alike.

The significance of self-consciousness and rationality

The philosophical defence of common morality on the ethics of animal research can find a number of bases for specific differences between human and animal subjects. The one that is made most of in the literature is the possession of self-consciousness and rationality by the human subject and their absence in the animal subject.

A standard account of self-consciousness is that offered by Jonathan Bennett in *Kant's Analytic* (1966, pp. 105–6; 116–17; expanded on in his *Rationality* 1964, pp. 80–94). Being self-conscious involves the ability to relate one's experiences to the idea of enduring self. That is, it entails being able to accompany a present experience or thought with the judgement 'This is how it is with me now' and to contrast that judgement with those about how it was with me in the past or will be with me in the future. The ability to accompany thoughts and experiences with the idea of an enduring self entails, then, the ability to make judgements about the past, the present and the future, which will in turn involve being able to distinguish between evidence that bears on the past, present and the future. Bennett would appear to be correct in saying that all this must be underpinned by: (a) the capacity for general knowledge and the ability to refer to and conceptualize facts not immediately present; (b) the possession of a complex symbolic system in which such judgements can be articulated and distinguished; and (c) the possession of a comparatively high degree of intelligence on which these various abilities can be based.

The experience of a temporally self-conscious creature is not one of pure immediacy, but is enmeshed with the idea of an enduring possessor of experiences which exists in a realm of past and present facts and future possibilities. Despite the evidence that some mammals (particularly primates) exhibit a rudimentary conception of the self and some capacity to reason in the present about how to get future goals, it seems reasonable to affirm that nothing like this kind of temporally self-conscious experience can be predicated of animals.

There is no reason for the approach articulated here to be dogmatic on this score, either on the point of whether any non-human species display abilities associated with language use or on whether forms of non-linguistic, non-symbolic ability might not provide sufficient evidence for the possession of concepts associated with temporal self-consciousness. The conceptual and judgmental abilities that need to be displayed by a self-conscious creature are fairly clear. The reason for tying these abilities to language-use is that in language-use we find the clearest expression of the possibility of making judgements that cannot be explained in simple stimulus/response terms, but might genuinely be about the non-immediate past and future. A being who possesses a language can make judgements which manifest a conception of the past and future and thus a conception of itself as something that endures through time. But we must recall that an individual human being grows into language-use and into a conception of self and that all these human endowments emerged gradually in the course of evolution. We might expect, therefore, that the origins of self-consciousness will be visible in forms of non-linguistic behaviour, which may indeed be found in some animals (see Chapter 4).

If there are primate species who display intelligent behaviour which points in the direction of the possession of self-consciousness, then they should be extended the protection of a research ethics closer to that used for human beings. (We might then refuse to interfere with their natural liberty in any way or use any research procedures on an individual primate that were not justified by expected benefit to *it*.) The evidence must be allowed to speak for itself.

The success of the appeal to self-consciousness as a ground of distinction between human and animal endowments needs no thought of an absolute gulf between the animal and the human without intermediate cases, as may be represented by some non-human primates or by the developmental process in a human child. What is needed is a perception of a clear difference that survives the awareness of gradation and development. This perception of clear difference is indeed confirmed by reflection. Full-blooded, self-conscious experience is only achieved by children out of infancy, presupposing as it does a stable hold on a sense of the past and future and the ability to compare these with the present. We can grasp these points if we ask what we presuppose in asking a human subject's consent to a proposed research procedure. If we cannot make sense of asking a laboratory animal to consider whether it would want a future in which it goes through some experiment or other, it is not simply that we cannot put the question to it in a form it will understand. We are asking the human subject to conceptualize a future unrelated to immediate stimuli and compare it with its past, its present and its *preferred* future. We make a nonsense of consent in the human case (with attendant worries about when a child can truly consent to medical and research procedures) if we make light of the differences between animal and human conceptions of the self and of the temporal and general order in which the self is set.

The human capacity for temporal self-consciousness is closely connected with the exercise of rational choice, thought of not merely as the ability to make choices between outcomes present in immediate stimuli, but as the ability to reason what to do and be in relation to non-immediate and general facts and possibilities.

The problem as we see it is not to make these differences between human and animal consciousness and attainments appear real, for their consequence could be drawn out endlessly. More pressing is the need to show how they establish a difference in moral status which will take us back to the unique restraints of justice with which we began. Do we make a difference in moral status merely because we happen to prefer our moral subjects to have temporal self-consciousness and to possess a rational nature?

The argument for grounding a relevant moral difference in these distinctions begins with noticing the unique harm that pain, distress and death represent to a self-conscious, rational creature.

(a) Suffering. This now takes on an extra dimension. It threatens the self-image of a self-conscious creature, through weakening his own sense of his worth. It disables and increases dependency thus undermining rational thought and action. If too great it will threaten self-consciousness itself by filling experience with total and inescapable immediacy. The extra dimension to pain in the human case gives the notion of torture a special sense and range of dimensions here. It is perhaps in an analogical sense that we describe someone as torturing an animal.

(b) Since the individual has a sense of its own future and value, death is a unique evil for a self-conscious creature, unless compensation is found in a relevant religious faith. There is something mysterious and unfathomable about death for such a being in consequence. Death is then normally contemplated as a unique evil, and a denial and negation of the individual's self-image and self-valuation.

All the facts highlighted in this section point to the idea of a human being as having an individual as well as a generic good, since the human being has an individual future, goals and self-value. We may speak of 'a human good' but this encompasses the thought that human individuals are capable of having their own unique conceptions of what they want to do and be. Each adult human is capable of having a plan of life, a set of interests, peculiarly his own. Infliction of pain or death upon a human being is a denial of his status as a source of one unique vision of the human good. In contrast the good of a non-self-conscious creature is simply that of a being of a certain species. A human life is valuable in the individual case and not just as an instance of generic human life. Human beings as instances of worthwhile life are not replaceable because they are not worthwhile merely as instances of a generic or general good. This is why to visit harm licitly on a human creature involves showing that the requirements of justice—of what is owing and due to this individual—have been met. All this we believe to be a matter of contrast with the standard animal case. The requirement of consent in research ethics bites in the case of a self-conscious, rational creature because we can ask if it wants its future to include the research activities we wish to subject it to. Non-maleficence gains real force for such a being also. We can ask if the future we propose for it really fits in with its sense of what is worthwhile and harmful for it. To harm it unduly for the sake of a general good is to deny its status as a source of a unique conception of the good and a unique locus of rational agency to the world.

Temporal self-consciousness and rational nature is a relevant moral difference for the shaping of research ethics because it: (a) creates the possibility of a clash between utility (the overall, general good) and justice; (b) suggests a real sense in which the subject who possesses these traits is inviolable and irreplaceable; (c) allows the concerns of justice for the circumstances and point of view of the individual to attach to something.

Speciesism and the moral significance of kinds

The charge of speciesism still has some force in it that needs to be considered. A critic of our use of animals in research might argue as follows. If mere membership of a kind cannot be a criterion of enhanced moral status and if reference must be made to morally enhancing properties of the kind, then those members of the kind 'human' who are not temporally self conscious and rational subjects should in consistency be considered as violable and governed by the kind of research ethics applying to animals. There are two ways in which this aspect of the speciesism charge can be put forward: as a plea for using infants, the senile and mentally handicapped in the manner in which we use animals, or as a plea to abandon our ethics of animal research (since we must be consistent and would not countenance stocking animal houses in laboratories with mentally handicapped humans). These approaches are represented by Frey (1983, pp. 115–16) and Singer (1980, p. 59) respectively.

In reply we must sort out two uses of species membership as a moral criterion. As a *necessary* condition of enhanced moral status it is objectionable on the ground that it denies status to possible or actual beings because of facts of classification which may be unrelated to endowment. There may be more than one life form capable in normal adult development of self-conscious, rational existence. Human beings can claim no monopoly over this mode of being independent of the facts about their and other species' attainments. On the other hand being of the human species may be a *sufficient* condition of being awarded enhanced moral status without detriment to the endowments of other beings (Byrne 1987, pp. 31–3). Possessing the *nature of a rational self-conscious creature* may be sufficient for being awarded this status even though this nature be impaired or underdeveloped in the individual case. *Humanity* in one's own and other persons may then rightly be valued in itself. For to be human is to be a creature possessing rational nature (Teichman 1985, p. 181–2). What it is to possess that nature is displayed fully by the normal, developed instances of the kind but the kind as a whole is valued. Two riders to this argument must be offered. First it would not deny awarding enhanced moral status and the protection of justice to any maverick individual of another 'lower' species who had attained what normal human beings attain. Second it espouses a principle that is consistently universal in its application. If on Mars there are races of beings who in normal development show the attainments which give humans enhanced status, then in consistency we should grant *all* members of that species that status. Being a member of a species could then be a morally significant fact and, suitably qualified, provide a relevant moral difference in how we treat individuals.

We offer this argument because we wish to maintain the difference in research ethics for animals and humans without flouting ordinary morality's feeling (enshrined in the best, intelligent practice of research) that all human beings are protected by a special ethics of research. There is no need to see common morality as hopelessly inconsistent. Speciesism is a label that conceals error as well as reveals truth. We should add that focusing on attainments as grounds for the enhanced moral status of humans is not to contend that humans are valued indirectly: as vehicles for the display of these attainments, not being of value then if they cannot now display these attainments. We value human beings and protect them by justice. Self-consciousness, etc. is the general basis for this valuing, but we value a substance: that is an enduring subject with a future and a past (Devine 1983, p. 515). That is why our ethics for research on infants and elderly senile human beings is an ethics for humanity and not for animals.

We offer a simple reflection on the philosophical attack on the customary research ethics for animal subjects: one of the errors concealed by talk of speciesism is that of attempting to question the standard of care given to some human beings. We feel that to do this can only dilute the resources of common morality, in the end to the detriment of research ethics in general.

The ethics of conflict

One kind of ethical argument that might support the Working Party's conclusion that in the present and in the foreseeable future some biomedical research using animal subjects is a necessary evil has now been outlined. This argument acknowledges that some of the treatment meted out to laboratory animals falls into the category of actions that ought not to be done, if taken in isolation.

In undertaking biomedical research on animals we ought to do what we ought not to do.

This paradox in the nature of the concept of 'necessary evil' can be explained following a standard analysis of reason in conduct in the philosophical literature (Edgely 1969, p. 132; Wertheimer 1972, pp. 81–2). To say that an act ought to be done is to declare that there are good reasons in favour of its being done. But 'ought' judgements can be entered at different stages in deliberation about whether an act is to be done or not. The judgement that biomedical research inflicting harm or distress on animals ought not to be done is one that arises in the initial stage of reflection on what range of reasons bears upon how we are to act. (It may be called a deliberative ought-judgement.) At this stage we register the presence of a clear set of reasons against treating animals thus. We also register reasons in favour of proceeding with research using animal subjects (research 'ought to proceed' in the light of these reasons). At this stage of reasoning we have a clash of 'oughts' (reasons) with no priority accorded to either. But deliberation on how to act presses to a conclusion about what is to be done, and a verdict is therefore forced upon us as to which set of reasons we judge the stronger and which is to shape the overall character of our practical policies. A verdictive judgement is then offered on the final balance of the varying sets of reasons that bear on what is to be done. If we say that, in general and in principle, biomedical research using animal subjects ought to proceed all things considered, this is a verdict on the overall pressure of reason after all deliberative 'oughts' have been entered and duly weighed. Its validity does not deny the reality of the reasons from animal welfare against research or the truth of the deliberative judgement that relative to these reasons research of this sort ought not to be done. The verdict that research ought to proceed is made in the belief that necessity (the presence of overwhelming reasons that tell us an act must be done) outweighs the deliberative judgement that research using animals ought not to take place. But the deliberative judgement remains with the reasons that lie behind it.

In other words, respect for the goals of biomedical research and animal welfare provide us with an example of moral conflict. But we do not regard it as the kind of conflict, distressing though it may otherwise be, in which any resolution would be as equally good (or bad) as any other. It is a rationally resolvable moral conflict because (speaking generally) the deliberative judgements in favour of pursuing medical advance and in favour of avoiding distress, pain and disruption to animals are of different strengths. Therefore, a clear verdict on how to act can be formulated and supported. Those engaged in biomedical research using animals may properly feel regret at having to use some procedures which inflict substantial harm and pain on sentient creatures, but they need feel no remorse.

Because deliberative judgements telling us that what we do ought not to be done remain in place in circumstances of moral conflict, we are under some pressure from moral reason to eliminate such conflicts so far as is possible. It remains true that even if we decide in favour of continuing research using whole animals, we are doing what in isolation would be wrong. We are flouting the force of one set of clear moral reasons. It is right and proper therefore that society should set itself the moral goal of eliminating research using whole animals as much as possible, and that any particular research project be faced with the question 'Can its goals be achieved without using whole animal subjects?' We ought not to be content to act and live in circumstances of moral conflict if this is avoidable. How far it will be possible to avoid this particular conflict in the future depends on the detailed assessment of present and future trends in the development of alternatives to whole animal work, and on the kinds of changes in the conduct of biomedical research that these would bring with them (see Chapter 6).

11.2.3 Conclusion

Two contrasts are central to the argument presented here. The first is between the ethics governing research using human subjects and the ethics governing research on animals. That contrast ultimately rests on a further one between individual, metaphysical visions as a basis for reasoning about our treatment of animals and common morality (as embodied in intelligent, reflective and rationally defensible practice) as a basis for such reasoning. Some metaphysical visions of the natural world and our relations to it will destroy the possibility of making the first contrast. For example, it is not difficult to imagine a religious perspective in which the common creaturehood of humans and animals is so stressed that all living things are seen as inviolable, and destruction or harm to any living thing, except in absolutely inescapable circumstances, viewed as an offence against the sacred. Secular ecological visions may produce similar results. We can respect such visions without conceding that they can be normative for general thought about the ethics of animal care. In the same way we can expect and cater for, but need not grant as normative for all, the intuitive horror many feel at *any* deliberate infliction of suffering on animals.

If there is no common morality, or at least common fund of moral notions, from which an argument to an ethics for our treatment of animals can be mounted, then all perspectives on animal–human relations are private visions with little chance of general acceptance.

The argument here, then, has been that a controlled, critical, questioning use of animals in biomedical research is warranted by common morality when supported by the requisite facts about animal and human consciousness. One important danger in this appeal to a common morality must be consciously avoided: that of assuming that common morality is fixed and static. A reading of Keith Thomas's *Man and the Natural World* demonstrates clearly that our nineteenth and twentieth century consensus that animal welfare matters, that animals have more than a use-value, is the product of dramatic change in moral consensus. If we think that common morality is established by agreement on and within intelligent practice, then we cannot see morality as following automatically from demonstrable first principles, guaranteed to be true eternally. It has evolved and we must expect it to evolve further.

Common morality on respect for non-human life continues to be on the move, as facts from ecology, evolutionary biology and animal behaviour spread through the moral community. Such facts enforce a stronger sense of the relationships, kinship and common interests binding animals and man. These facts are not such as to suggest to us that the contrast between appropriate ethics for research on animals and on humans should be abandoned. Philosophical scepticism about this contrast does not in our view demonstrate that the contrast rests on nothing other than a species prejudice comparable to racial prejudice. But scepticism on this score has its value. It is a stimulus to make common morality move further and faster in the direction of greater respect for animal welfare. It highlights the reality of the moral conflict and the acceptance of some evil in biomedical research using animal subjects, and invites us to eliminate quickly, so far as is possible, the circumstances that generate this moral conflict.

REFERENCES

Bennett, J. (1964). *Rationality*. Routledge, London.

Bennett, J. (1966). *Kant's analytic*. Cambridge University Press, Cambridge.

Bentham, J. (1789). Principles of morals and legislation. In *The Collected Works of Jeremy Bentham* (ed. J. H. Burns and H. L. A. Hart), Vol. 2.1, pp. 11–12. Athlone Press, London.

Byrne, P. (1987). The ethics of medical research. In *Medicine in Contemporary Society* (ed. P. Byrne) pp. 9–39. King's Fund, London.
Calabresi, G. and Bobbitt, P. (1978) *Tragic choices*. W. W. Norton and Company, New York.
Clark, S. R. L. (1984). *The moral status of animals*. Oxford University Press, Oxford.
Cornford, F. M. trs. (1941). *The Republic of Plato* (II, 376). Oxford University Press, London.
Devine, P. E. (1983). Abortion, infanticide, and contraception. *Philosophy* **58,** 513–20.
Dunstan, G. R. (1980). A limited dominion (the forty-eighth Stephen Paget Memorial Lecture, 29 October 1979). *Conquest* **170,** 1–8.
Edgely, R. (1969). *Reason in theory and practice*. Hutchinson, London.
Finnis, J. (1980). *Natural law and natural rights*. Clarendon Press, Oxford.
Frey, R. G. (1983). *Rights, killing and suffering*. Blackwell, Oxford.
Harrison, P. (1989). Theodicy and animal pain. *Philosophy* **64,** 79–92.
Infield, L. trs. (1963). Immanuel Kant: *Lectures on ethics* pp. 239–41. Harper and Row, New York.
Lucas, J. (1980). *On justice*. Clarendon Press, Oxford.
Rachels, J. (1990). *Created from animals*. Oxford University Press, Oxford.
Regan, T. (1983). *The case for animal rights*. Routledge and Kegan Paul, London.
Selby-Bigge, L. A. ed. (1888). David Hume: *A treatise on human nature* (I. III. XVI.). Clarendon Press, Oxford.
Singer, P. (1980). *Practical ethics*. Oxford University Press, Oxford.
Singer, P. (1983). *Animal liberation*. Thorsons, Wellingborough.
Sprigge, T. L. S. (1990). The ethics of animal use in biomedicine. In *The importance of animal experimentation for safety and biomedical research* (eds. S. Garattini and D. W. van Bekkum) pp. 17–28. Kluwer Academic Publishers, Dordrecht.
Teichman, J. (1985). The definition of 'person'. *Philosophy* **60,** 175–86.
Thomas, K. (1983). *Man and the natural world*. Allen Lane, London.
Wertheimer, R. (1972). *The significance of sense*. Cornell University Press, Ithaca, New York.
Wright, R. (1990). The human rights trap. *The Guardian,* 12 June 1990, p. 19.

12
Summary and conclusions

12.1 GENERAL CONCLUSION

Where there is no scientifically and morally acceptable alternative, some use of animals in biomedical research can be justified (albeit by different moral reasons for different people) as necessary to safeguard and improve the health, and to alleviate the suffering, of human beings and animals; as well as to advance fundamental scientific knowledge, upon which such therapeutic and practical benefits might depend. Such a justification, however, should be considered very carefully indeed.

In particular, a research project involving animal subjects should take place only when it can be shown:

(a) that the aim of the project is worthwhile;
(b) that the design of the project is such that there is the strong possibility that it will achieve the aim;
(c) that the aim could not be achieved using more morally and no less scientifically acceptable alternative subjects and procedures; and
(d) that the likely benefits of the project are substantial enough in relation to the suffering likely to be caused to the animals used (that is, the likely benefits of the research should be 'weighed' against the 'costs' to the animals involved).

12.2 BENEFITS OF BIOMEDICAL RESEARCH INVOLVING ANIMAL SUBJECTS

There can be no doubt that the use of animals in medical research in the past has proved worthwhile for human purposes, with consequent benefit to human and animal health. The benefits of animal use in the past, however, do not mean that the continued and unquestioning use of animals in biomedical research today is thereby also morally justified. In deciding what uses of animals in research might and might not be justified, there is thus a need to argue for, rather than to assume, the potential and likely benefits of the research (Chapter 3).

In particular, we recommend

(a) that judgements about the likely benefits of particular projects should be made by the scientific community in dialogue with informed public opinion;

(b) that the scientific community should not only seek to advance the public interest in research involving animal subjects, but should also be seen to do so: those who decide in such matters should be responsive to public attitudes;

(c) that any judgement that the use of animals is necessary should be regarded as an interim judgement: that is, it should be regarded as one which may change over time and with scientific advance, so that the need to use animals may diminish; and

(d) that the factors and interests taken into account in making such judgements should be well known and widely agreed to; since, if this is the case, there is likely to be more confidence in the soundness of the judgements.

The factors and interests which we believe should be taken into account in assessing the benefits of research are set out in a *Scheme for the assessment of potential and likely benefit of research involving animal subjects* (Chapter 7). It should be noted that the primary aim of the scheme is educational rather than legislative; and it can be seen as an aid to communication between the scientific community and public opinion. Its explicit and detailed character means that critics (save those who oppose all use of animals) will lack good reasons for opposing projects which can be shown, convincingly, to have a high rating; but where the rating is low, this will provide a structured argument for limiting animal use.

12.3 COSTS IMPOSED ON ANIMALS IN BIOMEDICAL RESEARCH

At present, human interpretation of what is observed in other animals can only be based on projections from humans, and there is fundamental uncertainty, of both a practical and conceptual kind, in assessing the thoughts and feelings of other species. An approach based on extrapolation from human experience can, nevertheless, lead to plausible inferences about animals' capacities for experiencing adverse states, such as pain, distress, and anxiety. Using such an approach, we have attempted to produce reasonably explicit criteria for assessing pain, stress and anxiety in animals; and we have reviewed scientific evidence concerning the application of these criteria to some selected vertebrates and invertebrates (Chapter 4).

Although the nature of such experiences may vary both in quality and quantity between the different animal species, the evidence suggests that the vertebrates, and perhaps some 'higher' invertebrates (such as *Octopus*) at least, may experience those states which humans find unpleasant or, in extremes, intolerable. Experiments can therefore impose costs, in terms of suffering, on these animals. For some of the animals, particularly the invertebrates, the evidence, although suggestive, is equivocal. Where such doubt exists, it is better to treat the animal as if it were sentient, rather than the other way round.

A variety of features of any given experimental context will contribute towards the overall harm or suffering caused to the animals, and each of these features should be considered in arriving at an overall assessment of the cost of a particular piece of research (Chapter 5). These features are summarized in a *Scheme for the assessment of cost to animals of research involving animal subjects* (Chapter 7). For each feature, it is also possible to identify strategies for reducing costs. Such strategies should be employed wherever possible.

12.3.1 Species of animal used

Species, strains and individuals with the lowest possible capacity for experiencing adverse effects, consistent with the aims of the experiment, should be used.

Invertebrates

Researchers should be alert to the possibility of suffering in cephalopods, and perhaps in other 'higher' invertebrates, and should give these species the benefit of the doubt in assessing the costs of involving them in scientific procedures. Furthermore, since there is no reason to assume that cephalopods are any less able to suffer than vertebrate fish and amphibia, there may indeed be a case for extending legal protection to cover these invertebrate animals.

Non-human primates

Non-human primates, in general, have a highly developed capacity for experiencing pain and distress, and have complex behavioural and psychological needs. Some species of primate are only available in the wild and their supply for research purposes imposes severe, additional, costs. The justification for any proposed use of primates must, therefore, be considered very carefully indeed.

Companion animals
The use of companion species, such as dogs and cats, in research might be seen as imposing greater costs, in terms of animal suffering, than the use of other mammalian species, such as rats, mice or pigs. Whether this greater cost is real is open to debate; nevertheless, the cost is *perceived* as being greater. Perceptions are sometimes important in moral analysis, and may be as legitimate a basis for giving these animals special protection as concerns based strictly on scientific evidence.

12.3.2 Animal supply

Purpose-bred, rather than wild-caught animals should be used wherever possible. If this is not possible, the preference should be for species which are easy to obtain, and which do not need to be transported over long distances or subjected to quarantine. The provision of UK and other legislation, which prohibits the use of stray animals in research, is welcomed. It should be noted that the use of animals with genetic defects (naturally-occurring or induced) may impose additional costs, since such abnormalities can, in themselves, give rise to animal suffering.

In all cases, and particularly where transgenic animals are concerned, there is a need to guard against the possibility that commercial exploitation might take precedence over concern for the welfare of the animals.

12.3.3 Animal housing and husbandry

The manner in which laboratory animals are housed and cared for is of utmost importance in determining the overall quality of their lives. Researchers should take particular care for the conditions under which the animals they use are housed and maintained; and housing and husbandry conditions should be taken into account when assessing the overall costs of a particular piece of research.

Where possible, more natural or enriched environments should be provided, so that the animals can carry out more of their natural behaviour patterns and fulfil their psychological needs. There is a need for further research aimed at improving the care and accommodation of laboratory animals.

12.3.4 Scientific procedures performed on the animals

Many different kinds of scientific procedure are used on animals in biomedical research. It is therefore difficult to make general statements concerning the costs, in terms of pain, distress and anxiety, caused to animals by each. Such predictions must take into account the precise details of the proposed experimental technique: the species of animal to be used; its age, sex, and previous experience; what, if any, anaesthetics and analgesics will be used; the quality of the facilities and the skill of those carrying out the procedure; and how the animal will be prepared for and cared for after the procedure.

The techniques used in the procedures should be refined, so that they cause the least pain, distress and anxiety to the animals. Such refinements might be suggested by examination of the relevant literature, through consultation with experienced colleagues and/or in the light of the researchers' own experience with the techniques. Consideration should be given to devising humane end points for scientific procedures. For example, more sophisticated measurements, such as clinical signs and biochemical analyses, rather than death should be used as end points in acute toxicity tests.

General or local anaesthesia is mandatory for surgical procedures and should be used in all other procedures, except when the giving of an anaesthetic would be more traumatic to the animal than the procedure itself, or when anaesthesia is incompatible with the object of the experiment. Where researchers have any doubts about the best method to be employed, or whether effective anaesthesia has indeed been achieved, they should seek veterinary advice. Attention should also be paid to standards of pre-and post-operative care, and analgesia should be provided whenever appropriate.

12.3.5 Number of animals

The minimum number of animals consistent with the scientific objectives should be used. Where appropriate, statistical advice should be sought, in order to find the minimum group sizes necessary to obtain meaningful results.

12.4 REPLACEMENT ALTERNATIVES TO ANIMAL EXPERIMENTS

In the context of animal experiments, the concept of alternatives embraces the 'Three Rs' of 'refinement, reduction and replacement' of animal use. The strategies described in sections 12.3.1–12.3.4 are all examples of refinements, which can help to diminish the amount of pain or distress suffered by animals in research; and the strategy in 12.3.5 is one of reduction of animal use.

Replacement alternatives can be said to be those which do not involve the use of 'protected' animals. Thus, in the UK, replacement alternatives encompass those methods which do not involve the use of vertebrate animals from halfway through gestation or incubation, or, where larval forms are concerned, when capable of independent feeding. In several countries there is a statutory, as well as moral, requirement for researchers to consider the possibility of using replacement alternatives to their proposed animal experiments (Chapter 6).

There can be certain advantages in the use of replacement alternatives. The principal advantages are that, scientifically, non-animal methods are sometimes the best methods available for tackling a particular problem or question; their use can help to reduce laboratory animal suffering, and they are often more economical than the equivalent animal procedures.

The use of such methods can, in itself, raise ethical questions. In general terms, it is essential that replacement methods be properly validated and evaluated: that is, it must be ascertained, from both a scientific and a practical point of view, that the methods are as good as, or better than, the animal procedures they are intended to replace. The various types of replacement alternative may also raise their own, special, ethical problems (section 6.4.2).

The development of replacement alternatives must be seen as an evolutionary process: it will be impossible to go from present practice to total replacement without many intermediate steps. Whilst replacement alternatives are evolving there will continue to be an interdependence of *in vivo* and non-animal methods. Some believe that animal research will always be necessary and that the potential contribution of alternatives will be limited. Others believe that total replacement will one day be possible. Nevertheless, in spite of these differences in long term vision, there is a moral as well as a legal duty for all concerned to recognize the concept of replacement alternatives, to inform themselves about the potential uses for particular replacement alternatives, actively to support the development of such alternatives, and to accept relevant and reliable, scientifically validated alternatives as replacements for animal procedures.

12.5 'WEIGHING' COSTS AND BENEFITS

Judgements about the likely costs (in terms of animal suffering) and benefits of research involving animal subjects (including whether these same benefits might be achieved by means which involve less animal suffering), can form the basis for a decision about the ethical acceptability of the work (Chapters 3, 4, 5, and 6). In order to draw conclusions about what ought or ought not to be done in a particular case, the costs and benefits must, somehow, be 'weighed', one against the other.

There can be no universally applicable rules for weighing costs and benefits, because the judgements involved are moral, or value, judgements. Nevertheless, these judgements need not be regarded as subjective, but can be inter-subjective: that is, they can be morally persuasive because they reflect consensus, not on the judgement *per se*, but on the procedures used to arrive at that judgement.

With this in mind, the Working Party has approached the question of weighing costs and benefits by examining a number of case studies. Seven cases have been used to show how the principles set out in the Working Party's assessment schemes can be applied (Chapter 7). In making its procedures explicit, the Working Party hopes to provide common ground for furthering the debate about what might and might not be judged acceptable practice.

Following on from these case studies, the lines of moral reasoning developed by the Working Party have been applied in two areas of special concern: the use of animals in toxicity testing and in education and training (neither of which uses is strictly 'research').

12.5.1 Use of animals in toxicity testing

Toxicity testing is concerned with identifying the potential of chemicals to cause adverse effects, that is, toxicity (Chapter 8). If society is to have the benefits of using new chemical products, some form of safety assessment of these products is required. Toxicity testing (by whatever method) provides a basis for the quantification of risks from exposure to chemicals, prior to human exposure, allowing government regulatory authorities and the general public to judge whether these risks are acceptable in the light of the benefits likely to accrue from their use. Testing fulfils practical, ethical and legal duties to try to ensure product safety and to improve knowledge of the risks posed by exposure to chemicals.

However, as practised, much toxicity testing requires adverse effects to be produced in animals, often leading to considerable pain and distress. This aim leads to a special ethical problem which does not arise in many other kinds of research in which suffering arises only indirectly, as an unwanted side effect of the experimental procedure used. A conflict thus arises between duties: on the one hand, to try to ensure the safety of chemicals for humans, animals and the general environment (for which reason testing is carried out) and, on the other hand, to try to safeguard the welfare of laboratory animals (which the testing, by its nature, harms).

Toxicity testing is also one of only a very few kinds of work in which there is a legal obligation to use animals. A conflict again arises because, whilst some laws and regulations demand the use of animals in testing, reflecting public concern about the safety of chemicals, others seek to reduce the use of animals in precisely this sort of activity, reflecting public concern about animal experiments.

In at least some instances, the moral conflicts could be avoided if, at the outset, we were to accept a popular criticism of present practice, which says that many products themselves are unnecessary, so that by implication the use of animals in testing them is unnecessary and therefore unjustified. However, whilst it is legitimate for individuals in society to decide that they do not wish certain products to be developed and produced if this requires animal testing, the question of the necessity of products is too uncertain and too complex to permit a general judgement to be made concerning the acceptability of animal use in testing products. The judgement has to be made case by case. In practice, it is best made by the consumers, who, given accurate information about the use of animals in testing, can exert their considerable buying power to influence the size of markets. Whenever society decides that it wishes to use certain products, there is a moral and legal responsibility on the provider of the products to do all possible to ensure their safety. Whilst safety assessment (in terms of toxicity) relies on animal toxicity tests, it is inevitable that those products will require to be tested in animals.

At present, therefore, the moral conflicts cannot be avoided altogether. Their resolution or diminution will depend on:

(1) reducing the harm caused to animals in testing;
(2) improving the validity, relevance and necessity of the tests in the assessment of risk;
(3) promoting the use of alternatives; and
(4) promoting debate about what products we really want, given the moral dilemma, and which we would prefer to do without.

In the first place, the harm caused to animals can be reduced by refining the test procedures and reducing the numbers of animals used in the tests. Strategies for reduction and refinement of practice include:

(1) employing good experimental design, so as to use the least number of animals consistent with the scientific objectives;
(2) refining the end-points used in the tests, and, in particular killing animals when evident signs of toxicity are shown, rather than continuing tests until the animals die;
(3) refining the techniques used in testing so that they cause fewer adverse effects in the animals (for example, using in-dwelling cannulae for body fluid collection);
(4) limiting the maximum dose of a chemical administered to the animals in testing;
(5) using the least sentient species possible;
(6) reducing the duration of the tests, where scientifically practical; and
(7) where possible, using a hierarchical approach in testing: that is, employing a series of non-animal screening methods, considered sequentially, to identify toxic chemicals at an early stage in the testing process (without using animals at all, or using only a very few animals).

The harm caused to animals in testing could be eliminated altogether if it were possible to replace the animals with non-animal alternatives. At present, non-animal methods are used mainly as screens or adjuncts in predominantly animal-based testing programmes. There are a few areas in which complete replacement of animal tests with *in vitro* alternatives is likely to occur in the foreseeable future (these areas are pyrogenicity testing, certain specific neurotoxicity studies, and eye and skin irritation tests). However, whether alternative methods can ever replace the full battery of animal tests remains controversial. Whatever the long-term vision, all concerned should share a commitment to try to implement change. There is a need for continued and increased funding for research aimed at improving understanding of the molecular and cellular mechanisms of toxicity and at developing, validating and evaluating alternative testing procedures.

On the second strategy for diminishing the moral conflicts which arise in testing, there are several means by which the validity, relevance and necessity of the tests can be improved:

(1) improving the validity of extrapolation from the test to the expected use, through further research into the physiology and mechanisms of toxicity in the various species used in testing, and also in the species which the tests are intended to protect;

(2) avoiding unnecessary duplication of tests by making data on toxic effects more widely available (whilst at the same time, having regard for the protection of the commercial interests of the companies involved);
(3) rationalizing regulatory guidelines and their interpretation, to allow for mutual acceptance, by the various authorities, of data from approved test procedures.

Rationalization of regulatory authority guidelines is especially important, since difficult moral and legal dilemmas are posed by the conflicting demands of the animal protection laws of some countries and some regulatory authority guidelines. Rationalization would help to reduce repetitive testing and ensure that toxicologists would not always need to employ the most demanding tests in terms of animal use. Instead of being forced into adopting a check list approach in determining the toxicological characteristics of a test compound, toxicologists would have more freedom to treat chemicals case by case, using the test procedures best suited to the characteristics and likely conditions of use of the substances under consideration, and causing the least possible animal suffering.

12.5.2 Use of animals in education and training

The use of animals in education and training is different in kind from the use of animals in biomedical research. In education and training, animals are used not, as in research, to try to discover, prove or develop novel facts, ideas or techniques, but to teach or demonstrate known facts, ideas or techniques (Chapter 9).

Some use of animals in teaching known fact and ideas can be justified, but this use should be limited and controlled. If any such use is to be justified, the animals should be kept in the best possible conditions and should not be subjected to any unnecessary stress. We recommend that schools keeping animals should be required to register with a local veterinary surgeon. The manner in which animals are treated in education and training may be important in shaping students' attitudes towards animals. It is therefore essential that animals used for these purposes are treated in a sensitive and humane manner, and that their use be demonstrably justifiable, so that animal life is not seen as being taken lightly.

In primary and secondary schools living animals might be observed, and in secondary schools dead animals might be dissected.

Keeping animals can help pupils to develop an interest in and respect for animals in general, and dissection may help pupils to understand how the body is organized. Some members of the Working Party believe that there should now be no animal dissection in schools; others agree that dissection should be permitted, but that students who object should be able to change to studying non-animal alternatives, without prejudice to their marks. Invasive studies of living vertebrates in schools are unjustifiable, since it is very unlikely that the benefits to pupils will outweigh the animal suffering caused by the studies.

In tertiary (undergraduate) education, some members believe that there is now no justification for invasive uses of animals, whilst others believe that there should be continued use of animals in biomedical degree courses. All are agreed that if animals are to be used, this use should be explicitly justified, carried out in a sensitive manner and well-supervised. Where surgery is involved, this should be under terminal anaesthesia, and animals should not be used in 'survival' surgery. The possibility of studying a non-animal alternative, instead of using animals, should be examined. The views of students who object to the use of animals in undergraduate studies should also be respected. Provided they could not have entertained their objections before starting the course, students should be allowed to opt out of using animals (studying non-animal systems instead), without prejudice to their marks or prospects. Course organizers should be willing to provide genuine prospective students with information about any proposed use of animals in their courses.

In pursuit of higher degrees, students embark on a research training, in which invasive uses of animals may be justified. There is a need for review of the checks and balances which operate to ensure that such students receive adequate supervision.

There is no need for pre-university students to use living animals in order to develop or practise manual skills. The same is true in most university courses. The use of animals in training for those wishing to carry out experiments on animals is more problematic, since these people must gain competence in the techniques they propose to use. However, provided new researchers are carefully supervised, they should be able to gain competence on the job, without compromising the well-being of the animals involved. Thus in the present political climate in the UK, we accept that it is reasonable to restrict the use of animals in gaining manual skills to the development of microsurgical skills by practising surgeons.

12.6 CONTROL OF LABORATORY ANIMAL USE

Several different kinds of control, taken together, can help to safeguard standards in biomedical research involving animal subjects (Chapter 10).

12.6.1 Statutory controls

In some countries, projects involving animals will require statutory authorization before the researchers can proceed. In the UK, the use of scientific procedures likely to cause pain, distress or lasting harm to living vertebrates is controlled by the Animals (Scientific Procedures) Act 1986. Such statutory control has not yet been established in all countries of the world. In many countries, laboratory animals are afforded only minimal legal protection, through general animal welfare or anti-cruelty statutes. In such countries, there may be *ad hoc*, voluntary, institutional or local systems of regulation, but these are patchy. Canada is the only country in the world to have established a national system of voluntary self-regulation of animal research.

There is a need for special laboratory animal protection laws, and funding to implement them, in most countries. In countries which rely on the provisions of more general statutes, there is difficulty in applying anti-cruelty legislation to the use of animals in the laboratory. Where voluntary systems of control are established, these may be open to criticism, not necessarily because they are not effective, but because the public may have less confidence in a voluntary than in a legal system of control. Even in Canada, where the CCAC system of voluntary self-regulation is reinforced by the threat of removal of funding for researchers who do not comply with the CCAC's requirements, there are calls for legislative back-up.

In countries which have special laboratory animal legislation, there is a general move away from authorizing people, or institutions in general, to perform experiments on animals, towards authorizing, or licensing, particular projects involving animals. This move means that the legislation can now control the design of each individual project, so as to encourage explicit consideration by researchers of the Three Rs and, in some cases, of the balance of the likely benefits of the work against the animal suffering involved. Such statutory provisions may exert a kind of ethical control over the use of animals in experiments, and are to be welcomed.

Prohibitions on areas of research

The controls might also include the prohibition of particular techniques which are considered to cause an unacceptable amount of animal suffering. There is some danger in the laws going further than this, by imposing bans on whole areas of research or testing. In particular, if a country with high standards of control over the use of animals in scientific procedures were to impose a total ban on work carried out for a particular purpose, the work might be exported to countries with lower standards of control. Most areas of research where a ban might be contemplated raise complex moral questions, which are not easily decided on a general basis. On balance, therefore, it would seem better for the animals if countries with stringent legislation were to retain control over all kinds of animal research, by reviewing every proposal on its merits, and not banning outright work carried out for particular purposes.

Effectiveness of legal controls

In practice the effectiveness of the various legislative controls, whatever they say on paper, depends on how well they are administered and enforced. Although all EC countries have established legislation which complies with the 1986 European Directive, the effectiveness of the laws varies considerably from country to country, because notification and authorization requirements and provision for inspection of establishments using animals are variable. In the USA, too, there is difficulty in ensuring compliance with the weak Animal Welfare Act, because of failure of research establishments to register under the Act, and difficulties in providing for sufficient inspection of research premises. Efficient administration and inspection are essential in monitoring compliance with laboratory animal laws.

Public reassurance and accountability

There is a need for the legislation not only to be effective, but for it to be seen to be effective. One way of assuring this would be for sanctions to be imposed more frequently or more seriously. Offences under the law should be matched with sufficiently serious penalties. For the most severe infringements the appropriate penalty would be prosecution, and administrative authorities should not be chary of prosecuting in cases of deliberate contravention of laboratory animal protection law. The withdrawal of authority to perform experiments on animals (in the UK, a personal or project licence) is also a serious sanction and should be applied rigorously, whenever appropriate.

Some form of external review of the system of administration of the law might also increase confidence in its provisions. In the UK, the statutory Animal Procedures Committee is an important means for such review. The possibility of instituting an animal ombudsman or other means of strictly independent review is also worth considering.

Lay involvement in the statutory review of applications to perform research involving animals is used by some countries as a means of ensuring public accountability in decisions taken under the law. In these countries, institutional, regional or national Research Review (or Ethics) Committees are responsible for reviewing and authorizing projects involving animals, and it is often a statutory requirement that such committees have lay members. Such a system can provide some local accountability and public reassurance. Research review committees may also play a complementary, more educational, role in countries where projects involving animals are authorized by some other authority. Indeed, this educational role might apply not only to institutional research review committees, but also to the other forms of statutory and non-statutory control described here.

12.6.2 Non-statutory controls

At the research proposal stage, before any work is carried out, a number of informal or more formal non-statutory controls may operate. The scientists writing the research proposal should consider the ethical aspects of their proposed work at this planning stage. The scientists can be helped in their planning by peer review within their institution (either informally, or in the more formal context of a Research Review Committee), by guidelines produced by the relevant learned society, and sometimes by review of the proposal by a committee of that society. If funding for the work is sought from a grant-giving body, the proposal might be subject to ethical review by a committee of that body. Indeed, it would be desirable for funding bodies to attempt to give more explicit consideration to animal interests when reviewing research proposals. This could be achieved by involving a laboratory animal scientist in the review.

Once the work is in progress, it might be subject to a number of ongoing reviews. The researchers themselves will evaluate the work as they go along, and should consider the effects on the animals as part of this evaluation (were the effects as expected?

might the techniques be improved, in the light of experience, so as to cause less harm to the animals? is the benefit still worth the harm caused to the animals?, and so on). Such review might be helped by discussion with scientific peers, including specialist laboratory animal scientists, technicians and others responsible for the care of the laboratory animals, the institution's veterinary surgeon and, in the UK, the Home Office Inspector. Where inspectors are appointed under the relevant laboratory animal legislation, they will be responsible for ongoing statutory review of the research, and should have the power to stop the research if, for example, it imposes much greater suffering than was expected, so making its justification questionable. In the UK, inspectors have this power and use it. All of the ongoing review might be influenced by guidelines produced by the learned society in the relevant specialist field.

It is envisaged that the Working Party's costs and benefits assessment schemes will prove helpful at all stages in this ethical review: at the research proposal stage, whilst the work is being carried out, and when the results are written up and submitted for publication.

In the last case, the review is retrospective. Although there is some doubt about whether editorial policies can influence research practice, policies which require stated ethical standards in work submitted for publication can fulfil an important educational role and might, in the longer term, lead to changes in practice. Usually it is only when the results of work are published (or reported at public conferences) that the details of that work become available to the general public. Some countries (including the UK and Netherlands) publish statistics of animal use, but these data refer to the overall scale of animal use, and do not give information about particular projects. Published material, therefore, represents the public face of research involving animals. Publication lays the research open to scrutiny by both scientific peers and the general public. For these reasons, it is important that careful consideration be given to the reporting requirements for animal experiments. Better reporting of the adverse effects on the animals (including any deaths) caused by the scientific procedures and the steps taken to alleviate these effects and to refine the procedures can help others to follow best practice. This reporting, together with the publishing of a clear, reasoned, justification for the work, may allay criticism of work which might otherwise be justified, and lead critics to focus their attention on work which indeed requires scrutiny.

12.7 PHILOSOPHICAL ARGUMENTS AND MORAL CONCERNS

The Working Party recognizes that its moral conclusions not only depend on information presently available, but also, like all moral conclusions, are philosophically contestable. Moral concern about the use of animals in biomedical research is expressed for a variety of reasons which command the Working Party's respect. It is not necessary to hold that animals have 'rights' in order to believe that it is wrong (other than in the direct interest of the animals concerned) to inflict pain and suffering on them, or to restrict their liberty. But the Working Party also respects the moral reasoning of those who argue that, at present, the use of animals is a necessary means to the end of preventing and relieving pain and suffering in both humans and animals. Again, it is not necessary to hold that humans have a 'right' to use animals in this way, in order to believe that it would be wrong to abandon this means in the absence of an alternative.

All members of the Working Party agree that at present it is morally unjustified to abandon the use of animals in biomedical research altogether. Within this agreement, however, some find it very difficult to resolve, in good conscience and in terms of rational arguments which they can accept, the conflict between this and their concern not to see pain or suffering inflicted on animals. They believe that the possession, by humans, of reason and self-consciousness provides a historical explanation (in terms of power) rather than a rational justification (in terms of justice) for our use of animals. Thus what essentially is involved, they argue, is a 'tragic conflict' to which there can be no satisfactory resolution, unless or until alternative means can be found (Chapter 11, part 1). They acknowledge that this position may be philosophically inconsistent. But they believe that it fits, better than any other so far advanced, with their sentiments, stronger than regret, if weaker than remorse, towards the use of animals in biomedical research. To acknowledge this inconsistency, they believe, is the most appropriate way to encourage the reduction, refinement and replacement of animal experiments–and so to reduce the moral conflict.

Other members of the Working Party believe that these conflicting moral concerns can be reconciled by logically consistent arguments based on a particular understanding of justice and common morality (Chapter 11, part 2). They argue that a different research ethics for humans and animals can be justified in terms of a morally-relevant difference between them. Sufficient reason for this difference is found in the argument that, typically, humans do, and animals do not, possess the nature of a rational self-conscious creature. This argument, they believe, suggests that the contrast between human and animal research ethics rests on more than a species prejudice; it rests on the obligations derived from an understanding of justice and of membership of a common moral community. At the same time, however, they acknowledge that all uses of laboratory animals fall into the category of actions which require to be justified as being of high scientific quality, having sufficient likely benefit, and imposing the minimum pain and distress compatible with the objectives of the work; and that, furthermore, within this category, there are some procedures which ought not to be done at all. In this argument, the use of animals in biomedical research is seen as a 'necessary evil'. Such a view, again, invites us to eliminate quickly, so far as is possible, the circumstances which generate the moral conflict it involves.

Index

abnormalities, congenital 93–4, 97
adjuvants 194
  Freund's 169–71
adverse effects 78–86, 111, 113, 117, 145–6, 150, 165, 183, 191, 222, 254, 261, 289, 310, 330, 335, 357; *see also* costs
aggression 108–9
alternatives 122–37, 234–7, 242, 252, 256, 262, 265, 269, 276, 294, 310, 312, 319, 326, 329, 333, 336, 339, 344; *see also* replacement
American Medical Association 26
Ames test 207
anaesthesia 11, 15, 16, 17, 21, 22, 26, 100, 111, 112–13, 116–17, 146, 149, 162, 164, 170, 229, 250, 252, 256, 259, 264–5, 267, 269, 275, 276, 283, 289, 292, 333
  terminal 100, 149, 154, 230, 236, 242, 292–4, 339
analgesia, analgesics 61, 62, 66, 85, 111, 112–13, 116, 146, 151–2, 154–5, 157, 256, 259, 264–5, 267, 269, 287, 294, 333
Anatomy Act 1832 26
anger 70
animal beneficiaries 39, 98, 141, 215, 308, 310, 315, 336
animal care/ethics committees 249, 262–3, 265, 267, 268–70, 280, 342; *see also* research review committees
Animals (Scientific Procedures) Act 1986 v, 5, 13–17, 23, 24, 25, 36, 41, 74, 79, 87–90, 91, 99, 100, 102, 104, 107, 114, 122–4, 130–1, 135, 139, 147, 163, 169, 176, 179, 213, 221, 224–5, 229, 231, 239, 254–6, 274, 277, 278, 279, 280, 284, 340
  infringements of 281–2, 283–4, 341–2
  sanctions under 279–81, 341–2
antibody production 114, 126, 129, 166, 168–72
anticipation 54, 62, 69, 71
anxiety 45–6, 68, 70–3, 78, 112, 116, 138–9, 145–6, 174, 195, 330
anxiolytics 72
Aristotle 140, 299, 300, 304, 313
ascites 169–71
Association for Science Education 228
Association for the Study of Animal Behaviour, guidelines of 286
awareness 4, 45–6, 52, 53, 69, 87, 91, 145, 307, 311–12
  of self 54–8, 300, 303; *see also* self-consciousness

Bacon, Francis 8
Balls, M. xv, 122, 123, 127, 134, 211
Barclay, R.J. 84–5
Bateson, P. xv, 90–1, 285
behaviour 45–6, 50–4, 69, 73, 80, 85, 103, 105–8, 110, 112, 117, 130, 173, 175, 213, 221–2, 241, 327
  non-linguistic 320
behavioural studies 282, 305–6
benefits of research 3, 4, 15, 24–44, 127, 245, 256, 261, 276, 305, 310–11, 329, 334, 345
  assessment of 36, 38, 39–42, 79, 138–82, 289
    scheme for 141–3, 330, 343
  see also costs; weighing
Bennett, J. 319
Bentham, J. 300
Bergman, P.G. 43
Bernard, Claude 3, 26, 183
Beynen, A.C. 84
Bobbit, P. 309–10
boredom 104
Boyle, Robert 2
brain 46–50, 51, 55, 59–61, 65–6, 68–9, 72; *see also* nervous system
breeding 252, 254–5, 260
  establishments 293
  purpose 90–3, 99, 100, 101, 117, 151, 255, 293
  selective 97–8
British Association for the Advancement of Science 249–50
British Psychological Society, guidelines of 286–7
British Toxicology Society (BTS), *see* fixed dose procedure
British Veterinary Association (BVA) vi, 253
Bryant, C.E. 109

Burch, R.L. 122
Byrne, P. xv, 317, 323
Byrne, R. 57–8

caging 89, 101–6, 108–110, 144; *see also* confinement; environment; housing
Calabresi, G. 309–10
Canada 102, 111, 125, 234, 257, 258, 268–71, 272–3, 275, 276, 279, 286, 340
cancer, carcinogens 96, 114, 126, 189, 193, 195–6, 229
captivity
  breeding in 89, 91–2, 132
  study in 91
  see also wild, capture from the
care, post-operative 113, 117, 151, 333
cell culture, tests 38, 132, 213; *see also in vitro* studies
cephalopods 47, 64, 66, 71, 87–8, 130, 161–6, 331; *see also* octopus
Chamove, A.S. 106, 108, 110
Clark, S.R.L. 302
clinical trials 153, 156–8; *see also* human subjects
Cochrane, W. 227
codes of practice 274
  housing and care 102–4, 254
  laboratory practice 177, 208
  see also guidelines; Home Office
cognitive ability 50–4
Committee for the Reform of Animal Experimentation (CRAE) 88–9, 253
companion animals 90, 251, 255, 264, 292–3, 403; *see also* pets
computer models, programmes 24, 116, 123, 125, 129, 211, 217, 228, 234, 259
confinement 68, 85; *see also* caging
conflict, ethics of 324–6, 327, 336, 344
conscientious objection 225–8, 237–8, 241–2, 339
controls, experimental 149–52, 154–6, 173, 192
  regulatory 5, 16, 179–81, 187–8, 206–8, 230, 245–84, 340–1
    international 258–9, 340; *see also* Council of Europe; European Commission
    self- (voluntary) 247, 248–50, 257–8, 265–6, 267, 268–71, 340, 342; *see also* peer review
cosmetics 24, 39, 124, 184, 188, 198, 200, 200n, 210, 273, 276, 277, 282
costs to animals 3, 4, 11, 21–2, 31, 36, 40, 78, 86–90, 92, 98, 101, 111, 187, 194–5, 275, 285, 289, 329, 333, 334
  assessment of 115–7, 138–82, 315, 343
    scheme for 144–6, 331; *see also* benefits; weighing
Council of Europe Convention 102, 108, 124, 222, 254, 258–60, 261
Council for International Organizations of Medical Sciences (CIOMS) 125
  guidelines of 258–9
cruelty, laws against 257–8, 275, 340
Cruelty to Animals Act 1876 v, 17, 21, 23, 79, 147, 169, 229, 230–1, 250–4, 278, 279, 279n, 292
  infringements of 281
curare, *see* neuromuscular blockade
curiosity 33, 34, 42; *see also* knowledge, pursuit of

Darwin, C. 303–6, 308
data banks, bases 206, 210
Dawkins, M.S. 58, 109–10
decline in animal use 22–4, 232
Denmark 246, 263, 272–3
Descartes, R. 299–300, 301, 312
design, scientific 252, 255–6, 269, 284, 329
  statistical 115, 117, 142, 159, 201, 333
developmental stages 87, 122, 123, 131–2, 213, 222, 293, 334
dietary restriction 158–60
discomfort 79, 150, 180, 252, 259, 315; *see also* costs
dissection 232, 235, 243
  in schools 225–8, 241, 338–9
    alternatives to 226–7
distress 46, 67, 78, 85, 113–14, 116, 122, 125, 138–9, 145–6, 157, 165, 175, 180, 184, 195, 198, 205, 236, 254, 259, 261, 262, 267, 294, 311, 315, 322, 330, 333, 345; *see also* stress
Dockerty, A. 227
dominion 299, 316
Draize test, *see* eye irritancy
Duncan, I.J.H. 103, 110
duplication 123, 128, 178, 180, 206, 208, 216, 249, 261, 292n, 294, 338
duties 302, 311, 312, 334, 336

earthworm 45, 46, 63–4, 71
editorial review 287–9; *see also* publication
education and training 5, 21, 41, 100, 129–30, 135, 142, 146, 173–6, 220–44, 255, 259, 260, 261, 265, 276, 293, 295, 335, 338
Egypt 257
emergency, ethics of 30

endpoints 111, 113–14, 117, 146, 152, 157, 188, 190, 201–2, 209, 216, 217, 333, 337
environment, enrichment of 108–9, 113, 117, 144–5, 151, 294, 332
Erasmus vii
European Commission 192, 193, 211, 212–13, 216
  Directives 78, 87–8, 91, 93, 102, 107, 112, 122, 124–5, 131, 135, 186, 188, 200n, 207, 209, 210 222, 230, 254, 259–60, 261, 293–5, 341
euthanasia, *see* killing
evil, lesser 30, 37
  necessary 308, 324–5, 327, 345
exercise 102–4, 265
experiment defined 293
eye irritancy tests 114, 128, 131, 177–8, 189, 192, 197, 200n, 203–4, 207, 213

fear 70, 72
fixed dose procedure (FDP) 191–2, 202, 209, 217
Fielder, R.J. 203–4
Frey, R.G. xv, 323
Fund for the Replacement of Animals in Medical Experiments (FRAME) vi, xv, 88–9, 126, 137, 253
funding of research 16, 34, 41, 171, 192, 214, 217, 257, 266–7, 270–1, 285, 337, 342
  peer review in 285–6

Galen 2
gene therapy 97, 99
  transfer 93–9
generic products 206; *see also* 'me-too'
genetic defects 116, 145, 332
Germany 191, 262, 276
Goodall, Jane 51
Greece 264
Griffiths, P.H.M. 80–2, 84, 85n, 86
guidelines 180, 207, 208, 215, 223, 245, 248–50, 254, 268, 272, 287, 342; *see also* codes of practice
  harmonization of 208
  see also Association for Study of Animal Behaviour; British Psychological Society; CIOMS; Home Office; learned societies; OECD; pain; refinement; RSPCA; UFAW

Hall, Marshall 248–9
Halsbury, earl of 253
handling 112, 117, 138, 144, 173–6, 223–4, 264
Harvey, William 2
Hollands, C. xvi, 288
Home Office 17, 24, 79, 88, 102, 108, 111–12, 114, 115, 170, 204, 229, 238–9, 275, 277, 282, 283
  Advisory Committee v, 252–3, 279
  Animal Procedures Committee 255, 273, 277, 279, 282, 283, 341
  Codes of Practice 102–4, 108, 112, 254
  Guidelines 114
  Inspectorate v, xiv, 16, 88, 111, 113, 177, 251–6, 262, 273, 281, 284, 285, 343
  statistics 195–8, 230, 279, 343
Houghton, Lord 253
housing of animals 14, 15, 91, 93, 101–10, 116–17, 138, 144–5, 254, 259, 260, 261, 264, 267, 294, 332; *see also* caging
  code of practice 102–4
human subjects, volunteers 11, 26, 28, 29–30, 38, 124, 126, 132–4, 153, 156, 160, 231, 245–7, 314–15, 326; *see also* clinical trials; research, ethics of; research ethics committees
Hume, D. 300
husbandry 91, 92, 101–10, 116–17, 139, 145, 223–4, 241, 275, 332
hybridomas 169–70

India 258, 264
industrial companies 176–81
information, pooling or withholding of 206, 216; *see also* publication
insecticides 176–81
'instincts' 51
intelligence 50, 320
intention 31, 34, 35, 36, 42
  in animals 51–2, 57–8
interests 301, 305, 315, 319, 344
International Society for the Study of Pain, guidelines of 286–7
invertebrates 87–8, 123, 129–30, 222, 224–5, 233, 251, 264, 274, 311–12, 331; *see also* cephalopods, octopus
in vitro studies 12, 24, 97, 123–6, 129, 131–3, 151, 159, 169, 170, 193, 203, 211–15, 217, 259, 274, 337; *see also* tissue culture

Japan 208
judgement
  deliberative 325–6
  verdictive 325
justice 313–16, 317, 318, 321, 322, 324, 344–5

Kant, I. 300–1, 314, 319
killing, humane 102, 113–14, 224–5, 227–8, 233, 240, 256, 259, 260, 267, 274, 292–4, 315
knowledge, advancement of 8–9, 21, 27, 31–4, 38, 39, 42–4, 141, 163, 166, 171, 180, 255, 259, 292, 310–11, 318, 329
Kurzok, R. 42–3

Laboratory Practice, Good, Code of 177, 208
language-use 320
law, enforcement of 279
  ethics and 275–80
  see also legislation; regulatory requirements
Lawlor, M. 105, 107
LC50 test 22
LD50 test 22, 177, 189–92, 197, 201, 205, 209, 211, 252
learned societies, ethical review by 286
  guidelines of 286–7, 342
learning 51, 52, 54, 62, 69
legislation 247–54, 340–1
  international 258–64, 294, 340–1
  see also Animals (Scientific Procedures) Act; cruelty, laws against; Cruelty to Animals Act; law; Martin's Act; Protection of Animals Act
'liberation', animal 317, 318–19
licences, personal 14, 16, 79, 238–9, 243, 254, 255–6, 260, 283, 341
  project 15, 16, 17, 18, 34, 41, 79, 88, 111, 113, 124, 207, 238, 255–6, 260, 262, 283, 340, 341
  revocation of, *see* Animals (Scientific Procedures) Act, infringments of
Lieb, C.C. 42–3
limit test 191–2, 197, 202, 208, 216, 337
Littlewood Committee 255
Lucas, J. 314, 317

Magnan, E. 248
Malaysia 257
Martin, P. 90–1
Martin's Act 1822 248, 248n, 257
Mather, J.A. 71
maximum tolerated dose (MTD) 177
medical education 232–3, 293
Mellor, David 253
'me-too' (generic) products 178, 180–1, 206
memory 69, 299
microsurgery, training in 111, 230–1, 240, 243, 276, 277, 282, 339
Moberg, G.P. 86
moral argument v, 3, 8, 25, 29–31, 35, 37–9, 41–2, 88, 90, 98, 115, 126–37, 139–41, 146–7, 180, 184, 198–210, 217, 220, 228–9, 246, 275, 277–8, 296–328, 335, 341
  conclusions of 310–11, 327, 344–5
moral claims 308
moral concern, reasons for 306–8, 332, 344
morality, common 316–17, 324, 326–7, 344–5
moral status of animals 316–18, 321; *see also* nature and status, animal and human
Morton, D.B. xvi, 80–4, 85n, 86, 223, 229

Named Person in Day to Day Care 79, 112, 280, 284
Named Veterinary Surgeon 79, 254, 280, 284
National Anti-vivisection Society 237
nature and status, animal and human vii, 42, 54, 55–6, 62, 65–7, 72, 78, 88, 98, 201, 205, 215, 299–305, 306, 309, 311, 316, 318–22, 323–4, 327, 344–5
Nazi experiments 28
necessity 4, 24, 36–9, 41, 143, 180, 199, 205, 208, 215, 236, 248n, 257, 260, 261, 293, 294, 308, 310, 315, 324, 325, 329, 336, 345
  interim, temporary 37, 306, 309, 330
nervous system 45, 47, 60, 71, 73; *see also* brain
Netherlands, The 17, 20, 21–2, 111, 233, 236, 237, 242, 263, 272–3, 343
neuromuscular blockade 114, 276, 287, 292
Nigeria 257
Norway 88, 261, 263

octopus 45, 47, 53, 54, 63–6, 71–3, 78, 83, 87, 161–6, 171, 331
ombudsman 282, 342
Organization for Economic Cooperation and Development (OECD) 181, 190–2, 216
  guidelines of 177, 179, 191, 201, 202, 204, 208–9

Paget, Stephen 26
pain 78–9, 112–14, 116, 122, 125, 131, 138–9, 145–6, 150, 157, 165, 180, 184, 195, 198, 252, 254, 259, 260, 261, 262, 263, 265, 267, 294, 303, 307, 312, 315, 322, 330, 333, 345
  anticipation of 62, 303

awareness of 4, 45–6, 58–67, 70, 73, 87, 259, 307–8, 311
experimental, guidelines for investigation of 287
memory of 62, 69
recognition of 79–81
responses to 61–2
painful experiments, restrictions on 292
Paracelsus 185
Paton, William 26, 28
peer review 16, 32, 40, 42, 271, 272, 274, 284–9; *see also* editorial review
Pepperberg, I.M. 53
pesticides 39, 176–81, 184, 187, 188
pets 223, 225; *see also* companion animals
pharmaceutical products 184, 188
philosophical argument 292–328, 344–5
Plato 299–300
Platt, Lord 253
Portugal 260, 264
practice, ethical analysis of 313–14, 316–17, 318, 325
primates 53, 57, 66, 320, 321
non-human 15, 20, 29, 30, 37, 45, 48, 49, 51, 56–7, 66, 88, 89, 91–2, 102, 104, 109, 112, 132, 148–53, 158–60, 197, 254, 264–5, 277, 282, 293, 305, 307, 321, 331
protected animals 87, 122, 129, 130–2, 213, 256, 264, 267, 268, 293, 331
Protection of Animals Act 1911 (1912) 222, 224, 279n, 281
psychological research 53, 282, 305
publication 16, 128, 289, 343; *see also* information
ethical review for 272, 274, 287–9, 343; *see also* peer review
purpose-breeding, *see* breeding
purposes, permissible 254–5, 260, 275, 276, 332

quarantine 91, 92, 117, 145, 149, 184, 332

Rachels, J. 303–6
radiation 113, 229
rationality 319–23, 323–4, 344
rationalization 209, 216, 338
recording
of animals used 89
of data 144, 265, 343
recreation 89
Reduction 4, 15, 38, 115, 117, 122, 134, 143, 152, 157, 165, 180–1, 184, 199, 201, 203, 208–9, 211, 216–17, 234, 261, 319, 333, 336–7, 344
Refinement 4, 15, 114, 116–17, 122, 143, 151–2, 157, 165, 170–2, 180, 201–2, 211, 216–17, 261, 275, 284, 289, 333, 337, 344
guidelines for 114
Regan, T. 301–3
regulatory authorities, requirements of 179–81, 184, 186, 187–8, 196, 198, 206–8, 211, 216, 294, 335; *see also* controls
harmonization of 208–9, 294, 338
rationalization of 209, 216–17, 338
repetitive testing, *see* duplication
Replacement 4, 12, 24, 29, 38, 122–37, 143, 180, 199, 211–14, 217, 234–7, 252, 261, 276, 333–4, 344
reporting, *see* publication; recording
research
assessing value of 141–3
ethics of 313–16, 317, 319, 321, 322–4, 327, 344–5
fundamental, *see* knowledge, advancement of
nature of, 8–24, 31–5
research advisory committees 89, 342
Research Defence Society 26
research ethics committees (medical) 29, 41, 151, 156–7, 246, 273
research review committees 271–5, 284, 342; *see also* animal care/ethics committees
re-use of animals 261, 294
rights 300–3, 305, 317, 344
risk 185–6, 199, 205, 210, 211, 214, 217
Rowan, A.N. 71
Royal Society for the Prevention of Cruelty to Animals (RSPCA) vi, 223–4, 227, 248, 248n, 250, 253
guidelines of 224
Russell, W.M.S. 122

St Andrew Animal Fund vi, xiv, 253
safety 128, 179, 186, 199, 200, 210, 214, 335–6
Sales, G.D. 107
Sanford, J. 80–1, 83, 85, 86
self-consciousness, *see* awareness
temporal 319–23, 323–4, 327, 344
sensitivity 4, 87, 311, 312
sentience 4, 74, 202, 256, 260, 275–6, 301, 312, 319, 331, 337
severity, assessment of 4, 16, 21, 22, 78–9, 83, 89, 111, 116, 140, 143, 145–6, 157, 205, 230, 256, 260
Silver, I.A. 26
Singer, P. 301, 303, 323
skills, manual, acquisition of 111, 227, 229–30, 232–3, 240–2, 250, 276, 282, 293, 339

skin irritancy tests 192, 197, 200, 203–4, 213
social contact 89, 103–4, 109, 151, 221
Socrates 298
South Africa 257
Spain 260, 264
species
  choice of 91, 117, 216, 249, 252, 259, 275, 293, 331, 337
  endangered 89, 92, 126, 145
  used 15, 18–20, 87–90, 111, 256, 333
'speciesism' 7, 301, 304–5, 317, 323–4, 327, 345
Sprigge, T. 302
stocking density 105
stray animals 91, 99–100, 332
stress 4, 45, 67–70, 73, 79, 85, 92, 102, 105, 112–13, 139, 146, 174, 195; *see also* distress
suffering 54, 138–40, 151, 205, 249, 252, 261, 307, 322; *see also* anxiety; distress; pain; stress
supply of animals 90–100, 117, 149, 252, 254, 260, 332; *see also* breeding; wild, capture from
surgically altered animals 99
Sweden 261, 262, 272–4
Switzerland 263

Taylor, G.T. 110
technicians, animals 89, 112, 221, 224, 280, 284, 343
temperature 106–7
theology 297, 298–300, 304, 312, 316, 326
Thomas, Keith 316, 327
Tigerstedt, R. 43
tissue culture 24, 123, 125; *see also in vitro* studies
tobacco 276, 277, 282
toolmaking and use 51
toxicity testing, toxicology 5, 10, 17, 26, 113, 123, 126, 129–32, 134, 146, 176–82, 183–219, 333, 335
  legal obligation for 184; *see also* regulatory authorities
'tragic conflict' 309–10, 344
training 89, 111, 116–17, 143, 144, 220–44, 227–8, 229, 231–2, 239–40, 243; *see also* education
transgenic animals 93–9, 332

ultrasound 107
United States of America (USA) 18, 20, 99, 102, 111, 122, 125, 168, 188, 210, 212, 213, 223, 225, 228–9, 232–3, 236, 242, 257–8, 260, 264–8, 270, 272, 275, 276, 278, 286, 341
Universities Federation for Animal Welfare (UFAW) vi, 112, 113, 224–6
  guidelines of 112

vaccine testing 113, 184
validation 127, 188, 192, 199, 206–7, 209, 211, 217, 334, 337
Vane, John 26

Walker, S.F. 65
Wallace, M.E. 82, 86, 103, 111
washing powders 276
weapons, bacterial and chemical 5
Weaver, A.D. 233
weighing benefits and costs 138–82, 198, 215, 229, 241, 245, 249, 255, 275, 311, 315, 329, 334, 340
Weiss, J. 110
welfare of animals 14, 42, 74, 89, 126, 143, 215, 221, 224, 252, 254, 294, 311, 325, 327, 336; *see also* anaesthesia; analgesia; caging; costs; environment; exercise; handling; recreation; social contact, stocking density; suffering
Whiten, A. 57–8
Wigglesworth, V.B. 64, 74
wild, capture from the 90, 92–3, 116–17, 126, 131–2, 145, 150–1, 331

Yamauchi, C. 106–7

Zbinden, G. 195, 207–8